NEWTONIAN DYNAMICS

Ralph Baierlein
Charlotte Ayres Professor of Physics
Wesleyan University

McGraw-Hill Book Company
New York St. Louis San Francisco Auckland Bogotá Hamburg
Johannesburg London Madrid Mexico Montreal New Delhi
Panama Paris São Paulo Singapore Sydney Tokyo Toronto

This book was set in Times Roman by A Graphic Method Inc.
The editors were John J. Corrigan and James S. Amar;
the production supervisor was Leroy A. Young.
The drawings were done by Allyn-Mason, Inc.
The cover was designed by Robin Hessel.
Halliday Lithograph Corporation was printer and binder.

NEWTONIAN DYNAMICS

Copyright © 1983 by McGraw-Hill, Inc. All rights reserved. Printed in the United States of America. Except as permitted under the United States Copyright Act of 1976, no part of this publication may be reproduced or distributed in any form or by any means, or stored in a data base or retrieval system, without the prior written permission of the publisher.

1234567890 HALHAL 89876543

ISBN 0-07-003016-2

Library of Congress Cataloging in Publication Data

Baierlein, Ralph.
   Newtonian dynamics.

   Includes index.
   1. Dynamics. I. Title.
QA845.B33 1983   531'.11   82-17944
ISBN 0-07-003016-2

TO MY FAMILY
Jean, Eric, and Jeffrey

CONTENTS

	Preface	vii
	A Note to the Problem Solver	ix

Chapter 1   A Review of Some Basics   1

1.1	Qualitative Reasoning	1
1.2	Vectors and the Scalar Product	7
1.3	Tossing Rocks	8
1.4	A Consequence of Newton's Second Law: Energy	15
1.5	Gravitational Potential Energy	20
1.6	Potential Energy and Stability	23
1.7	Some Implications of Invariance	27
1.8	Newton's Laws: Structure and Meaning	30
1.9	Reviewing the Review	34

Chapter 2   The Harmonic Oscillator   45

2.1	Damped Harmonic Oscillator	45
2.2	Phase Space: An Introduction	51
2.3	Harmonic Oscillation in Two Dimensions	54
2.4	Sinusoidally Driven Oscillator	56
2.5	Quality Factor Q	61
2.6	A Guide to the Major Ideas	63

Chapter 3   Nonlinear Oscillators   71

3.1	A Nonlinear Oscillator	71
3.2	Amplitude Jumps and Hysteresis	81
3.3	van der Pol's Equation: The Limit Cycle	88
3.4	More Uses for the Averaging Method	93
3.5	Series Expansions	97
3.6	About the Methods	104

Chapter 4   Lagrangian Formulation   116

4.1	Fermat's Principle	116
4.2	Calculus of Variations	118
4.3	Newton II as an Extremal Principle	123
4.4	Lagrangians and Constraints	128
4.5	Another Instance of Constrained Motion	131
4.6	Conversion to First-Order Equations: Hamilton's Equations	136
4.7	Liouville's Theorem	140
4.8	The Lagrangian and Quantum Mechanics	144
4.9	A Sense of Perspective	146

Chapter 5   Two-Body Problem   156

- 5.1 Reduction to Motion in a Plane   156
- 5.2 Effective Potential Energy   161
- 5.3 Orbit Shape   164
- 5.4 Orbits around the Spherical Sun   168
- 5.5 The Oblate Sun   175
- 5.6 Stability of Circular Orbits   181
- 5.7 The Orbit in Time   187
- 5.8 Compendium on Central-Force Motion   189

Chapter 6   Rotating Frames of Reference   202

- 6.1 Vectors in a Rotating Frame of Reference   202
- 6.2 Physics on a Rotating Table   209
- 6.3 The Rotating Earth   210
- 6.4 Foucault Pendulum   217
- 6.5 The Figure of the Earth   223
- 6.6 A Perspective on Rotating Frames   228

Chapter 7   Extended Bodies in Rotation   235

- 7.1 Equations for Location and Orientation   235
- 7.2 Simple Precession   240
- 7.3 How $\mathbf{L}$ Is Related to ω 244
- 7.4 A Novel Pendulum   247
- 7.5 $\mathbf{L}$ Is Not Necessarily Parallel to ω 249
- 7.6 Diagonal Form for the Inertia Tensor   251
- 7.7 Euler's Equations for a Rigid Body   254
- 7.8 Axisymmetric and Torque-Free   255
- 7.9 Chandler Wobble   260
- 7.10 An Interlude on Kinetic Energy   264
- 7.11 The Symmetric, Supported Top   265
- 7.12 Precession of the Equinoxes   272
- 7.13 Survey of Critical Notions   275

Chapter 8   Cross Sections   286

- 8.1 Scattering Effectiveness: The Idea Behind the Cross Section   286
- 8.2 A Capture Cross Section   289
- 8.3 A Differential Cross Section   291
- 8.4 Rutherford Scattering   295
- 8.5 Major Ideas   300

Appendixes   308

- A Expansions, Identities, and Miscellany   308
- B Vector Product   311
- C The Averaging Method   317
- D The Craft of the Physicist   321

Index   324

PREFACE

A book on classical mechanics for juniors and seniors is more than just that. The topics chosen and the methods used reveal the author's attitude toward the subject. When I look back at what I emphasized in class and then in writing the book, I find a stress on analytical methods, on techniques that are useful throughout physics and engineering—things such as expansions, dimensional analysis, sketches (rather than just equations), and stability analysis. These are the stock in trade of an astrophysicist, say, and of anyone who is working on the back of an envelope.

I have written with the aim to free more of class time for doing problems, for discussion, and for gloss. To this end, the text develops the theory with enough words to supply motivation and context. A difficult piece of theory is immediately followed by an illustration of how the principle works in practice. Thus one can ask students to read sections before the topics arise in class. While this may be uncommon, the benefits are great: both a richer use of class time and a maturing ability to understand science directly from the printed page.

Some less central theoretical developments have been left to the homework problems, for two reasons. First, a text can become overburdened with information. Then readers have an unnecessarily difficult time assimilating the more basic parts of the subject. Second, nothing sinks in like a problem tackled and solved. Those homework problems that extend the theory in any notable way have been flagged in the margin with a superscript T, like this:T

Equations that are especially notable, either as starting points or as results, are starred in the margin: ★

Something should be said about the sections that are essential for continuity. Within a given chapter, the progression and prerequisites are, I hope, clear. It is the continuity from one chapter to another that merits an outline. Chapter

1, as a whole, is essential. Then come sections 2.1, 2.2, and 3.1. These sections introduce the harmonic oscillator, phase space, and motion near an equilibrium point. They provide the basis for stability analysis and "small oscillations" theory, and thus they lay the foundation for frequent applications in the succeeding chapters. No subsequent material depends in an essential fashion on Chapter 4. Portions of Chapter 5 are needed in the last chapter, on cross sections, specifically Sections 5.1 through 5.4. Section 6.1 is a prerequisite for Chapter 7. Even after some picking and choosing, there is enough to occupy a class productively for a semester.

An equation with the structure $\ddot{x} = -\omega_0^2 x + f(x, \dot{x})$, where $f(x, \dot{x})$ incorporates all the complicated terms, arises surprisingly often in mechanics and elsewhere in science. When f has only a small influence on the motion during any single time interval of order $2\pi/\omega_0$, a general method for solving the equation can be given. The method, called here the averaging method, is derived in Appendix C and is used in three contexts: nonlinear oscillators in Chapter 3, perturbed planetary motion in Chapter 5, and the Foucault pendulum in Chapter 6. (Homework problems illustrate its usefulness still elsewhere, for example, in laser physics.) One can pick up the method the first time one needs it.

Every instructor uses personal experience as a guide in selecting some topics from a text and omitting others. Still, it may be useful to ask, If I were strapped for time, what would I omit? Formulating a reply is painful. Nonetheless, let me specify massive omissions (so that there may be space to reinstate a few topics). I would omit Chapter 3 on nonlinear oscillators, except for Section 3.1; Sections 4.6 and 4.7 on Hamilton's equations and Liouville's theorem; Section 5.6 on the stability of circular orbits; Sections 6.4 and 6.5 on the Foucault pendulum and the figure of the earth; Section 7.11 on the symmetric, supported top; and Sections 8.3 and 8.4 on differential cross sections.

My thanks for help and advice go to many people: to my classes at Wesleyan; to Janet Morgan, George Zepko, and David Todd in the Wesleyan Computer Center; to my colleagues Robert Behringer, Richard Lindquist, Ronald Ruby, Peter Scott, and Alan Spero. Appendix D is an adapted version of an article that Ron Ruby and I wrote; without our conversations, the appendix might never have been thought of. Peter Scott allowed me to include some of his homework problems, for which I tender my special thanks. Alan Spero offered to teach from a preliminary edition. His experience was a rich source of comments; I have incorporated many of Alan's suggestions and express here my appreciation.

A solutions manual for instructors is available from the publisher upon request.

Often readers hesitate to write the author. Please do not hesitate. I will be grateful for corrections (typographical or otherwise), for suggestions, and for comments.

Ralph Baierlein

A NOTE TO THE PROBLEM SOLVER

There are two basic approaches to solving homework problems. You can read the chapter from epigraph through the end of the assigned sections, look at the worked problems, and then tackle the homework problems. Or you can start with the homework problems and flip through the chapter for relevant equations or sections. This book is designed for the first approach.

A colleague taught from a preliminary edition; here are some of his comments, made at the semester's end.

> This is a book for reading—each chapter should be read in its entirety and ideally in one sitting. The students who realize that will benefit the most.... In the preface, tell your readers that they must read each chapter through completely first. Then they can go back and read sections, and parts of sections, when doing the problems. Maybe even advise them to make a flowchart or compendium of ideas as they read.

An entire chapter at one sitting is severe—I usually assign half a chapter at a time—but the general drift makes excellent advice: *read first*.

CHAPTER ONE

A REVIEW OF SOME BASICS

1.1 Qualitative reasoning
1.2 Vectors and the scalar product
1.3 Tossing rocks
1.4 A consequence of Newton's second law: Energy
1.5 Gravitational potential energy
1.6 Potential energy and stability
1.7 Some implications of invariance
1.8 Newton's laws: structure and meaning
1.9 Reviewing the review

If we pass in review the period in which the development of dynamics fell—a period inaugurated by Galileo, continued by Huygens, and brought to a close by Newton—its main result will be found to be the perception that bodies mutually determine in each other accelerations *dependent on definite spatial and material circumstances and that there are* masses.

Ernst Mach
The Science of Mechanics

1.1 QUALITATIVE REASONING

Suppose you drop a small rubber ball from the highest building accessible to you. At what rate does it fall, and how does the distance fallen depend on time? To answer these questions, we need to know the forces that act on the ball. As sketched in figure 1.1-1, they are gravity and air resistance. Here, of course, m denotes the ball's mass and g denotes the force per unit mass exerted gravita-

2 A REVIEW OF SOME BASICS

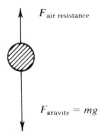

Figure 1.1-1 The forces on the ball.

tionally by the earth. The forces go into Newton's second law, which here asserts

$$\frac{d}{dt}(mv) = mg - F_{\text{air resistance}}. \tag{1.1-1}$$

The downward direction has been taken as positive.

We need to assess the force due to air resistance. Look at the issue from the point of view of momentum gain and loss. The moving ball sets air into motion, imparts momentum to the air it disturbs. At what rate does the ball transfer momentum to the ambient air?

In a time Δt, the ball sweeps out a cylindrical volume in space equal to $\pi R^2 v \, \Delta t$, as sketched in figure 1.1-2, and so the ball affects an amount of air approximately equal to $\rho_{\text{air}} \pi R^2 v \, \Delta t$. (Here R denotes the ball's radius, and ρ_{air} denotes the density of air.) To the mass of air, the ball gives a velocity of order v, and so

$$\text{Momentum given to air in } \Delta t \simeq \rho_{\text{air}} \pi R^2 v^2 \, \Delta t.$$

The gain in momentum by the air comes at the cost of momentum lost by the ball. Thus the temporal rate of momentum transfer goes into Newton's second law as $F_{\text{air resistance}}$:

$$\frac{d}{dt}(mv) = mg - \#\rho_{\text{air}} \pi R^2 v^2, \tag{1.1-2}$$

where # denotes a dimensionless constant of order 1.

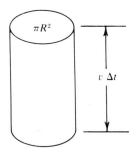

Figure 1.1-2 The volume of space swept out by the ball.

Equation (1.1-2) is our basic working differential equation. Rather than try to solve it immediately, however, we should use it first to infer the qualitative behavior of the motion. A dropped ball starts from rest and hence moves slowly at first. Because the air resistance term is quadratic in the velocity v, it is insignificant at small v; the speed must build up approximately as $v \simeq gt$. But as v grows, so does the air resistance. The net force, given by the right-hand side of equation (1.1-2), decreases toward zero; thus the rate of growth of v decreases. There is a characteristic speed for which the net force would be precisely zero:

$$v_t \equiv \left(\frac{mg}{\#\rho_{\text{air}} \pi R^2} \right)^{1/2}. \qquad (1.1\text{-}3)$$

As v approaches v_t, the net force approaches zero, implying that v will never exceed (or even reach) the "terminal velocity" v_t. The evolution of v is depicted in figure 1.1-3.

Once we know how v depends on time, at least qualitatively, we can compute the distance fallen. In the beginning, while $v \simeq gt$ holds, the distance will increase approximately as $\frac{1}{2}gt^2$, a quadratic increase with time. The situation is quite different when v gets close to its terminal value v_t and hence changes little with time. Then, with $v \simeq v_t$, distance will increase only linearly with time, though at a high rate. The full run of behavior is indicated in figure 1.1-4.

Now let us go back and solve equation (1.1-2) analytically. Once characteristic quantities have emerged, it is wise to display them prominently; so we should rewrite equation (1.1-2) in terms of v_t, the terminal velocity:

$$\frac{d}{dt}(mv) = \frac{mg}{v_t^2}(v_t^2 - v^2). \qquad (1.1\text{-}4)$$

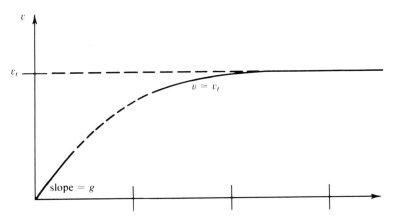

Figure 1.1-3 The evolution of v, as predicted by qualitative reasoning. The dashed, interpolating section can be inferred from the continuous decrease in slope, predicted by equation (1.1-2), as v grows toward v_t.

4 A REVIEW OF SOME BASICS

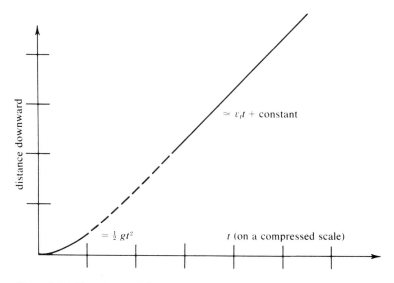

Figure 1.1-4 The distance fallen, as inferred from qualitative reasoning.

Only velocity, time, and constants appear: no function of position per se. Hence we can try to solve immediately for $v(t)$. If we transfer all dependence on v to the left-hand side, we have

$$\frac{dv}{v_t^2 - v^2} = \frac{g}{v_t^2} \, dt. \tag{1.1-5}$$

To integrate the left-hand side, we can expand in partial fractions,

$$\frac{1}{v_t^2 - v^2} = \frac{1/(2v_t)}{v_t + v} + \frac{1/(2v_t)}{v_t - v},$$

and then integrate each term to get a logarithm. Upon combining the logarithms, we arrive at

$$\frac{1}{2v_t} \ln\left(\frac{v_t + v}{v_t - v}\right) = \frac{gt}{v_t^2} + \text{constant}.$$

Because $t = 0$ is intended to imply $v = 0$ (and because $\ln 1 = 0$), the constant is actually zero. Finally, we can solve for v by transferring $2v_t$ to the right-hand side, taking anti-logarithms, and rearranging terms:

$$v = \frac{e^{2gt/v_t} - 1}{e^{2gt/v_t} + 1} v_t. \tag{1.1-6}$$

We can check this result. When $t \gg v_t/g$, the numerator and denominator are dominated by the exponential terms, and so $v \simeq v_t$. To deduce the behavior at early times, when $t \ll v_t/g$, we need the expansion of an exponential:

$$e^\alpha = 1 + \alpha + \cdots \quad \text{when } |\alpha| \ll 1.$$

(Greater detail about this expansion—and others—is provided in appendix A.)
Applying the expansion to equation (1.1-6) yields

$$v = \frac{1 + 2gt/v_t + \cdots - 1}{1 + 2gt/v_t + \cdots + 1} v_t \simeq gt.$$

Thus in both limits—long times and short—the analytic solution and our results from qualitative reasoning agree.

From equation (1.1-6) and our analysis of it we can infer the characteristic time of the physical problem:

$$\text{Characteristic time} = v_t/g \equiv \tau. \tag{1.1-7}$$

A time t is long or short according to whether $t \gg v_t/g$ or $t \ll v_t/g$ holds. One could quibble over a factor of 2 in how one should specify the characteristic time. A characteristic time is seldom uniquely specified by the problem down to factors of 2 or π; what matters is the order of magnitude, for that determines the sense of the strong inequality when one asks, Is t long or short relative to the characteristic time?

Now that we have the speed securely in hand, we can compute the distance fallen:

$$\text{Distance fallen at time } t = \int_{t'=0}^{t'=t} v(t') \, dt'$$

$$= v_t \int \frac{e^{2t'/\tau} - 1}{e^{2t'/\tau} + 1} \, dt'$$

$$= v_t \tau \int \frac{d(e^{t'/\tau} + e^{-t'/\tau})}{e^{t'/\tau} + e^{-t'/\tau}}$$

$$= v_t \tau \ln (e^{t'/\tau} + e^{-t'/\tau}) \Big|_{t'=0}^{t'=t}$$

$$= v_t \tau [\ln (e^{t/\tau} + e^{-t/\tau}) - \ln 2]. \tag{1.1-8}$$

The step from the second line to the third is not obvious: we multiply numerator and denominator by $e^{-t'/\tau}$; then we can recognize the numerator as the differential of the denominator, except for a factor of $1/\tau$ that must be supplied (and then compensated for).

Again a check is in order. When $t \gg \tau$, the argument of the logarithm is dominated by the first term, $e^{t/\tau}$; we may neglect the second term. Moreover, the first term will be much larger than 2; so we may neglect the ln 2 term. Thus the distance fallen will be approximately $v_t \tau \ln e^{t/\tau}$, which is just $v_t t$. And that must be right: when $t \gg \tau$, the ball has spent *most* of the time traveling at speeds close to the terminal velocity, and so, of course, the distance traveled is approximately $v_t t$.

6 A REVIEW OF SOME BASICS

How much of our analysis is really germane to a rubber ball dropped from a building of realistic height? That depends on the characteristic time for the ball. We can express τ as

$$\tau = \frac{v_t}{g} = \frac{[mg/(\#\rho_{air}\pi R^2)]^{1/2}}{g}$$

$$= g^{-1/2}\left[\frac{(4\pi/3)\rho_{ball}R^3}{\#\rho_{air}\pi R^2}\right]^{1/2}$$

$$\simeq g^{-1/2}\left(\frac{\rho_{ball}}{\rho_{air}}\right)^{1/2} R^{1/2}. \quad (1.1\text{-}9)$$

In a solid rubber ball the molecules are essentially in constant contact. The average distance between air molecules, however, is about 10 times the diameter of such a molecule. To speak loosely, air is like rubber expanded by a factor of 10 in all three directions; thus we can estimate the density ratio as $\rho_{ball}/\rho_{air} \simeq 10^3$. The value of g is 9.8 newtons/kilogram. (One often refers to g as "the acceleration due to gravity" and describes its units as meters per second squared. That can be misleading. Only near the start does the ball have an acceleration of about 9.8 meters/second^2; as air resistance grows, the acceleration decreases toward zero, but the gravitational force per unit mass remains 9.8 newtons/kilogram.) Thus our estimate is

$$\tau \simeq (9.8)^{-1/2}(10^3)^{1/2}R^{1/2}$$

$$\simeq 10(R \text{ in meters})^{1/2} \text{ seconds}$$

$$\simeq 2 \text{ seconds if } R = 0.03 \text{ meter},$$

to take a typical small ball.

Now let's reason this way. If there were no air resistance, in time τ the ball would fall a distance $\frac{1}{2}g\tau^2$, which amounts to about 20 meters. The actual time to go 20 meters will be somewhat longer, of course, because of the resistance. If the building's height is at least 20 meters, then the ball will spend a time longer than τ on its way to the ground, and its motion will enter the domain where resistance is significant. Whenever the height exceeds a modest number of stories, about four, an analysis like ours is not merely germane but actually mandatory.

To be useful in our analysis, figures 1.1-3 and 1.1-4 need only have been qualitatively correct sketches. But, in fact, the curves were computed from the analytic expressions in equations (1.1-6) and (1.1-8). The horizontal scale interval is τ, and the vertical scale interval in figure 1.1-4 is $v_t\tau$.

What are the working principles that one can extract from this introductory section? First, sketches are helpful. They aid the mind in seeing how far one has come, in thinking out the next steps, and—sometimes—in preventing error. Second, a physical problem usually has characteristic quantities. Sometimes they just emerge in the course of calculation; often they can be inferred at the beginning by dimensional analysis. Using the characteristic quantities (such

1.2 VECTORS AND THE SCALAR PRODUCT

The falling ball is a problem in one-dimensional motion. A more typical dynamical problem, such as the orbit of a space-craft or the irregularity in the earth's rotation, takes place in three dimensions. For such problems, vectors are an indispensable tool; a review is in order.

What Is a Vector?

To qualify as a mathematical vector, a quantity must satisfy two conditions:

1. It must have both a direction and a magnitude, and these must be independent of any specific choice of coordinate axes.
2. It must satisfy the parallelogram law of addition, as sketched in figure 1.2-1.

Vector addition is commutative and associative:

$$\mathbf{A} + \mathbf{B} = \mathbf{B} + \mathbf{A},$$

$$\mathbf{A} + (\mathbf{B} + \mathbf{C}) = (\mathbf{A} + \mathbf{B}) + \mathbf{C}.$$

The Scalar Product

The next step is to define the product of two vectors. The guiding principle is that the product should be independent of any specific choice of coordinate axes.

The *scalar product* generates, from two vectors, a scalar:

Figure 1.2-1 The parallelogram law for vector addition. The law amounts to a tail-to-head construction.

8 A REVIEW OF SOME BASICS

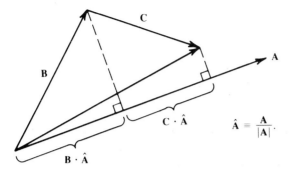

Figure 1.2-2 A start toward proving the distributive law for the scalar product. The situation is special: **A** lies in the plane defined by **B** and **C**. The dashed auxiliary lines are constructed perpendicular to **A**. The length denoted by $\mathbf{B} \cdot \hat{\mathbf{A}}$ is $|\mathbf{B}||\hat{\mathbf{A}}| \cos(\mathbf{B}, \hat{\mathbf{A}})$, that is, $|\mathbf{B}| \cos(\mathbf{B}, \hat{\mathbf{A}})$. The sum of the lengths $\mathbf{B} \cdot \hat{\mathbf{A}}$ and $\mathbf{C} \cdot \hat{\mathbf{A}}$ is equal to the length $(\mathbf{B} + \mathbf{C}) \cdot \hat{\mathbf{A}}$, which we get by projecting the vector $\mathbf{B} + \mathbf{C}$ onto **A**. This establishes distributivity in the present special case; the geometric reasoning can be extended to the general situation in three dimensions.

$$\mathbf{A} \cdot \mathbf{B} \equiv |\mathbf{A}|\,|\mathbf{B}| \cos(\mathbf{A}, \mathbf{B}), \qquad (1.2\text{-}1)$$

where $|\mathbf{A}|$ denotes the magnitude of vector **A** and $\cos(\mathbf{A}, \mathbf{B})$ denotes the cosine of the angle between **A** and **B**. By virtue of its geometric definition, the scalar product is independent of coordinate axes.

Commutativity of the scalar product follows by inspection of the definition, for it is symmetric in **A** and **B**. The product is also distributive:

$$\mathbf{A} \cdot (\mathbf{B} + \mathbf{C}) = \mathbf{A} \cdot \mathbf{B} + \mathbf{A} \cdot \mathbf{C}.$$

The start of a geometric proof is sketched in figure 1.2-2.

When two vectors are perpendicular to each other, their scalar product is zero (because the cosine of a right angle is zero). We call such vectors *orthogonal*.

A second product, the *vector product*, is reviewed in appendix B.

Unit Vectors

Vectors of unit length provide a useful basis for representing other more general vectors. Unit vectors usually are denoted by boldface characters with a circumflex. For example, a unit vector along the direction of **A** is denoted by $\hat{\mathbf{A}}$; one along the direction of the x coordinate axis, by $\hat{\mathbf{x}}$.

At times we need to work with a triplet of mutually orthogonal unit vectors. To be concise, let us simply number them 1, 2, 3 and write them as $\mathbf{e}_1, \mathbf{e}_2, \mathbf{e}_3$. With these unit vectors a circumflex would be superfluous.

1.3 TOSSING ROCKS

It is time to get some practice with vectors. A projectile problem in two dimensions is likely to be familiar and provides a good start. Suppose we are stand-

1.3 TOSSING ROCKS

ing at the edge of a pond and tossing rocks. At the time of release, the rock's speed is v_0. What angle (relative to the horizontal) will give us the most distant splash? And what is that maximum range? The situation is sketched in figure 1.3-1.

Suppose the arc of the toss goes so high that we can ignore the thrower's height in comparison: the rock starts from the coordinate origin, $x = 0$, $y = 0$, situated at the pond's edge. Moreover, let us ignore all forms of air resistance. That may not be realistic, but right now we are looking for practice, not realism. Then Newton's second law asserts simply

$$\frac{d}{dt}(m\mathbf{v}) = m\mathbf{g}. \tag{1.3-1}$$

We can represent the rock's position $\mathbf{r}(t)$ in terms of horizontal and vertical unit vectors, $\hat{\mathbf{x}}$ and $\hat{\mathbf{y}}$.

$$\mathbf{r}(t) = x(t)\hat{\mathbf{x}} + y(t)\hat{\mathbf{y}}. \tag{1.3-2}$$

This is illustrated in figure 1.3-2.

Next, figure 1.3-3 shows how to construct the velocity vector from two adjacent locations on the path and then a limit:

$$\mathbf{v} = \lim_{\Delta t \to 0} \frac{\Delta x\, \hat{\mathbf{x}} + \Delta y\, \hat{\mathbf{y}}}{\Delta t}$$

$$= \lim_{\Delta t \to 0} \left(\frac{\Delta x}{\Delta t}\hat{\mathbf{x}} + \frac{\Delta y}{\Delta t}\hat{\mathbf{y}}\right)$$

$$= \frac{dx}{dt}\hat{\mathbf{x}} + \frac{dy}{dt}\hat{\mathbf{y}}.$$

Because the numerator in the first line points along the path, the construction shows that $\mathbf{v}$ is tangent to the path.

We can streamline the route to the velocity. The numerator in the original

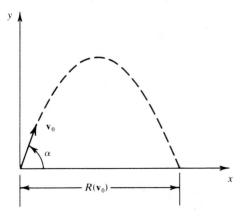

Figure 1.3-1 Tossing rocks into a pond. The range $R(\mathbf{v}_0)$ is a function of the initial velocity $\mathbf{v}_0$, which makes an angle α with the horizontal.

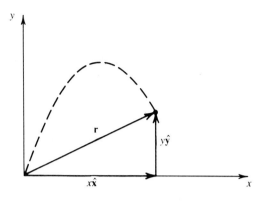

Figure 1.3-2 Representation of the rock's position.

definition is the change in location between times t and $t + \Delta t$. Thus we may write

$$\mathbf{v} = \lim_{\Delta t \to 0} \frac{\mathbf{r}(t + \Delta t) - \mathbf{r}(t)}{\Delta t}.$$

The limit is the derivative of the position vector with respect to time. The position vector for the rock has the form shown in equation (1.3-2). Since the cartesian unit vectors are constants, differentiating $\mathbf{r}$ with respect to time yields simply

$$\mathbf{v} = \frac{d\mathbf{r}}{dt} = \dot{x}\hat{\mathbf{x}} + \dot{y}\hat{\mathbf{y}}, \qquad (1.3\text{-}3)$$

where the dot denotes a time derivative: $\dot{x} = dx/dt$. (This, of course, reproduces what we found a few lines above.)

For Newton's second law, we need the derivative of the velocity. Differen-

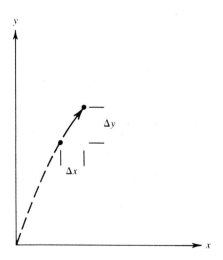

Figure 1.3-3 The two dots indicate the rock's location at times t and $t + \Delta t$.

tiating the position vector a second time and inserting it into equation (1.3-1) gives

$$m(\ddot{x}\hat{\mathbf{x}} + \ddot{y}\hat{\mathbf{y}}) = m\mathbf{g}. \tag{1.3-4}$$

From this relationship we can extract equations for x and y by taking scalar products with unit vectors. Specifically, taking the scalar product with $\hat{\mathbf{x}}$ yields

$$m(\ddot{x}\hat{\mathbf{x}} \cdot \hat{\mathbf{x}} + \ddot{y}\hat{\mathbf{x}} \cdot \hat{\mathbf{y}}) = m\hat{\mathbf{x}} \cdot \mathbf{g},$$

which reduces to

$$\ddot{x} = 0$$

when we note that $\hat{\mathbf{x}}$ is orthogonal to $\hat{\mathbf{y}}$ and $\mathbf{g}$. This tells us that the horizontal motion occurs at constant velocity; thus we can write

$$x(t) = (v_0 \cos \alpha)t \tag{1.3-5}$$

since the velocity component along $\hat{\mathbf{x}}$ is $\hat{\mathbf{x}} \cdot \mathbf{v}_0 = v_0 \cos \alpha$.

To determine the range from equation (1.3-5), we need the time of flight t_F, the time when the rock has $y = 0$ a second time. To get an equation for y, we need only take the scalar product of equation (1.3-4) with $\hat{\mathbf{y}}$:

$$m(\ddot{x}\hat{\mathbf{y}} \cdot \hat{\mathbf{x}} + \ddot{y}\hat{\mathbf{y}} \cdot \hat{\mathbf{y}}) = m\hat{\mathbf{y}} \cdot \mathbf{g},$$

which reduces to

$$\ddot{y} = -g$$

because $\mathbf{g}$ is antiparallel to $\hat{\mathbf{y}}$. One integration with respect to t yields

$$\dot{y}(t) = v_0 \sin \alpha - gt$$

since the initial velocity component along $\hat{\mathbf{y}}$ is $\hat{\mathbf{y}} \cdot \mathbf{v}_0 = v_0 \sin \alpha$. A second integration produces

$$y(t) = (v_0 \sin \alpha)t - \tfrac{1}{2}gt^2. \tag{1.3-6}$$

Setting the left-hand side equal to zero and solving for t_F yields

$$t_F = \frac{2v_0 \sin \alpha}{g}.$$

Inserting t_F into equation (1.3-5) gives the range $R(\mathbf{v}_0)$:

$$R(\mathbf{v}_0) = x(t_F) = 2\frac{v_0^2}{g} \sin \alpha \cos \alpha. \tag{1.3-7}$$

Several trajectories are plotted in figure 1.3-4. In equation (1.3-7) we have the range as a function of the tossing angle. To determine the optimum angle, we differentiate with respect to α and look for a zero slope, indicating an extreme value:

$$\frac{\partial}{\partial \alpha} R(\mathbf{v}_0) = 2\frac{v_0^2}{g} (\cos^2 \alpha - \sin^2 \alpha) = 0,$$

12 A REVIEW OF SOME BASICS

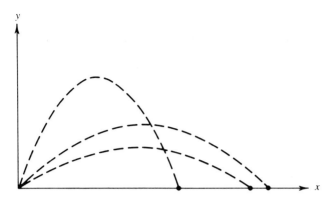

Figure 1.3-4 Several trajectories, all with the same v_0. The angles of toss are 35°, 45°, and 70°.

which implies that $\alpha = \pi/4$ (or 45°) is the optimum angle. Inserting the optimum α into the equation (1.3-7) yields

$$\text{Maximum range} = \frac{v_0^2}{g}. \tag{1.3-8}$$

Tossing up the Hill

Tossing rocks into the pond was hardly a challenge for the scalar product. Let us try tossing up a hill and ask similar questions about optimum angle and maximum range. The hill has a smooth slope, as indicated in figure 1.3-5, and rises at an angle β with respect to the horizontal.

There are two natural directions in the problem: along the hillside and the vertical, associated with **g**. They are not orthogonal directions, however, and so we should pick one and then manufacture its orthogonal companion to use in representing the rock's position vector. Since our goal is to determine the maximum range, we take a unit vector $\hat{x}$ along the hillside and a companion $\hat{y}$ perpendicular to it, as indicated in figure 1.3-5.

The position vector **r** and Newton's second law have the *same structure* as previously:

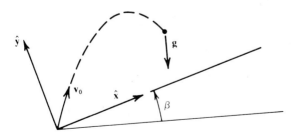

Figure 1.3-5 Tossing up the hill. The hill rises uniformly at angle β.

$$\mathbf{r} = x(t)\hat{\mathbf{x}} + y(t)\hat{\mathbf{y}},$$

$$m(\ddot{x}\hat{\mathbf{x}} + \ddot{y}\hat{\mathbf{y}}) = m\mathbf{g}. \tag{1.3-9}$$

A difference appears when we take scalar products to get equations for the new x and y. The scalar product with $\hat{\mathbf{x}}$ yields

$$m\ddot{x} = m\hat{\mathbf{x}} \cdot \mathbf{g} \neq 0, \tag{1.3-10}$$

for $\hat{\mathbf{x}}$ is no longer perpendicular to $\mathbf{g}$. Nonetheless, we have merely constant acceleration in the direction down the hillside, and so we can integrate equation (1.3-10) to get the distance:

$$x(t) = \hat{\mathbf{x}} \cdot \mathbf{v}_0 t + \tfrac{1}{2}\hat{\mathbf{x}} \cdot \mathbf{g} t^2. \tag{1.3-11}$$

Again we need the time of flight t_F, which is still the time until $y = 0$ a second time. Taking the scalar product of equation (1.3-9) with $\hat{\mathbf{y}}$ yields

$$m\ddot{y} = m\hat{\mathbf{y}} \cdot \mathbf{g},$$

which integrates to

$$y(t) = \hat{\mathbf{y}} \cdot \mathbf{v}_0 t + \tfrac{1}{2}\hat{\mathbf{y}} \cdot \mathbf{g} t^2. \tag{1.3-12}$$

The coordinate perpendicular to the hillside is zero for a second time at

$$t_F = \frac{-2\hat{\mathbf{y}} \cdot \mathbf{v}_0}{\hat{\mathbf{y}} \cdot \mathbf{g}}.$$

Inserting this time into equation (1.3-11) yields the range:

$$R(\mathbf{v}_0) = \hat{\mathbf{x}} \cdot \mathbf{v}_0 \frac{-2\hat{\mathbf{y}} \cdot \mathbf{v}_0}{\hat{\mathbf{y}} \cdot \mathbf{g}} + \tfrac{1}{2}\hat{\mathbf{x}} \cdot \mathbf{g} \left(\frac{-2\hat{\mathbf{y}} \cdot \mathbf{v}_0}{\hat{\mathbf{y}} \cdot \mathbf{g}}\right)^2. \tag{1.3-13}$$

As it stands, the expression for the range is correct, but inscrutable. At this stage we should convert to explicit trigonometric form. Figure 1.3-6 illustrates the angle γ between $\mathbf{v}_0$ and the hillside and displays all the vectors that enter into the range expression. The trigonometric identity

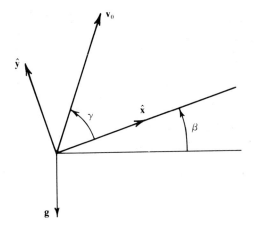

Figure 1.3-6 The four vectors that enter into the range expression.

14 A REVIEW OF SOME BASICS

$$\cos(a+b) = \cos a \cos b - \sin a \sin b$$

will come in handy. Indeed, using the figure and identity in conjunction, we can evaluate the scalar products as follows:

$$\hat{\mathbf{x}} \cdot \mathbf{v}_0 = v_0 \cos \gamma,$$

$$\hat{\mathbf{x}} \cdot \mathbf{g} = g \cos\left(\beta + \frac{\pi}{2}\right) = -g \sin \beta,$$

$$\hat{\mathbf{y}} \cdot \mathbf{v}_0 = v_0 \cos\left(\frac{\pi}{2} - \gamma\right) = v_0 \sin \gamma,$$

$$\hat{\mathbf{y}} \cdot \mathbf{g} = g \cos(\pi - \beta) = -g \cos \beta.$$

Inserting these into equation (1.3-13) yields

$$R(\mathbf{v}_0) = \frac{2v_0^2}{g}\left(\frac{\cos \gamma \sin \gamma}{\cos \beta} - \frac{\sin \beta \sin^2 \gamma}{\cos^2 \beta}\right). \tag{1.3-14}$$

As before, to determine the optimum angle, we differentiate with respect to a tossing angle and look for a zero derivative. Here we can use the angle γ:

$$\frac{\partial R}{\partial \gamma} = \frac{2v_0^2}{g}\left(\frac{\cos^2 \gamma - \sin^2 \gamma}{\cos \beta} - \frac{\sin \beta \cdot 2 \sin \gamma \cos \gamma}{\cos^2 \beta}\right)$$

$$= \frac{2v_0^2}{g}\left(\frac{\cos 2\gamma \cos \beta}{\cos^2 \beta} - \frac{\sin 2\gamma \sin \beta}{\cos^2 \beta}\right)$$

$$= \frac{2v_0^2}{g} \frac{\cos(2\gamma + \beta)}{\cos^2 \beta}.$$

The second and third lines follow from the trigonometric identities (A.2-2), (A.2-4), and (A.2-5) in appendix A. The derivative will be zero if $2\gamma + \beta = \pi/2$, and thus the optimum angle γ is provided by $\gamma = \pi/4 - \beta/2$. More meaningful is the total tossing angle relative to the horizontal:

$$\gamma + \beta = \frac{\pi}{4} + \frac{\beta}{2}. \tag{1.3-15}$$

This optimum total angle exceeds $\pi/4$, the optimum for level ground; as might be expected, one has to toss farther upward to maximize the distance up a hill.

With the optimum angle known, we can go back to equation (1.3-14) and compute the maximum range. If we use the trigonometric identities in appendix A several times, we can cast the result into the form

$$\text{Maximum range up the hill} = \frac{v_0^2/g}{1 + \sin \beta}. \tag{1.3-16}$$

The present maximum is smaller, by the divisor $1 + \sin \beta$, than the maximum for level ground.

What has this calculation to teach us? The intermediate expressions were far from transparent. But we set up the calculation easily: after we recognized the natural directions and chose orthogonal unit vectors accordingly, the next few steps followed naturally and readily. Choosing the unit vectors astutely can simplify a problem, and taking scalar products can neatly separate a vector equation into natural, manageable pieces.

1.4 A CONSEQUENCE OF NEWTON'S SECOND LAW: ENERGY

Newton's second law,

$$\frac{d}{dt}(m\mathbf{v}) = \mathbf{F}, \qquad (1.4\text{-}1)$$

is the cornerstone of our subject. Because we will refer to it often, let's agree on an abbreviation: Newton II.

Once one has Newton II, one can neatly derive general consequences by applying the same vector operations to both sides. For example, suppose we know a portion of the trajectory followed by the mass m. Let us break up the path into short directed segments $\Delta\mathbf{r}_i$, as indicated in figure 1.4-1. For each section, take the scalar product of $\Delta\mathbf{r}_i$ with Newton II, as evaluated in that section, and then sum over all N segments:

$$\sum_{i=1}^{N}\left[\frac{d(m\mathbf{v})}{dt}\right]_{(i)} \cdot \Delta\mathbf{r}_i = \sum_{i=1}^{N}[\mathbf{F}]_{(i)} \cdot \Delta\mathbf{r}_i. \qquad (1.4\text{-}2)$$

Next, multiply numerator and denominator of each term on the left side by Δt_i, the time it took the particle to move along $\Delta\mathbf{r}_i$. Now take the limit as each segment is reduced in length, while further short segments are added to continue covering the trajectory between points $\mathbf{r}_A$ and $\mathbf{r}_B$. On the left-hand side

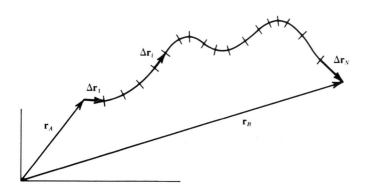

Figure 1.4-1 The mass's trajectory between points $\mathbf{r}_A$ and $\mathbf{r}_B$ is broken up into N short directed segments.

16 A REVIEW OF SOME BASICS

$$\lim_{N \to \infty} \sum_{i=1}^{N} \left[m \frac{d\mathbf{v}}{dt} \right]_{(i)} \cdot \frac{\Delta \mathbf{r}_i}{\Delta t_i} \Delta t_i = \int_{t_A}^{t_B} m \frac{d\mathbf{v}}{dt} \cdot \mathbf{v} \, dt$$

$$= \int_{t_A}^{t_B} \frac{d}{dt} \left(\frac{1}{2} m\mathbf{v} \cdot \mathbf{v} \right) dt = \frac{1}{2} mv^2 \Big|_{t_A}^{t_B}.$$

The limit of the right-hand side of equation (1.4-2) defines the *line integral* of the force—the actual force acting on the mass—taken along the actual path. Thus

$$\tfrac{1}{2} m v_B^2 - \tfrac{1}{2} m v_A^2 = \int_{r_A, \text{ actual path}}^{r_B} \mathbf{F} \cdot d\mathbf{r}. \qquad (1.4\text{-}3)\bigstar$$

The words used to describe this result are, no doubt, familiar: the change in kinetic energy $\tfrac{1}{2} m v^2$ is equal to the work done by the force actually acting on the mass. (Note what the scalar product on the right implies: only the portion of **F** along the direction of $d\mathbf{r}$ does work and affects the kinetic energy.)

The force **F** may be complicated. In addition to depending on position, the force may depend on the velocity of the mass, as with magnetic or viscous forces. Or the force may vary, at a fixed location, with time if, say, someone moves the end of a spring. Often, however, the situation is simpler than the most general case. Let us suppose, and specify, the following:

1. The force **F** is a function of position only.
2. The line integral of $\mathbf{F}(\mathbf{r})$ between any two specified points is independent of the path chosen to connect those two points.

The second condition is illustrated in figure 1.4-2.

Given conditions 1 and 2, we can define unambiguously a scalar function of position

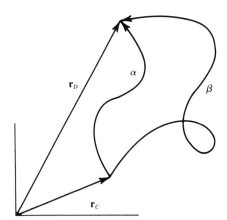

Figure 1.4-2 The line integral of **F** between points $\mathbf{r}_C$ and $\mathbf{r}_D$ is the same for path α as for path β and, indeed, for any path connecting the two points.

1.4 A CONSEQUENCE OF NEWTON'S SECOND LAW: ENERGY

$$U(\mathbf{r}) \equiv -\int_{r_A}^{r} \mathbf{F}(\mathbf{r}') \cdot d\mathbf{r}' + U(\mathbf{r}_A), \qquad (1.4\text{-}4)\bigstar$$

such that the right-hand side of equation (1.4-3) may be written

$$\int_{r_A,\text{ actual path}}^{r_B} \mathbf{F} \cdot d\mathbf{r} = -[U(\mathbf{r}_B) - U(\mathbf{r}_A)]. \qquad (1.4\text{-}5)$$

Because equations (1.4-3) and (1.4-5) now imply

$$\tfrac{1}{2} m v_B^2 + U(\mathbf{r}_B) = \tfrac{1}{2} m v_A^2 + U(\mathbf{r}_A),$$

based on faith or trust

we call the scalar $U(\mathbf{r})$ the potential energy. The fiducial point $\mathbf{r}_A$ and $U(\mathbf{r}_A)$ may be chosen freely. Once we have chosen them, however, $U(\mathbf{r})$ is firmly determined.

The first worked problem at the chapter's end provides an example of how the line integral in equation (1.4-4) can be evaluated to produce U.

F and grad U

To establish a further connection between U and $\mathbf{F}$, we can look at the contours of constant potential energy. For ease in representation, specify that the motion is in two dimensions. Then we might have contours like those in figure 1.4-3. At every point, the force $\mathbf{F}$ is perpendicular to the contour of constant potential energy passing through the point. The proof is by contradiction. Suppose $\mathbf{F}$ had a component along the contour; then, by equation (1.4-4), U would change if $d\mathbf{r}'$ were taken along that component, which is contrary to assumption, namely, that the contour is one of constant potential energy.

Next we need the geometric definition of the *gradient* of a scalar like U. In the neighborhood of a given point, we search for the direction that gives the

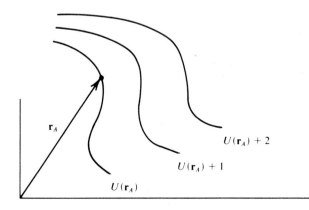

Figure 1.4-3 Contours of constant potential energy might look like this (in two dimensions).

maximum spatial rate of change in U. We might denote the rate itself by $(dU/ds)_{\max}$, where ds denotes the infinitesimal distance to a nearby point and dU is the associated change in U. Then the gradient of U, abbreviated grad U, is defined by

$$\text{grad } U \equiv \begin{array}{c} \text{the vector whose magnitude is } (dU/ds)_{\max} \\ \text{and whose direction is that of the} \\ \text{maximum spatial rate of change.} \end{array}$$

Note that the vector grad U is also perpendicular to the contours of constant potential energy. (A proof is sketched in figure 1.4-4.) Since both **F** and grad U are perpendicular to the contours, they must be parallel to each other or antiparallel. To compare magnitudes, let us take the integration vector $d\mathbf{r}'$ in equation (1.4-4) to be parallel to **F**. Then that equation implies, in a loose notation, that

$$\Delta U = -|\mathbf{F}| \, |\Delta \mathbf{r}|.$$

Because the right-hand side is negative, U decreases in the direction of **F**, and so grad U must be antiparallel to **F**. If we take the absolute value of both sides and divide by $|\Delta \mathbf{r}|$, we deduce that $|\text{grad } U| = |\mathbf{F}|$. The upshot of the analysis is that

$$\mathbf{F} = -\text{grad } U. \qquad (1.4\text{-}6)\bigstar$$

There is a way to visualize the result. Take two-dimensional motion, but imagine plotting U vertically, above the plane of motion, in the fashion of a relief map. The associated contours are sketched in figure 1.4-5. In the vicinity of a potential "hill," grad U will point toward the top of the hill. The force **F** will point in the opposite direction—downhill—telling us that the actual push is toward the valley.

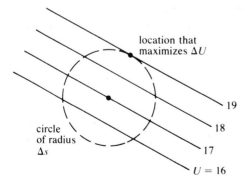

Figure 1.4-4 On a sufficiently fine scale, the contours of constant potential energy are essentially straight. Around the point in question, draw a circle of fixed small radius Δs, and ask where on the circle you should go to maximize ΔU. As the sketch shows, the direction of that location, relative to the central point, is perpendicular to the contours, and hence so is grad U.

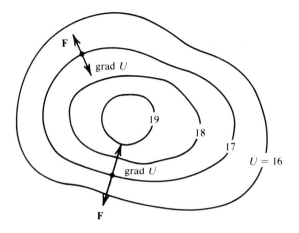

Figure 1.4-5 Looking down on two-dimensional motion: an illustration of $\mathbf{F} = -\text{grad}\ U$. (The name *gradient* for our vector function comes from this kind of picture. The steepness of the hill—the gradient of the ascent—determines grad U and gives the function its name.)

Cartesian Representation

Without much trouble we can generate an expression for grad U in terms of partial derivatives and cartesian unit vectors. Because that may be more familar, the connection is worth making. The first step is to express the spatial rate of change of U in an arbitrarily chosen direction $\hat{\mathbf{e}}$ in terms of grad U. With the aid of figure 1.4-6, we can reason that, for a fixed change ΔU, we must go farther along $\hat{\mathbf{e}}$ than along grad U; the factor is precisely $1/\cos \theta$. Thus we can say that

$$\text{Rate of change in } U \text{ along } \hat{\mathbf{e}} = \frac{|\text{grad}\ U|}{1/\cos\theta} = \hat{\mathbf{e}} \cdot \text{grad}\ U. \quad (1.4\text{-}7)$$

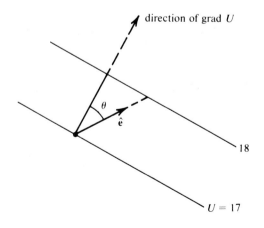

Figure 1.4-6 The distance from $U = 17$ to $U = 18$ is farther along $\hat{\mathbf{e}}$ than along grad U by the ratio $1:\cos \theta$.

If we take $\hat{\mathbf{e}}$ to be a unit vector along the x axis in a cartesian coordinate system, then the left-hand side in equation (1.4-7) is just $\partial U/\partial x$, and so

$$\frac{\partial U}{\partial x} = \hat{\mathbf{x}} \cdot \text{grad } U.$$

Similar expressions hold for $\hat{\mathbf{y}}$ and $\hat{\mathbf{z}}$. Thus we must be able to write

$$\text{grad } U = \frac{\partial U}{\partial x} \hat{\mathbf{x}} + \frac{\partial U}{\partial y} \hat{\mathbf{y}} + \frac{\partial U}{\partial z} \hat{\mathbf{z}}. \qquad (1.4\text{-}8)$$

1.5 GRAVITATIONAL POTENTIAL ENERGY

Some examples of potential energy computations are certainly in order. Gravity provides a context that is familiar and often tractable; we will focus on it.

The properties of a *spherically symmetric* mass distribution are probably familiar; the essentials are simply enumerated here.

1. *Point mass M.* The gravitational force exerted by a point mass M, located at the origin, on another point mass m is radially inward and inversely proportional to the square of the distance:

$$\mathbf{F} = -\frac{GMm}{r^2} \hat{\mathbf{r}}, \qquad (1.5\text{-}1)$$

where G is the newtonian gravitational constant, equal to 6.67×10^{-11} meter³/(kilogram · second²) or newton · meter²/kilogram.

The (mutual) potential energy is

$$U(r) = -\frac{GMm}{r}, \qquad (1.5\text{-}2)$$

provided we adopt the convention that U should go to zero as the separation r goes to infinity.

2. *An extended mass.* Let the point mass m be located (at least in our imagination) at a point $\mathbf{r}$ inside the extended mass M, as in figure 1.5-1. The net force on m arises solely from the mass distribution at radial distances less than or equal to r, and that mass acts as though it were concentrated at the center. The mass arranged at radial distances greater than r produces forces that mutually cancel at point $\mathbf{r}$ and hence may be ignored. We may describe the situation symbolically by writing

$$\mathbf{F}(r) = -\frac{G\mathfrak{M}(r)m}{r^2} \hat{\mathbf{r}}, \qquad (1.5\text{-}3)$$

where $\mathfrak{M}(r)$ denotes the amount of mass at radial distances less than or equal to r.

The potential energy can be calculated as an integral; we need only in-

1.5 GRAVITATIONAL POTENTIAL ENERGY

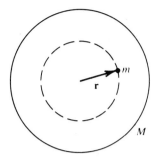

Figure 1.5-1 The mass m is at a point r inside the extended mass M.

voke the general expression (1.4-4) and insert **F** from equation (1.5-3):

$$U(r) = -\int_{r'=\infty}^{r'=r} \mathbf{F}(\mathbf{r}') \cdot d\mathbf{r}' + U(\infty)$$

$$= +\int_{\infty}^{r} \frac{G\mathfrak{M}(r')m}{r'^2} dr' + 0$$

$$= -\int_{r}^{\infty} \frac{G\mathfrak{M}(r')m}{r'^2} dr', \qquad (1.5\text{-}4)$$

where the step to the third line is just a rearrangement of the integration limits.

3. *A sphere of constant density.* We need an example to bring to life the abstractions. A sphere of constant density provides a tractable extended mass: there is constant density ρ out to a boundary radius R and then vacuum beyond. Then the amount of mass $\mathfrak{M}(r)$ inside radius r is given by

$$\mathfrak{M}(r) = \begin{cases} \dfrac{4\pi}{3} r^3 \rho & \text{if } r \leq R, \\[2mm] \dfrac{4\pi}{3} R^3 \rho \equiv M & \text{if } r \geq R. \end{cases}$$

This permits us to evaluate $\mathbf{F}(r)$ via equation (1.5-3); the result is illustrated in figure 1.5-2.

The computation for the potential energy $U(r)$ depends, in its details, on whether $r < R$ or $r > R$ pertains. Take the former first. Because $\mathfrak{M}(r)$ changes functional form at the boundary, we split the range of integration into two pieces: the interval out to R and then the interval from R to infinity. Thus

$$U(r) = -\int_{r}^{R} \frac{G(4\pi/3)r'^3\rho m}{r'^2} dr' - \int_{R}^{\infty} \frac{GMm}{r'^2} dr'$$

$$= -\frac{4\pi}{3} G\rho m \cdot \frac{1}{2}(R^2 - r^2) - \frac{GMm}{R}.$$

22 A REVIEW OF SOME BASICS

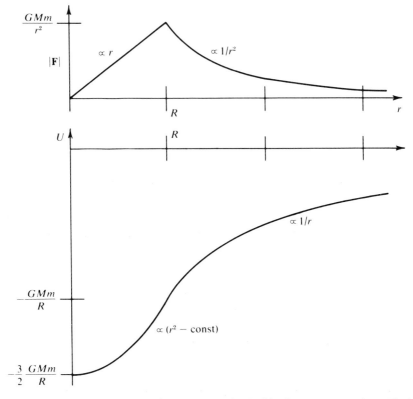

Figure 1.5-2 The force and potential energy associated with a homogeneous sphere. The boundary is at radius R.

The potential energy grows as r^2 as long as $r < R$. Its value at the boundary is simply $-GMm/R$, and so $U(r)$ matches smoothly to the functional form that holds for $r \geq R$, namely, $-GMm/r$, which follows from

$$U(r) = -\int_r^\infty \frac{GMm}{r'^2}\, dr' \qquad \text{if } r \geq R$$

$$= -\frac{GMm}{r}.$$

The full run of $U(r)$ is shown in figure 1.5-2. The force, remember, is always radially inward. The closer m is to the center, the larger the energy that must be expended to pull m out to infinity. The increasing depth of the potential energy "well" reflects this fact.

In this section all the computations are restricted to spherically symmetric mass distributions. To be sure, in part 2 the radial dependence of the density was unrestricted—the density could vary wildly as a function of r—but spherical symmetry was presupposed. That extreme symmetry is relaxed in the next section.

1.6 POTENTIAL ENERGY AND STABILITY

The new context is illustrated in figure 1.6-1. The massive ring has a radius a and a thickness much less than a, so that the ring is drawn simply as a circle. Interacting with the ring is a small mass m, located at a point with position vector $\mathbf{r}$. We may vary the location where we imagine m is placed—and with no velocity, at least initially.

If we place m at the ring's center, there is no *net* gravitational force on it, by the symmetry of the situation. After all, in which direction could a nonzero $\mathbf{F}$ choose to point?

What if we displace m a little, so that $r > 0$ but still $r \ll a$? Will a force act? Will it tend to return m to the center? For all possible small displacements? This is the issue of *stability* for the central location.

We can try some qualitative reasoning. Suppose we displace m upward along the symmetry axis $\mathbf{e}_3$. Every mass element in the ring will exert on m an attractive force that is partly downward and partly radially outward. When we sum over all those forces, the radial parts cancel (by symmetry), but the downward components add constructively. The net result will be a downward force: a restoring force. Good.

Now try displacing m in the plane of the ring, as sketched in figure 1.6-2. The mass elements that lie above the dashed line through m tend to restore m to the center. The elements that lie below tend to increase the displacement. Which elements win? The elements that tend to restore are larger in number

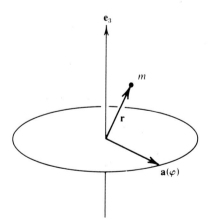

Figure 1.6-1 A point mass m interacting with a ring of radius a and mass M. The position vector $\mathbf{r}$ originates from the ring's center.

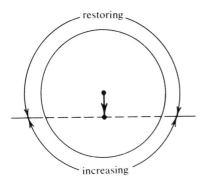

Figure 1.6-2 The view looking down, when mass m is displaced in the plane of the ring.

(and hence in aggregate mass), but they are farther away. It is hard to say. (We could compare the ring context with a hollow sphere, where the gravitational force is zero everywhere inside, but it is time to move on.)

To answer the stability question definitively, we can calculate the potential energy $U(\mathbf{r})$, at least for $r \ll a$. Since we do not know the force as a function of position, we can not compute $U(\mathbf{r})$ by equation (1.4-4). Another route, however, is available. The gravitational force between m and any little section of the ring is like that between two point masses. The associated potential energy must have the form

$$-\frac{G \text{ (mass of ring section)} m}{\text{Distance between them}}.$$

All we need to do is to add all the contributions from around the ring. That is accomplished by an integration.

We are appealing here to the *structure* of equation (1.4-4). If the force $\mathbf{F}$ is the sum of two other forces, $\mathbf{F}_1$ and $\mathbf{F}_2$, say, then the integral can be split into two integrals and the constant can be split also. Each pair (an integral plus a constant) gives the potential energy, U_1 or U_2, say, associated with the corresponding force. The full potential energy is the sum, $U_1 + U_2$. In short, we may add potential energies. Since an energy has only a magnitude and a sign, not a direction in space, adding energies is easier than adding forces, which are vectors.

For the integration, we need some way of specifying location around the ring; we can use an angle φ, measured from some fiducial line in the plane. The mass associated with an angular increment $d\varphi$ is the fraction $d\varphi/(2\pi)$ of the total mass M. The location of the mass element, relative to the origin, can be denoted by a vector $\mathbf{a}(\varphi)$; the vector has magnitude a and points toward the mass element, as in figure 1.6-1. Each mass element contributes to $U(\mathbf{r})$ in inverse proportion to its distance from m. Thus we may write

$$U(\mathbf{r}) = -\int_{\varphi=0}^{2\pi} d\varphi \; G \; \frac{M}{2\pi} \; m \; \frac{1}{|\mathbf{a}(\varphi) - \mathbf{r}|}. \quad (1.6\text{-}1)$$

1.6 POTENTIAL ENERGY AND STABILITY

To make the integration tractable, we need to work out the relative separation:

$$|\mathbf{a}(\varphi) - \mathbf{r}| = [(\mathbf{a} - \mathbf{r}) \cdot (\mathbf{a} - \mathbf{r})]^{1/2}$$
$$= (a^2 - 2\mathbf{a} \cdot \mathbf{r} + r^2)^{1/2}$$
$$= a\left(1 - 2\hat{\mathbf{a}} \cdot \frac{\mathbf{r}}{a} + \frac{r^2}{a^2}\right)^{1/2}. \tag{1.6-2}$$

The last step is anticipatory: we need the potential energy only when $r \ll a$, and so we can afford to approximate, using $r/a \ll 1$ as the basis.

The binomial expansion, with exponent $-\frac{1}{2}$, is

$$(1 + \alpha)^{-1/2} = 1 - \tfrac{1}{2}\alpha + \tfrac{3}{8}\alpha^2 + \cdots,$$

provided $|\alpha| < 1$. (More about such an expansion appears in appendix A.) We may regard

$$-2\hat{\mathbf{a}} \cdot \frac{\mathbf{r}}{a} + \frac{r^2}{a^2}$$

in equation (1.6-2) as α. To get all terms in the integrand that are of order r^2/a^2, we need to go to quadratic order in "α":

$$U(\mathbf{r}) = -\frac{GMm}{2\pi} \int d\varphi \, \frac{1}{a}\left\{1 - \frac{1}{2}\left[-2\hat{\mathbf{a}} \cdot \frac{\mathbf{r}}{a} + \frac{r^2}{a^2}\right] + \frac{3}{8}\left[-2\hat{\mathbf{a}} \cdot \frac{\mathbf{r}}{a} + \frac{r^2}{a^2}\right]^2 + \cdots\right\}$$
$$= -\frac{GMm}{2\pi a} \int d\varphi \left\{1 + \frac{r}{a}\hat{\mathbf{a}} \cdot \hat{\mathbf{r}} + \frac{r^2}{2a^2}[-1 + 3(\hat{\mathbf{a}} \cdot \hat{\mathbf{r}})^2] + \cdots\right\}. \tag{1.6-3}$$

We are free to choose orthogonal unit vectors $\mathbf{e}_1$ and $\mathbf{e}_2$ in the plane as we wish. To simplify the integrations, we choose $\mathbf{e}_1$ to lie along the projection of $\mathbf{r}$ onto the plane, as shown in figure 1.6-3. Then we have

$$\mathbf{r} = x\mathbf{e}_1 + (0)\mathbf{e}_2 + z\mathbf{e}_3. \tag{1.6-4}$$

If we take the fiducial line for the angle φ to be $\mathbf{e}_1$, then

$$\mathbf{a}(\varphi) = a(\cos\varphi \, \mathbf{e}_1 + \sin\varphi \, \mathbf{e}_2). \tag{1.6-5}$$

The scalar product that we need reduces to

$$\hat{\mathbf{a}} \cdot \hat{\mathbf{r}} = \frac{x}{r}\cos\varphi.$$

And the integrals are

$$\int_0^{2\pi} d\varphi \, \hat{\mathbf{a}} \cdot \hat{\mathbf{r}} = \frac{x}{r}\int_0^{2\pi} d\varphi \cos\varphi = 0,$$

$$\int_0^{2\pi} d\varphi (\hat{\mathbf{a}} \cdot \hat{\mathbf{r}})^2 = \frac{x^2}{r^2}\int_0^{2\pi} d\varphi \cos^2\varphi = \frac{2\pi}{2}\frac{x^2}{r^2}.$$

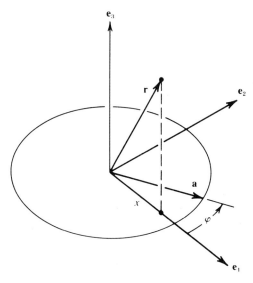

Figure 1.6-3 A good choice for the direction of e_1 is along the projection of **r** onto the plane.

According to equation (1.6-4), $x^2 = r^2 - z^2$. Moreover, $z = \mathbf{e}_3 \cdot \mathbf{r}$. Thus we can write

$$\frac{x^2}{r^2} = \frac{r^2 - (\mathbf{e}_3 \cdot \mathbf{r})^2}{r^2} = 1 - (\mathbf{e}_3 \cdot \hat{\mathbf{r}})^2$$

and remove every trace of the special choice we made for axes in the plane. Inserting the various results into equation (1.6-3) and collecting terms produces

$$U(\mathbf{r}) = -\frac{GMm}{a}\left\{1 + \frac{r^2}{4a^2}[1 - 3(\mathbf{e}_3 \cdot \hat{\mathbf{r}})^2] + \cdots\right\}. \qquad (1.6\text{-}6)$$

With the potential energy in hand, we can return to the stability question. Let us reconsider the two displacements that we examined previously. A displacement upward along the symmetry axis implies $\mathbf{e}_3 \cdot \hat{\mathbf{r}} = +1$. Then

$$U = -\frac{GMm}{a} + \frac{GMm}{2a^3} r^2 + \cdots,$$

and so the potential energy increases. The associated force is directed back to the center—a restoring force, just as we reasoned.

Next, suppose the displacement is in the plane. Then $\mathbf{e}_3 \cdot \hat{\mathbf{r}} = 0$ and

$$U = -\frac{GMm}{a} - \frac{GMm}{4a^3} r^2 + \cdots.$$

Now the potential energy decreases; the associated force is directed radially outward, tending to increase the displacement.

The question of stability is usually understood to ask, Is there a restoring force for all possible small displacements? We have already found that for all displacements in the plane, there is no restoring force, indeed, quite the contrary. Thus we can answer our primary question: no, the center is not a stable location.

Moreover, we can see the general principle at work here. If a location is to be a stable equilibrium point, the potential energy U must have a minimum there. Equilibrium by itself requires merely that the force be zero and hence that $-\text{grad } U = 0$, which implies that the first derivative of U in any direction is zero. Stability requires that every small displacement lead to a restoring force. Only a rise in U in every direction will guarantee that, whence the minimum property is the succinct requirement for a stable equilibrium.

There is a point worth noting before we go on. If the mass m were far from the ring, then we would need an expansion for $U(\mathbf{r})$ that was valid when $r \gg a$. We could go back and adapt our method, using the inequality $r \gg a$ as the basis for an expansion. Such a potential energy could describe the *extra* potential energy resulting from the bulge if we approximated the earth's mass distribution by a sphere plus a ring at the equator. It could do the same for the (barely perceptible) solar bulge. In chapter 5, when we study planetary orbits, we will use the expansion for that purpose.

1.7 SOME IMPLICATIONS OF INVARIANCE

Let us write Newton II as

$$\frac{d}{dt}(m\mathbf{v}) = -\text{grad } U, \qquad (1.7\text{-}1)$$

making the explicit restriction (in this section) to forces that can be derived from a scalar potential energy via $\mathbf{F} = -\text{grad } U$. What changes from problem to problem is the function U; therein reside the details. Nonetheless, we can make some statements of substantial generality because U often has one or more invariance properties.

That proposition, of course, provokes questions. What is an invariance property? And what is the dynamical implication of each specific invariance property? The best route to answers is by working out some examples.

Translational Invariance

Suppose that, in some specific cartesian coordinate system, the function U does not depend on y, that is, $\partial U/\partial y = 0$ everywhere. We can say that U is invariant under *translations* along y, the direction of the y axis.

To extract a dynamical implication, we can take the scalar product of Newton II with $\hat{\mathbf{y}}$:

$$\hat{\mathbf{y}} \cdot \frac{d(m\mathbf{v})}{dt} = -\hat{\mathbf{y}} \cdot \text{grad } U.$$

The scalar product of $\hat{\mathbf{y}}$ with grad U isolates $\partial U/\partial y$, as a look back at equation (1.4-8) will confirm, and so we find

$$\frac{d}{dt}(\hat{\mathbf{y}} \cdot m\mathbf{v}) = -\frac{\partial U}{\partial y} = 0,$$

that is,

$$\hat{\mathbf{y}} \cdot m\mathbf{v} = \text{constant}. \qquad (1.7\text{-}2)$$

The component of linear momentum along the translation direction is constant, is conserved in time. We can put the implication succinctly: *translation invariance implies momentum conservation.*

The most common example would be U for a ball near the surface of the earth: $U(\mathbf{r}) = mgz$, with z being distance from the ground and $U(0) = 0$ by a revised convention. This U is invariant under translation in any horizontal direction, implying that any initial horizontal momentum component is conserved in time.

Rotational Invariance

Suppose that U does not change as one goes around the $\hat{\mathbf{z}}$ axis at any fixed height and radial distance. This is illustrated in figure 1.7-1. The tangent to such a circle is proportional to $\hat{\mathbf{z}} \times \mathbf{r}$. Thus

$$(\hat{\mathbf{z}} \times \mathbf{r}) \cdot \text{grad } U = 0. \qquad (1.7\text{-}3)$$

We can say that U is invariant under *rotations* about the $\hat{\mathbf{z}}$ axis.

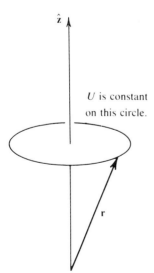

Figure 1.7-1 The function U is constant on the circle shown. On any other similar circle that is generated by rotating the position vector $\mathbf{r}$ around $\hat{\mathbf{z}}$, the value of U is also assumed to be constant. The numerical value of the constant may differ from circle to circle, however.

As a preliminary to extracting the dynamical implications, we can rearrange the triple product:

$$(\hat{\mathbf{z}} \times \mathbf{r}) \cdot \text{grad } U = \hat{\mathbf{z}} \cdot (\mathbf{r} \times \text{grad } U). \tag{1.7-4}$$

significant

(The salient properties of the vector product are reviewed in appendix B.) The $\mathbf{r} \times$ form in equation (1.7-4) suggests that we take the vector product of Newton II with $\mathbf{r}$:

$$\mathbf{r} \times \frac{d}{dt}(m\mathbf{v}) = \mathbf{r} \times (-\text{grad } U);$$

$$\frac{d}{dt}\left[\mathbf{r} \times m\mathbf{v}\right] - \frac{d\mathbf{r}}{dt} \times m\mathbf{v} = \mathbf{r} \times (-\text{grad } U). \tag{1.7-5}$$

The second term on the left-hand side, equivalent to $\mathbf{v} \times m\mathbf{v}$, is identically zero. The remaining terms say that the time derivative of the angular momentum $\mathbf{r} \times m\mathbf{v}$ is equal to the applied torque. If we take the scalar product with $\hat{\mathbf{z}}$ and then heed equations (1.7-3) and (1.7-4), we find

$$\frac{d}{dt}[\hat{\mathbf{z}} \cdot (\mathbf{r} \times m\mathbf{v})] = 0,$$

that is,

$$\hat{\mathbf{z}} \cdot (\mathbf{r} \times m\mathbf{v}) = \text{constant}. \tag{1.7-6}$$

Thus the component of angular momentum along the rotation axis is conserved in time. Again, we can put the implication succinctly: *rotational invariance implies conservation of angular momentum.*

A potential energy with rotational invariance appeared already in section 1.6: the $U(\mathbf{r})$ for the ring is invariant under rotations about the axis $\mathbf{e}_3$; the symmetry evident in figure 1.6-1 says it must be, as well as the explicit expression in equation (1.6-6). Electrostatics provides another example. A long, straight insulating rod, charged uniformly along its length and oriented along $\hat{\mathbf{z}}$, produces a potential energy of the form $-aF_0 \ln[(x^2 + y^2)^{1/2}/a]$. Here a denotes the rod's radius, and the constant F_0 accounts for a product of charges. The dependence on x and y through only the combination $x^2 + y^2$ is what ensures the rotational invariance about $\hat{\mathbf{z}}$. Finally, the gravitational potential energies of section 1.5, arising from spherically symmetric mass distributions, depend on position through only the magnitude r of $\mathbf{r}$. Thus they are invariant under rotations about any axis; consequently, all components of the angular momentum vector are conserved.

Invariance under Time Translation

Suppose that U does not depend on time explicitly: $\partial U/\partial t = 0$. A suitable example is $U = mgz$.

A potential energy that would *not* meet the supposition is $U = -xF_0 \cos \omega t$.

30 A REVIEW OF SOME BASICS

The negative gradient of that function yields a force $F_0 \cos \omega t \, \hat{x}$, which might describe the electric force produced by an oscillating electric field. The function $-xF_0 \cos \omega t$ is acceptable as a potential energy in the sense that its negative gradient yields the relevant force, which explicitly depends on time. Here we relax the conditions imposed in section 1.4, where U was introduced via a line integral of the actual force. We are saying, we will call a function the potential energy as long as its negative gradient produces the desired force; no further demands will be made. In the context of that generalization, it is readily possible for $\partial U/\partial t$ not to be zero, as in the example of this paragraph. But back to our supposition.

If $\partial U/\partial t = 0$, we can say that U is invariant under *translation in time*.

To extract the dynamical implication, it will suffice to work in one dimension. Let $x(t)$ denote the position as it evolves according to Newton II. As for U, we do not yet specify that U has no explicit time dependence; rather, we take the general form $U(x, t)$, with the sole requirement that $-\partial U/\partial x$ yield the actual force.

If we differentiate the sum of kinetic and potential energies, we find

$$\frac{d}{dt}\left[\frac{1}{2} m\dot{x}^2 + U(x, t)\right] = m\dot{x}\frac{d\dot{x}}{dt} + \frac{\partial U}{\partial x}\frac{dx}{dt} + \frac{\partial U}{\partial t}$$

$$= \left(m\frac{d\dot{x}}{dt} + \frac{\partial U}{\partial x}\right)\dot{x} + \frac{\partial U}{\partial t}$$

$$= \frac{\partial U}{\partial t}. \qquad (1.7\text{-}7)$$

By virtue of Newton II, the factor multiplying $\dot{x}$ in the second line is zero, and so the step to the last line follows. If $\partial U/\partial t$ is indeed zero, then the sum of the kinetic and potential energies is constant in time. Succinctly, *invariance under time translation implies energy conservation.*

1.8 NEWTON'S LAWS: STRUCTURE AND MEANING

We have been able to do several calculations and to develop some theorems without ever examining the foundations of newtonian dynamics. That is typical of physics. One can know how to do a calculation without knowing—in clear and concise fashion—the logical underpinnings. Indeed, progress in physics would be slow if the situation were otherwise.

It is time, however, to examine the foundations. They are Newton's three laws, which we may take in the following form.

Newton I When no external forces act on a body, its velocity—as measured in an inertial frame of reference—remains constant.

1.8 NEWTON'S LAWS: STRUCTURE AND MEANING

Newton II In an inertial frame of reference, the time rate of change of momentum $m\mathbf{v}$ is equal to the net force on the body:

$$\frac{d}{dt}(m\mathbf{v}) = \mathbf{F}.$$

Newton III Whenever two bodies interact, they exert on each other forces that are equal in magnitude but opposite in direction:

$$\mathbf{F}_{2 \text{ on } 1} = -\mathbf{F}_{1 \text{ on } 2}.$$

The phrase *inertial frame of reference* appears prominently. How does one define an inertial frame of reference? How does one give the phrase operational meaning?

Einstein and Infeld asked those questions, too. Let us see how they proceeded in *The Evolution of Physics*.*

> There still remains one point to be cleared up. One of the most fundamental questions has not been settled as yet: does an inertial system exist? ... We have the laws but do not know the frame to which to refer them.
>
> In order to be more aware of this difficulty, let us interview the classical physicist and ask him some simple questions:
> "What is an inertial system?"
> "It is a CS [Coordinate System] in which the laws of mechanics are valid. A body on which no external forces are acting moves uniformly in such a CS. This property thus enables us to distinguish an inertial CS from any other."
> "But what does it mean to say that no forces are acting on a body?"
> "It simply means that the body moves uniformly in an inertial CS."
> Here we could once more put the question: "What is an inertial CS?" But since there is little hope of obtaining an answer differing from the above, ...

Here we are presented with a neat logical circle:

Inertial frame: An inertial frame is a frame in which a body moves with constant velocity if no force acts on that body.
No force (or no net force): No force acts on a body if it moves with constant velocity in an inertial frame.

Is there no way to break the circle? Perhaps we can define a special case of no net force without reference to an inertial frame. The prescription might be

---
*Albert Einstein and Leopold Infeld, *The Evolution of Physics* (Simon and Schuster, New York, 1961).

this: Get far away from all nearby bodies, and locate yourself symmetrically relative to the large-scale universe. Our experience is that physical forces diminish with distance; that plus symmetry may suffice to establish a "no net force" situation. Then we may turn Newton I into a definition of an inertial frame.

If we can establish one frame as an inertial frame, we are all set. We can test any other frame by asking, Does it move with constant velocity relative to the established inertial frame? If yes, then it, too, is an inertial frame. If no, then it is not.

Whether this strategy is logically sound is debatable. Certainly it is not practical. What, indeed, does the practical person say?

Structural engineers say that the earth is an inertial frame, as far as they are concerned. Meteorologists say no, the trade winds and a hurricane are understandable only if we recognize that the earth rotates relative to an ideal inertial frame. A frame fixed with respect to the nearby stars will, however, serve as an inertial frame for meteorologists. But radio astronomers, mapping the Milky Way by the radio emission of hydrogen or carbon monoxide, find that the galaxy rotates relative to an ideal inertial frame. For them, the local stars do not suffice as an inertial frame, but a grid based on the nearby galaxies would be sufficiently inertial.

Here we have various approximations to an ideal inertial frame, indeed, a hierarchy of them. The practical person merely picks a frame of suitable quality and continues working. Whether the ideal inertial frame even exists is a question that the practical person can defer.

We defer the question also. The issues have been raised. The notion of an inertial frame is a topic more for discussion than for definitive exposition, and that is true for all three of Newton's laws. Within the community of physicists and philosophers, there is no unanimity on how the foundations should be conceived.

Now we go on to Newton's second law. We suppose we have at least a practical approximation to an inertial frame. Then Newton II describes the effect of a physical force when it is observed in an inertial frame. If we want to, we may regard Newton II as defining the force $\mathbf{F}$. The usefulness of Newton II then lies in the empirical property that forces can be constructed out of positions and velocities, with no need for higher derivatives; and after the forces have been constructed in one context, they may be used in another.

We have, however, a natural sense of a force as a push or a pull. We push a swing or pull a stake out of the ground and know that we are exerting a force. The import of Newton II is clearer, then. The physical force determines $d(m\mathbf{v})/dt$, rather than some other kinematic quantity, such as $d^3(mv^4\mathbf{v})/dt^3$, which is abstractly conceivable.

Newton's third law was an excellent approximation for the physics of his day. When objects are sharply accelerated or move at high speed, however, the approximation breaks down. An electron shot close to an atomic nucleus will

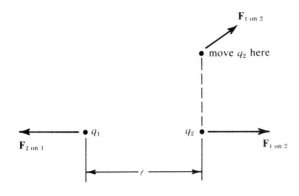

Figure 1.8-1 A violation of Newton III (for a time of order ℓ/c).

be abruptly accelerated and will emit electromagnetic radiation, a photon, say. Momentum is conserved, and the photon itself carries off some momentum. The sum of the electron and nuclear momenta must change, but that could not occur if Newton III were literally true.

The fields—electromagnetic and gravitational—may absorb and contribute momentum, and thereby they invalidate Newton III, which considers only the masses as carriers of momentum.

Figure 1.8-1 shows another instance in which Newton III breaks down. The two positive charges q_1 and q_2 have long been at rest and exert electric forces on each other that are equal but opposite. Now move q_2 quickly to the new location. Since the electric field produced by q_1 points radially away from q_1, the new force on q_2 differs from the old. The charge q_1, however, cannot "know" anything about q_2's motion until a time ℓ/c has elapsed, because changes in the electric field propagate with a finite speed, the speed of light c. For an interval ℓ/c, the force on q_1 "due to q_2" remains unchanged, but the force on q_2 "due to q_1" has been changing. Obviously Newton III fails here. The quotation marks suggest why: We really have three entities—the two charges and the electric field.

Whenever fields are changing, the amount of momentum they carry may change, and Newton III may be violated. The effect is vital; without it, we would have no electromagnetic communication and could not even see. Nonetheless, Newton III is an excellent approximation in most terrestrial and astrophysical contexts where one is at all inclined to invoke it.

The questions raised in this section beg to be followed up. Here are several routes. Surely it is proper to start with the author of the three laws: Sir Isaac Newton, *The Mathematical Principles of Natural Philosophy*, as translated by Motte from the Latin original. (The original, with the title *Philosophiae Naturalis Principia Mathematica*, was published in 1687, while Samuel Pepys was president of the Royal Society and only after Newton had been prodded into publishing by Edmund Halley.) Critiques of Newtonian mechanics that are now classic were written by Ernst Mach, *The Science of Mechanics* (Open

34 A REVIEW OF SOME BASICS

Court Publishing, La Salle, Ill., 1960), and by Henri Poincaré, *Science and Hypothesis* (Dover, New York, 1952). The law of inertia, another name for Newton I, is the focus of a paper by G. J. Whitrow, *Brit. J. Phil. Sci.*, **1**, 92 (1950). Whitrow provides a good sense of how many different views and interpretations a single law can engender. Some reassurance may be found in the paper "Laws of Classical Motion: What's F? What's m? What's a?" by Robert Weinstock, *Am. J. Phys.*, **29**, 698 (1961). Finally, Mary Hesse provides an annotated bibliography in "Resource Letter on Philosophical Foundations of Classical Mechanics," *Am. J. Phys.*, **32**, 1 (1964).

1.9 REVIEWING THE REVIEW

A review as long and complex as this chapter deserves a review itself. Here is a recapitulation of the major points.

There are two geometrically defined ways to multiply vectors: the scalar and the vector product. With the scalar product we can decompose a vector equation into a set of equations in the components; we take the scalar product with unit vectors such as $\hat{\mathbf{x}}, \hat{\mathbf{y}},$ and $\hat{\mathbf{z}}$. In general, we can determine the amount of a vector ($\mathbf{A}$, say) along a specific direction ($\hat{\mathbf{B}}$, say) by forming a scalar product: $\mathbf{A} \cdot \hat{\mathbf{B}} = |\mathbf{A}| \cos (\mathbf{A}, \hat{\mathbf{B}})$. The vector product figures prominently in the definition of angular momentum, $\mathbf{r} \times m\mathbf{v}$, and, in general, whenever the notion of rotation arises.

Vectors provide a natural language for writing Newton's second law:

$$\frac{d}{dt}(m\mathbf{v}) = \mathbf{F}.$$

The derivative of a vector with respect to time means "vector at $t + \Delta t$ minus vector at t, all divided by Δt, with the limit $\Delta t \to 0$ then taken." (Because we take the difference of two vectors, which itself is a vector, the result of the limit process is also a vector.) Sometimes we stay at this relatively abstract level, as when writing $\mathbf{v} = d\mathbf{r}/dt$. At other times, we express the vector in terms of unit vectors and components, apply the definition to that expression, and emerge with derivatives of the components and of the unit vectors (which need not be temporally constant).

Taking the scalar product of Newton II with an infinitesimal displacement and then summing along the object's path leads to the ideas of kinetic energy and work. Thus, it *follows* from Newton II that the change in kinetic energy is equal to the work done by the force actually acting on the mass.

Often the work can be computed as a change in a potential energy U. Then we arrive at energy conservation: The sum of the kinetic and potential energies remains constant. Beyond that, we can compute the force from the gradient of U: $\mathbf{F} = -\text{grad } U$. By definition, grad U is a vector formed by finding the direction in which U changes the fastest. The vector grad U has that direction, and its magnitude is that maximum spatial rate of change.

The potential energy may be computed in two ways. They are not logically distinct; the second method follows from the first, but is often preferred because it is easier to implement.

1. We may compute U from the definition, as -1 times a line integral of the force, taken from a fiducial point to the location where we want U. This is displayed in equation (1.4-4).
2. We may split the force **F** into contributions for which the integral is easy or known, compute the corresponding contributions to the total potential energy, and then add those contributions. The addition is just the addition of scalars, not vectors, and so is often easily done. (We used this method in section 1.6.)

We touched on the idea of a stable equilibrium. (A systematic study comes later, in section 3.1, and instances recur throughout the book.) The question of stability is usually understood to ask, Is there a restoring force for all possible small displacements? The answer, we found, is yes if—and only if—the potential energy U has a minimum at the point in question.

The conservation laws are intimately connected with invariance properties. If the potential energy does not change numerical value under some mentally conceived motion—a translation or a rotation—then the corresponding "momentum" component is conserved: a component of linear momentum for a translation, a component of angular momentum for a rotation.

Last, but not least, come Newton's three laws. They can be stated succinctly, but a thoughtful inspection raises questions. How do we avoid circularity in the definitions? How do we select an inertial frame of reference? When will Newton's third law be valid or at least an adequate approximation? And other questions arise. For some of these questions, section 1.8 provides an answer or a partial answer. Other questions remain as food for thought or subjects for debate.

The chapter also illustrated some problem-solving techniques and commented on them (as in the final paragraphs of sections 1.1 and 1.3). The topic of how best to tackle a problem is a wide one, but there are some techniques to bear in mind or to turn to when in despair. A number of those techniques are collected in appendix D. We have not met all of them yet, but now is a good time for a first reading of that appendix.

WORKED PROBLEMS

WP1-1 The force **F** is specified to satisfy the conditions under which a potential energy exists. In component form, the force is

$$\mathbf{F} = F_x\hat{\mathbf{x}} + F_y\hat{\mathbf{y}} \tag{1}$$

with
$$F_x = F_0(x + by)$$

and
$$F_y = bF_0(x + by),$$

where F_0 and b are constants.

Problem. Determine the potential energy function $U(\mathbf{r})$. Then sketch contours of constant U.

According to equation (1.4-4), we need to evaluate an integral of the form

$$U(\mathbf{r}) = -\int_{(0,0)}^{(x,y)} \mathbf{F}(\mathbf{r}') \cdot d\mathbf{r}' + U(0). \tag{2}$$

Because the integral is specified to be independent of the path we take between the origin and the point (x, y), we can simplify our calculation by choosing a path such that $d\mathbf{r}$ runs parallel to the cartesian coordinate axes. Figure WP1-1a shows a good choice for the two legs that comprise the total path. On the first leg we have

$$d\mathbf{r}' = \hat{\mathbf{x}} \, dx',$$

and so

$$-\int_{(0,0)}^{(x,0)} \mathbf{F}(\mathbf{r}') \cdot d\mathbf{r}' = -\int_{(0,0)}^{(x,0)} F_0(x' + b \cdot 0) \, dx'$$

$$= -F_0 \frac{x^2}{2}.$$

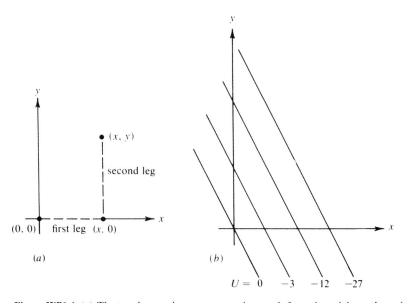

Figure WP1-1 (a) The two legs make up a convenient path from the origin to the point (x, y). (b) Lines of constant U. The numerical constants were taken to be $F_0 = 6$, $b = \frac{1}{2}$ and $U(0) = 0$.

On the second leg we have

$$d\mathbf{r}' = \hat{\mathbf{y}}\, dy',$$

and so

$$-\int_{(x,0)}^{(x,y)} \mathbf{F}(\mathbf{r}') \cdot d\mathbf{r}' = -\int_{(x,0)}^{(x,y)} bF_0(x + by')\, dy'$$
$$= -bF_0(xy + \tfrac{1}{2} by^2).$$

Adding these two contributions gives the integral in equation (2); thus

$$U(\mathbf{r}) = -\tfrac{1}{2} F_0(x^2 + 2bxy + b^2 y^2) + U(0)$$
$$= -\tfrac{1}{2} F_0(x + by)^2 + U(0). \tag{3}$$

Nothing in the statement of the problem determines $U(0)$; we are free to choose its value.

Equation (3) tells us that contours of constant U are given by

$$x + by = \text{constant},$$

i.e., by the lines

$$y = -\frac{1}{b} x + \frac{\text{constant}}{b}. \tag{4}$$

Such contours are sketched in figure WP1-1b.

WP1-2 The potential energy $U(\mathbf{r})$ is that given by equation (3) in WP1-1.

Problem. Compute the associated force $\mathbf{F}$. Is it perpendicular to lines of constant U?

Equation (1.4-6) tells us that we need to evaluate $-\text{grad}\, U$. Because we are given the potential energy in cartesian coordinates, we can use the prescription in equation (1.4-8). The partial derivatives that we need are

$$\frac{\partial U}{\partial x} = -\frac{1}{2} F_0 [2(x + by)]$$

and

$$\frac{\partial U}{\partial y} = -\frac{1}{2} F_0 [2(x + by)b].$$

Upon remembering the minus sign, we have

$$\mathbf{F} = F_0(x + by)\hat{\mathbf{x}} + bF_0(x + by)\hat{\mathbf{y}}.$$

This expression agrees with equation (1) of WP1-1. Good.

We can readily construct a unit vector in the direction of $\mathbf{F}$:

$$\frac{\mathbf{F}}{F} = \frac{\hat{\mathbf{x}} + b\hat{\mathbf{y}}}{(1 + b^2)^{1/2}}.$$

According to equation (4) of WP1-1, a line of constant U has a slope

$$\frac{dy}{dx} = -\frac{1}{b}.$$

Therefore a vector tangent to such a line is proportional to

$$\Delta x\, \hat{\mathbf{x}} + \Delta y\, \hat{\mathbf{y}} = \Delta x\left[\hat{\mathbf{x}} + \left(-\frac{1}{b}\right)\hat{\mathbf{y}}\right].$$

The scalar product of this vector with $\mathbf{F}/F$ is zero, and so $\mathbf{F}$ is perpendicular to a line of constant U.

PROBLEMS

1-1 In some inertial frame, a particle of mass m is acted on by forces (unknown to us) such that the *velocity* as a function of time is

$$\mathbf{v}(t) = \sqrt{3}t^2\hat{\mathbf{x}} + 1\hat{\mathbf{y}} + 0\hat{\mathbf{z}},$$

where $\hat{\mathbf{x}}, \hat{\mathbf{y}}, \hat{\mathbf{z}}$ are orthogonal unit vectors. You may leave the answers to (*a*) and (*c*)—but not (*b*)—in component form.

    (*a*) What is the force $\mathbf{F}$ on the particle at $t = 7$ seconds?

    (*b*) What is the angle between the velocity vector and the x axis at $t = 1$ second?

    (*c*) If the particle's position at $t = 0$ was

$$x(0) = 13 \quad y(0) = 19 \quad z(0) = 0,$$

what is the position of the particle at $t = 10$ seconds?

1-2 Jack and Jill together hold up a bucket of water. If Jill pulls with a force of 10 newtons at an angle of 45° with respect to the vertical and if Jack has to pull (in some directon) twice as hard as Jill, how much does the bucket weigh? (You may want to start with a sketch.)

1-3 The context is section 1.1. Denote the ball's vertical displacement from its initial position by s, whence $v = ds/dt$.

    (*a*) Can you convert equation (1.1-5) to an equation connecting dv and ds? What is v as a function of s for arbitrary initial downward velocity v_0?

    (*b*) Suppose the ball were thrown vertically upward, with $v_0 < 0$ but $|v_0| < v_t$. How high would it go? When it passed the tossing point on the return trip, what would be the ratio of the new speed to the initial speed (as a function of v_0 and v_t)? Is this consistent with energy conservation?

You will need to be careful with signs on the upward journey. Equation (1.1-5) presumes downward motion and is not applicable. Remembering that air resistance always opposes the motion will help you to insert it with the correct sign.

1-4 *Dimensional analysis and characteristic quantities.* This problem harks back to section 1.1. The air resistance on the ball may depend on the ball's size (specified by the radius R), on the ball's speed v through the air, and on the density ρ_{air} of the air. Suppose these are the only quantities on which $F_{air\ resistance}$ may depend. How must R, v, and ρ_{air} be combined so that the result has the same dimensions as $F_{air\ resistance}$ and hence is a candidate for that force? Is the combination unique (aside from numerical factors)?

When we go on to characteristic velocity and distance fallen, there may reasonably be a dependence on mass m and on g. We must, however, delete the instantaneous velocity v from the list of potential ingredients. Which combinations of R, ρ_{air}, m, and g have the dimensions of velocity or length and hence are candidates for a characteristic velocity or distance? Are those combinations unique?

You might wonder whether molecular viscosity should be included in the list. It may be ignored (as an explicit parameter) once the ball is well underway, specifically, as soon as vR substantially exceeds the product of the molecular thermal speed and the molecular mean free path. More about the subtle aspects of falling balls may be found in B. L. Coulter and C. G. Adler, *Am. J. Phys.*, **47**, 841 (1979).

1-5 The context is section 1.1. Does the distance fallen, equation (1.1-8), have the behavior you expect when $0 < t \ll \tau$?

1-6 A tiny object moves (with initial velocity $\mathbf{v}_0$) through a viscous fluid that exerts a drag force $-\gamma\,\mathbf{v}$, with γ being a positive constant. The viscous drag is the sole force on mass m. (The object might be one of Millikan's charged oil drops, moving through air and in an electric field that cancels the gravitational force.) Can you find an exponential solution for the velocity as a function of time? How does the distance traveled grow with time?

1-7 A tiny object moves upward (with initial velocity $\mathbf{v}_0$) against a constant downward force $\mathbf{F}_0$. In addition, viscosity produces a variable resisting force $-\gamma\,\mathbf{v}$, where the constant γ is positive. (The object might be one of Millikan's charged oil drops, moving through air against the combined gravitational and electric forces.) Calculate the time required for mass m to reach its maximum height. Compare this with the time required if $\gamma = 0$. Establish the specific sense of the inequality: which time is shorter? Evaluate the ratio of the two times numerically for the following three situations: $v_0 = v_t$, $10v_t$, and $0.1v_t$, where v_t is the terminal speed for *downward* motion in the *present* context of a drag that is linear in the velocity.

1-8 *Non-orthogonal unit vectors.* The vectors $\mathbf{e}_1$ and $\mathbf{e}_2$ are unit vectors, but they are not orthogonal:

$$\mathbf{e}_1 \cdot \mathbf{e}_2 = \cos\alpha \neq 0,$$

as indicated in figure P1-8. Vector $\mathbf{A}$ lies in the plane defined by $\mathbf{e}_1$ and $\mathbf{e}_2$. Therefore, we may write

$$\mathbf{A} = A_1\mathbf{e}_1 + A_2\mathbf{e}_2;$$

40 A REVIEW OF SOME BASICS

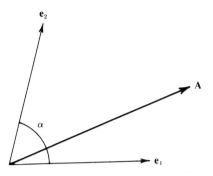

Figure P1-8

that is, we may express the general vector **A** in terms of the two unit vectors.

(a) Determine coefficients A_1 and A_2 in terms of $\mathbf{e}_1 \cdot \mathbf{A}$, $\mathbf{e}_2 \cdot \mathbf{A}$, and $\cos \alpha$.

(b) Do the coefficients reduce to something familiar in the limit as $\alpha \to \pi/2$?

(c) Do you prefer orthogonal or non-orthogonal unit vectors?

1-9 A mass m moves in the xy plane, acted on by an oscillating electric force $qE_0 \cos \omega t \, \hat{\mathbf{y}}$. At $t = 0$, $\mathbf{v}(0) = \mathbf{v}_0$ and $\mathbf{r}(0) = 0$. Compute the velocity and position at subsequent times. Describe the motion in words and sketch it, too.

1-10 Suppose you are in a hole, tossing rocks onto the level ground a distance h_0 above your outstretched arm. The speed of the rock on release is a fixed v_0 (determined by the kinetic energy you can give the rock). At what angle should you toss so that the rock will land as far from the hole as possible? And what is that maximum distance? (You may ignore air resistance, and you need not worry about hitting the edge of the hole as the rock moves on its upward arc.) A calculation correct through first order in the dimensionless ratio $mgh_0/(mv_0^2/2)$ will suffice.

The shot-putter faces the complementary problem: maximizing the range from an outstretched arm that is *above* ground level. A realistic analysis is given by D. B. Lichtenberg and J. G. Wills, *Am. J. Phys.*, **46**, 546 (1978).

1-11 This is a one-dimensional problem. The potential energy for the particle of mass m has the symmetric form $U(x) = U_0 b^2/(x^2 + b^2)$, where U_0 and b are positive constants.

(a) Sketch the potential energy as a function of x.

(b) If the particle started from far to the left (really, at $x = -\infty$) with an initial speed to the right of $[U_0/(5m)]^{1/2}$, will the particle be able to get to positive values of x? If yes, why? If no, how far to the right will it get?

(c) What are the magnitude and direction of the force on the particle when $x = -7$?

1-12 Suppose we think just of the earth and moon, dropping the sun and other planets from consideration. We put them at rest for the moment and look at a small object—a powerless spaceship, say, of mass m—located somewhere

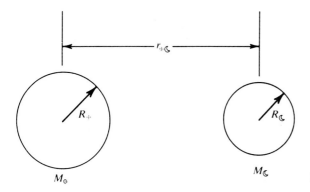

Figure P1-12

along the line connecting the earth and moon. Note the sketch in figure P1-12, where the subscript $\oplus$ denotes the earth and $\mathbb{C}$, the moon.

(a) Is there any position along the line at which the spaceship would feel zero gravitational force? Why?

(b) If the gravitational potential energy is taken to be zero at infinity, is the net potential energy equal to zero anywhere along the line between the earth and moon? Why? Make a rough sketch of the potential energy as a function of position along the line.

(c) If we wanted to toss the spaceship from the earth's surface to the moon, what would be the minimum initial speed of the "toss"? You may answer this in terms of $M_\oplus$, $M_\mathbb{C}$, m, $R_\oplus$, $R_\mathbb{C}$, $r_{\oplus\mathbb{C}}$, G, and the distances specified in (a) and (b) if you said those parts had "yes" answers. Do not bother to compute numerical values.

Note: Part (c) is not sneaky, but it is not trivial either. (The "activation energy" idea in chemical kinetics arises in a similar context.)

1-13 The force $F(x)$ in one-dimensional motion has the profile indicated in figure P1-13. Calculate the associated potential energy $U(x)$ and sketch its profile.

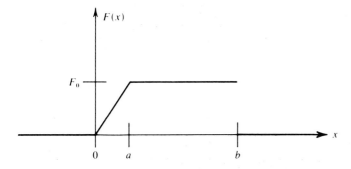

Figure P1-13

42 A REVIEW OF SOME BASICS

1-14 For a small mass m, calculate the gravitational potential energy along the symmetry axis of a disk with total mass M, radius R, and thickness much, much less than R. The tiny thickness-to-radius ratio means that you may integrate as though the disk had vanishing thickness. Determine the gravitational force, too. Examine your potential energy at axial distances much larger than R. Does it approach a familiar (nonzero) expression?

T1-15* Several times we have taken Newton's second law, applied the same operation to both sides, and obtained a *useful* consequence. By no means have we exhausted the possibilities. Try taking first the *scalar* product of both sides of

$$\frac{d}{dt}(m\mathbf{v}) = \mathbf{F}$$

with $\mathbf{v}$, then the *vector* product with $\mathbf{v}$. After a little manipulation, one product gives a mathematical result that you can interpret physically; it *is* useful. The other product leads to no useful result. Which is which?

T1-16 When does a potential energy exist? Conditions 1 and 2 in section 1.4 are sufficient, but we would like an efficient way to implement the test in condition 2. If the line integral is independent of path, then the integral over a closed contour will be zero: the homeward trip cancels the outward trip between any two points on the contour. The zero value implies that curl $\mathbf{F} = 0$. Indeed, curl $\mathbf{F} = 0$ is necessary and sufficient for condition 2 to be satisfied.

For each of the following forces, test for the existence of a potential-energy function. If such a function exists, construct it (or just exhibit it, having arrived at it by inspired guesswork).

(a) $\mathbf{F} = C(x\hat{\mathbf{x}} + y\hat{\mathbf{y}}) \exp[-\alpha(x^2 + y^2)]$.
(b) $\mathbf{F} = C(y\hat{\mathbf{x}} - x\hat{\mathbf{y}}) \exp[-\alpha(x^2 + y^2)]$.
(c) $\mathbf{F} = C(x\hat{\mathbf{x}} + y\hat{\mathbf{y}}) \exp[-\beta(x^2 + y^2)^{1/2}]$.
(d) $\mathbf{F} = C(x^3 y\hat{\mathbf{x}} + xy^2\hat{\mathbf{y}} + z\hat{\mathbf{z}})$.

1-17 Plane polar coordinates, r and θ, and the associated unit vectors, $\hat{\mathbf{r}}$ and $\hat{\boldsymbol{\theta}}$, are defined in figures 4.3-2 and 4.3-3. Can you use equation (1.4-7) to show that

$$\text{grad } U = \frac{\partial U}{\partial r}\hat{\mathbf{r}} + \frac{1}{r}\frac{\partial U}{\partial \theta}\hat{\boldsymbol{\theta}}$$

in plane polar coordinates?

T1-18 *Adapting a previous result.* In section 1.6 we worked out the gravitational potential energy $U(\mathbf{r})$ of a mass m interacting with a ring of radius a and total mass M. Although we wrote down a formal integral expression valid for any r, we evaluated the integral explicitly only for $r \ll a$. Adapt our method to the op-

---

*The superscript T denotes a homework problem that extends the theory in some notable way.

posite situation, $r \gg a$, and work out $U(\mathbf{r})$ correct through terms of order $(GMm/r)(a^2/r^2)$. We will need this in chapter 5.

1-19 Can you adapt the results of problem 1-18 to calculate the gravitational potential energy of problem 1-14 for points off the symmetry axis as well as on it, provided the points are at distances large relative to the disk's radius?

1-20 For the potential energy $U = C(x^2 + 4y^2)$, with C a positive constant, sketch some contours of constant U. Then compute the associated force $\mathbf{F}$ and sketch it at several locations. Is there a location of stable equilibrium?

1-21 For each of the following potential energies, use invariance arguments to deduce the conserved quantities, be they momentum, angular momentum, or energy. The mass may move in the full three-dimensional space: its position vector is $\mathbf{r} = x\hat{\mathbf{x}} + y\hat{\mathbf{y}} + z\hat{\mathbf{z}}$.

(a) $U = \frac{1}{2}k(x^2 + y^2)$.

(b) $U = \frac{\alpha}{r}$.

(c) $U = \beta(\hat{\mathbf{z}} \cdot \mathbf{r})^2$.

(d) $U = \frac{\alpha}{r} + \beta(\hat{\mathbf{z}} \cdot \mathbf{r})^2$.

1-22 An object of mass m, initially at rest at location $x = x_0$, is attracted to the origin by a force of magnitude C/x^3. Determine the time it takes the object to reach the origin. (You do *not* need to solve directly a second-order differential equation. See what energy conservation will do for you. Also, try dimensional analysis: what combination of m, C, and x_0 has the dimensions of time?)

T**1-23** *The virial theorem.* This problem continues the approach described in problem 1-15: apply the same operation to both sides of Newton II. This time, try the scalar product with the position vector $\mathbf{r}$. (We did the vector product in section 1.7.) See what you can say about the *time average* of the terms if the motion is periodic or at least bounded in r and $|\mathbf{v}|$.

Next, see what special consequences arise if the force $\mathbf{F}$ can be derived from a potential energy that depends only on some power of r: $U(r) = (\text{const})r^n$, for some number n.

Last, apply your relationship to a planet in orbit around the sun. What can you conclude, regardless of whether the orbit is circular or elliptical?

The results that emerge are various aspects of the *virial theorem*, derived by Clausius in 1870. The word *virial* is derived from the Latin *vis* [plural, *vires*] meaning "force."

1-24 Even the design of a lawn sprinkler can pose a nice problem in applied physics. The lawn sprinkler is to consist of a spherical cap ($\alpha_0 = 45°$) provided with a large number of equal holes through which water is ejected with speed v_0; this is sketched in figure P1-24. The lawn is not uniformly sprinkled if these holes are uniformly spaced over the surface of the cap. How must $n(\alpha)$, the number of holes per unit area, be chosen to achieve uniform sprinkling of a circular area? Assume the size of the cap to be small compared with the area to

44 A REVIEW OF SOME BASICS

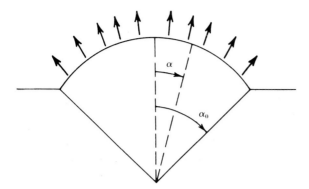

Figure P1-24

be sprinkled, the surface of the cap to be level with the lawn, and air resistance to be ignorable. Sketch a graph of $n(\alpha)$ versus α.

(I find it useful to ask, Where does the water ejected by holes in the angular range α to $\alpha + d\alpha$ land on the lawn? A certain range of distances emerges, and I can ask about the "rain" per unit area out there.)

CHAPTER TWO

THE HARMONIC OSCILLATOR

2.1 Damped harmonic oscillator
2.2 Phase space: An introduction
2.3 Harmonic oscillation in two dimensions
2.4 Sinusoidally driven oscillator
2.5 Quality factor Q
2.6 A guide to the major ideas

You can do everything with a bayonet except sit on it.

Napoleon

2.1 DAMPED HARMONIC OSCILLATOR

Imagine a small object moving slowly through some viscous automotive oil. The mass is attached to the end of a spring, as indicated in figure 2.1-1. The spring exerts a restoring force proportional to the displacement x from the point where the spring is relaxed. We can write that force as $-kx$, where k is the (positive) spring constant and the minus sign ensures that we have a *restoring* force.

46 THE HARMONIC OSCILLATOR

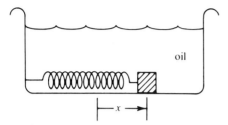

Figure 2.1-1 A damped harmonic oscillator. The coordinate x is the object's location relative to the point where the spring is relaxed (neither extended nor compressed).

If this were the only force, then the mass, when set into motion, would oscillate endlessly. Every time it went to one side, it would be pulled back, would overshoot the zero-spring-force position, would be pulled back the other way, would overshoot.... Motion through the viscous oil, however, saps energy from the mass. The oscillations will die out, sooner or later. A typical graph of position versus time would look like figure 2.1-2.

For the one-dimensional motion that we have here, Newton II would be

$$m\ddot{x} = -kx - \gamma\dot{x}. \qquad (2.1\text{-}1)\bigstar$$

The positive constant γ specifies the strength of the viscous damping, which is proportional to the velocity in magnitude but opposite in direction. (Under the stipulation of slow motion in a very viscous fluid, the resistance is approximately linear in the velocity, unlike the quadratic dependence that we studied in section 1.1.)

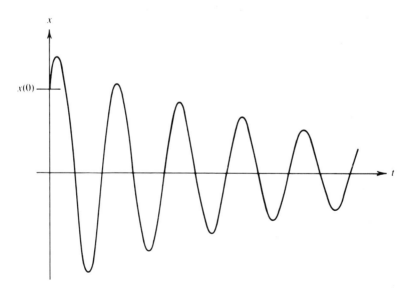

Figure 2.1-2 Motion of a damped oscillator.

2.1 DAMPED HARMONIC OSCILLATOR

Surely this is an artificial problem. Why should we bother with it? The reason lies in the *structure* of equation (2.1-1). An electron tethered to the atomic nucleus can be described—for some purposes—by an equation like (2.1-1); the damping term arises because the oscillating electron radiates energy, and some electromagnetic force of reaction must act on the electron, to balance the energy budget. Better yet, atoms in a molecule vibrate about their equilibrium locations; the damping is again associated with radiation. Almost any stable system, when displaced a little from its equilibrium configuration, will respond as does a harmonic oscillator. Moreover, there are analogies in electric circuit theory, in particular, the series connection of a capacitor, an inductor, and a resistor. The instances of similar structure are legion, and the harmonic oscillator is a versatile model as one gropes to understand a puzzling, new situation. For these reasons we study the object immersed in oil.

Equation (2.1-1) is a linear, homogeneous equation: the dependent variable x or a derivative of it enters each term, but only linearly. The coefficients are merely constants. This suggests that, in searching for a solution, we try the form

$$x = A e^{pt}, \quad (2.1\text{-}2)$$

because differentiation of an exponential just reproduces the exponential, times the constant p. When we substitute the trial form into the differential equation, the latter reduces to merely an algebraic equation in the unknown constant p. In particular, the substitution leads to

$$mp^2 = -k - \gamma p,$$

after common factors have been deleted. The solution for p is

$$p = -\frac{\gamma}{2m} \pm \left[\left(\frac{\gamma}{2m}\right)^2 - \frac{k}{m} \right]^{1/2}. \quad (2.1\text{-}3)$$

In general, there are two distinct algebraic solutions. If $\gamma/(2m) > (k/m)^{1/2}$, then both solutions give $p < 0$. If we were to construct a general solution to equation (2.1-1) as a linear combination of the two exponential solutions, each p would produce a monotonically decreasing contribution. The mass soon would be in the neighborhood of $x = 0$ for good. The physical situation, of course, is heavy damping. More about it can be found in problems 2-5 and 2-6; for now, let us move on to light damping.

Damped Oscillations

If $\gamma/(2m) < (k/m)^{1/2}$, the argument of the square root operation is negative. If we first factor out $-k/m$, we can write

$$p = -\frac{\gamma}{2m} \pm i\left(\frac{k}{m}\right)^{1/2} \left[1 - \left(\frac{\gamma}{2m}\right)^2 \frac{m}{k} \right]^{1/2}. \quad (2.1\text{-}4)$$

48 THE HARMONIC OSCILLATOR

To understand this result, we need the identity

$$e^{a+ib} = e^a(\cos b + i \sin b), \qquad (2.1\text{-}5)$$

valid for any real quantities a and b. If we adopt the abbreviations

$$\omega_0 \equiv \left(\frac{k}{m}\right)^{1/2} \qquad (2.1\text{-}6)$$

and

$$\tau \equiv \frac{m}{\gamma}, \qquad (2.1\text{-}7)$$

we can use identity (2.1-5) to write

$$e^{pt} = e^{-t/(2\tau)}\left\{\cos \omega_0\left[1 - \left(\frac{1}{2\omega_0\tau}\right)^2\right]^{1/2} t \pm i \sin \omega_0\left[1 - \left(\frac{1}{2\omega_0\tau}\right)^2\right]^{1/2} t\right\}.$$

The first factor on the right tells us that the time τ is a characteristic damping time. The larger γ, the shorter that characteristic time is. The trigonometric functions give us oscillation, sinusoidal in the general sense of that word. If γ were zero, so that τ were infinite, the oscillation would be at the angular frequency ω_0. For this reason, ω_0 is called the *natural (angular) frequency*. (An angular frequency is always 2π times the number of oscillations or revolutions per unit time.)

So far we have found solutions to the mathematical problem, but the presence of i, equal to $\sqrt{-1}$, tells us that we have not yet solved the physical problem: how does the object's location change with time? We complete the solution in two steps. First, we construct the general solution to equation (2.1-1). If we denote the two algebraic solutions for p by

$$p_\pm = -\frac{1}{2\tau} \pm i\omega_0',$$

where

$$\omega_0' \equiv \omega_0\left[1 - \left(\frac{1}{2\omega_0\tau}\right)^2\right]^{1/2}, \qquad (2.1\text{-}8)$$

then we can write the general solution as

$$x = A_+ e^{p_+ t} + A_- e^{p_- t}$$
$$= e^{-t/(2\tau)}[(A_+ + A_-)\cos \omega_0' t + i(A_+ - A_-)\sin \omega_0' t]. \qquad (2.1\text{-}9)$$

Here A_+ and A_- are two complex constants that we are free to choose. Indeed, our second step is to choose those constants so that the initial conditions of the motion will be met. Evaluating equation (2.1-9) at $t = 0$ and then doing the same for its derivative yields

$$x(0) = A_+ + A_-,$$
$$\dot{x}(0) = -\frac{1}{2\tau}(A_+ + A_-) + i(A_+ - A_-)\omega_0'.$$

2.1 DAMPED HARMONIC OSCILLATOR

These equations give us immediately the sum and difference that we need in equation (2.1-9); the result is

$$x = e^{-t/(2\tau)}\left\{x(0)\cos\omega_0't + \frac{1}{\omega_0'}\left[\dot{x}(0) + \frac{x(0)}{2\tau}\right]\sin\omega_0't\right\}. \quad (2.1\text{-}10) \bigstar$$

The right-hand side is now entirely real, meets the initial conditions, and (as one can confirm by substitution) satisfies Newton II. It describes damped oscillatory motion, the angular frequency being ω_0'. The smaller the damping, the closer ω_0' is to the natural frequency ω_0. Figure 2.1-2 illustrates the motion. There the parameters τ and ω_0 have the ratio $\tau/(2\pi/\omega_0) = 2$. The initial velocity was chosen to be $\dot{x}(0) = \omega_0 x(0)$.

Energy

The restoring force $-kx$ meets the conditions, enumerated in section 1.4, that guarantee the existence of a potential energy function. Adapting equation (1.4-4) to our one-dimensional context, we can write

$$U(x) = -\int_{x_A}^{x}(-kx')\,dx' + U(x_A)$$
$$= \tfrac{1}{2}kx^2 \quad (2.1\text{-}11)$$

if we choose $x_A = 0$ and $U(0) = 0$.

The other force in the problem, $-\gamma\dot{x}$, describes the viscous damping, a process that transfers energy from the object to the fluid. There is no associated potential energy. Rather, we are led to construct the "energy" of the mass as

$$E = \tfrac{1}{2}m\dot{x}^2 + \tfrac{1}{2}kx^2 \quad (2.1\text{-}12)$$

and to see how the damping reduces its numerical value with time. Differentiating the energy expression with respect to time and then invoking equation (2.1-1) yields

$$\frac{dE}{dt} = m\dot{x}\ddot{x} + kx\dot{x} = -\gamma\dot{x}^2. \quad (2.1\text{-}13)$$

The right-hand side is never positive: the damping term always extracts energy, never returns any.

Equation (2.1-13) gives the instantaneous rate of energy loss. Often the characteristic damping time τ is long relative to the natural period of the oscillator. The natural period is, of course, the time required for the product $\omega_0 t$ to increase by 2π; hence that period is $2\pi/\omega_0$. (We could introduce a separate symbol for the period, $P_0 \equiv 2\pi/\omega_0$, say, but already the chapter's essential symbols make a long list.) To say that the damping time τ is long relative to the natural period means that the strong inequality

$$\tau \gg 2\pi/\omega_0$$

holds. In this context, the amplitude of the motion changes little during one period of oscillation. The mass makes many oscillations before the cumulative effect of damping becomes substantial. Averages over one oscillation period, then, may be sufficient for one's purposes, and their simpler structure makes them appealing. To compute the average rate of energy loss, we need only determine the average of $\dot{x}^2$.

An interlude, however, is in order. Because the solution in equation (2.1-10) has two trigonometric terms, forming products and averages with it is a cumbersome business. We can compress the solution into a single term by using the trigonometric identity

$$\mathcal{A} \cos (\omega_0' t + \varphi) = \mathcal{A} \cos \varphi \cos \omega_0' t - \mathcal{A} \sin \varphi \sin \omega_0' t.$$

If we compare the right-hand side here with equation (2.1-10) and equate coefficients of the corresponding terms, we find

$$\mathcal{A} \cos \varphi = x(0) e^{-t/(2\tau)}$$

and

$$\mathcal{A} \sin \varphi = -\frac{1}{\omega_0'} \left[\dot{x}(0) + \frac{x(0)}{2\tau} \right] e^{-t/(2\tau)}.$$

Forming the ratio of these two equations gives $\tan \varphi =$ known constant, which we can solve for φ. Then either equation (or the sum of their squares) gives

$$\mathcal{A} = \text{(known constant)} \ e^{-t/(2\tau)}.$$

The upshot is that we can compress our solution in equation (2.1-10) to the form

$$x = A e^{-t/(2\tau)} \cos (\omega_0' t + \varphi), \tag{2.1-14}$$

where the constants A and φ are determined by the initial conditions. The principle of this process is quite general; we will refer to it again.

Now we can return to the averages. We had had light damping in mind, which we quantified as $\omega_0 \tau \gg 2\pi$. Then $\omega_0' \simeq \omega_0$; moreover, when we differentiate x as displayed in equation (2.1-14), we may ignore the derivative of the exponential (which would describe the slow change in amplitude), and so we find

$$\dot{x} \simeq -\omega_0 A e^{-t/(2\tau)} \sin (\omega_0 t + \varphi).$$

The average of equation (2.1-13) becomes

$$\left\langle \frac{dE}{dt} \right\rangle \simeq -\gamma \omega_0^2 A^2 e^{-t/\tau} \left\langle \sin^2 (\omega_0 t + \varphi) \right\rangle;$$

the exponential varies so slowly that it may be factored out of the averaging integral. To evaluate the remaining average, note that we average over one full period, that the average of $\cos^2 (\omega_0 t + \varphi)$ must have the same value over the interval, and that the sum of the averages must be 1 because $\sin^2 a + \cos^2 a = 1$ for any a. Thus the average we seek must be $\frac{1}{2}$. Therefore

$$\left\langle \frac{dE}{dt} \right\rangle \simeq -\tfrac{1}{2}\gamma\omega_0^2 A^2 e^{-t/\tau}. \tag{2.1-15}$$

This is correct but not yet particularly instructive. We can express the right-hand side in terms of the energy averaged over one period:

$$\begin{aligned} \langle E \rangle &= \left\langle \tfrac{1}{2}m\dot{x}^2 \right\rangle + \left\langle \tfrac{1}{2}kx^2 \right\rangle \\ &\simeq \tfrac{1}{4}m\omega_0^2 A^2 e^{-t/\tau} + \tfrac{1}{4}kA^2 e^{-t/\tau} \\ &\simeq \tfrac{1}{2}m\omega_0^2 A^2 e^{-t/\tau}, \end{aligned} \tag{2.1-16}$$

after we eliminate k in terms of ω_0^2 and m. Inserting this into equation (2.1-15) yields

$$\left\langle \frac{dE}{dt} \right\rangle \simeq -\frac{1}{\tau} \langle E \rangle. \tag{2.1-17} \bigstar$$

The first implication of equation (2.1-17) is that the energy decays on a time scale set by τ, a conclusion that could be extracted also from the exponential in equation (2.1-16). Beyond that, we can compute the energy dissipated per period and compare it with the energy itself. We can write

$$\text{Energy dissipated per period} = \left| \left\langle \frac{dE}{dt} \right\rangle \right| (\text{period})$$

$$= \frac{1}{\tau} \langle E \rangle \frac{2\pi}{\omega_0}. \tag{2.1-18}$$

Comparison with the energy still stored takes the form

$$\frac{\text{Energy dissipated per period}}{\langle E \rangle} = \frac{2\pi}{\omega_0 \tau}. \tag{2.1-19}$$

When we look at the fraction on the right as

$$\frac{2\pi/\omega_0}{\tau} = \frac{\text{natural period}}{\text{damping time}} \tag{2.1-20}$$

and remember that we are dealing with light damping, we see that the ratio is small indeed.

2.2 PHASE SPACE: AN INTRODUCTION

We have worked out the damped harmonic oscillator in complete analytic detail. Although that has cost us a lot of effort, let us suppose we do not know the solution, and let us see what we can learn by qualitative reasoning. This will be an introduction to methods that more difficult problems force on us.

To begin, we set the damping to zero and study the equation

$$m\frac{d^2x}{dt^2} = -kx. \tag{2.2-1}$$

52 THE HARMONIC OSCILLATOR

To specify the motion unambiguously, we would need to specify two initial conditions: x and $\dot{x}$ at $t = 0$. To see that this is both necessary and sufficient, we could write two equations that determine the subsequent values of x and $\dot{x}$:

$$\frac{d\dot{x}}{dt} = -\frac{k}{m}x, \quad (2.2\text{-}2a)$$

$$\frac{dx}{dt} = \dot{x}. \quad (2.2\text{-}2b)$$

Equation (2.2-2a) is just equation (2.2-1); equation (2.2-2b) follows by definition but is neither vacuous nor a swindle. What equations (2.2-2a and b) do for us is this: They replace a single second-order differential equation, namely, equation (2.2-1), by a pair of coupled first-order equations, provided we think of x and $\dot{x}$ as *separate dependent variables*. We can use the equations to integrate forward in time. For a short step Δt, they assert that

$$\dot{x}(t + \Delta t) = \dot{x}(t) - \frac{k}{m}x(t)\,\Delta t, \quad (2.2\text{-}3a)$$

$$x(t + \Delta t) = x(t) + \dot{x}(t)\,\Delta t. \quad (2.2\text{-}3b)$$

If we start at $t = 0$, then we need to know $x(0)$ and $\dot{x}(0)$ to evaluate the right-hand side and thus determine $x(\Delta t)$ and $\dot{x}(\Delta t)$. We can substitute them, in turn, on the right-hand side, thereby determining $x(2\,\Delta t)$ and $\dot{x}(2\,\Delta t)$. Step by step, we can march forward in time, though the tedium is best left to a digital computer.

For the initial conditions $x(0) = x_0$ and $\dot{x}(0) = 0$, the first few integration steps are shown in figure 2.2-1. By equations (2.2-3a and b), the first step yields

$$\dot{x}(\Delta t) = 0 - \omega_0^2 x_0\,\Delta t,$$

$$x(\Delta t) = x_0 + 0;$$

the next step produces

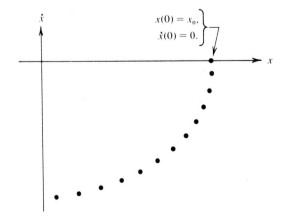

Figure 2.2-1 Some points along the trajectory in phase space. The time interval is $\Delta t = 0.02(2\pi/\omega_0)$.

2.2 PHASE SPACE: AN INTRODUCTION

$$\dot{x}(2\,\Delta t) = -\omega_0^2 x_0\,\Delta t - \omega_0^2 x_0\,\Delta t,$$

$$x(2\,\Delta t) = x_0 + (-\omega_0^2 x_0\,\Delta t)\Delta t.$$

Because the object was placed at rest but where a restoring force acts toward the left, the object acquires a negative velocity and begins to move inward.

The "space" of the diagram, consisting of the $(x, \dot{x})$ plane, is an instance of a *phase space*, in which the instantaneous state of the physical system is represented by the *system point*. As time goes on, the system point, which here has coordinates $[x(t), \dot{x}(t)]$, moves around in the phase space; we speak of the system's *trajectory in phase space*.

Integration, step by step, can always be used to determine the trajectory. Sometimes, however, there are shortcuts; energy conservation, when it holds, supplies one. In the present context,

$$\tfrac{1}{2}m\dot{x}^2 + \tfrac{1}{2}kx^2 = E \tag{2.2-4}$$

is constant in time. Every point along the trajectory must satisfy this equation. We can solve for $\dot{x}$ as a function of x and then plot the trajectory. Or we may recognize equation (2.2-4) as the equation of an ellipse in variables x and $\dot{x}$. Either way, the outcome is the set of trajectories in figure 2.2-2, a distinct trajectory for each different value of energy E. That the motion is bounded and periodic becomes apparent immediately.

Let us go on now and include the damping force $-\gamma\dot{x}$ in Newton II. The corresponding pair of first-order equations is

$$\frac{d\dot{x}}{dt} = -\omega_0^2 x - \frac{1}{\tau}\dot{x}, \tag{2.2-5a}$$

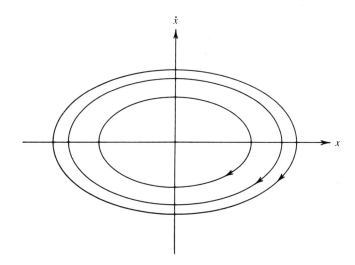

Figure 2.2-2 Trajectories for the harmonic oscillator. The larger the energy, the larger the ellipse.

54 THE HARMONIC OSCILLATOR

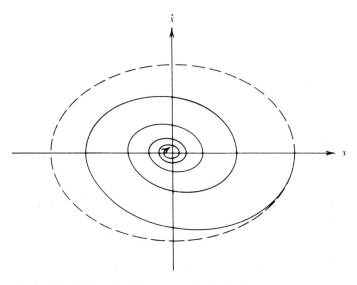

Figure 2.2-3 When a damping force is included, the trajectory spirals in toward the origin. The characteristic damping time τ here is $\tau = 0.7(2\pi/\omega_0)$.

$$\frac{dx}{dt} = \dot{x}. \qquad (2.2\text{-}5b)$$

Suppose we start the mass as before: $x(0) = x_0$ and $\dot{x}(0) = 0$. The trajectory will head into the fourth quadrant of the phase space, just as previously. What is the effect of the additional term? In the fourth quadrant, $\dot{x}$ is negative, and so $-\dot{x}/\tau$ is positive. Thus equation (2.2-5a) says that $\dot{x}$ does not grow through negative values as rapidly as before. The tangent to the trajectory points downward less than before, and so the trajectory lies inside the ellipse that would apply if there were no damping. Figure 2.2-3 illustrates this. A similar analysis can be made for each quadrant. The result is that the trajectory spirals in toward the origin. This geometric description is the analogue, in phase space, of the exponential damping that we found analytically in equation (2.1-10).

2.3 HARMONIC OSCILLATION IN TWO DIMENSIONS

Now we let our object move in two dimensions, but we remove the oil and contrive to make the restoring force proportional to distance from the origin in a symmetric fashion, so that Newton II reads

$$\frac{d}{dt}(m\mathbf{v}) = -k\mathbf{r}. \qquad (2.3\text{-}1)$$

In terms of cartesian unit vectors, the position vector **r** is

2.3 HARMONIC OSCILLATION IN TWO DIMENSIONS

$$\mathbf{r} = x\hat{\mathbf{x}} + y\hat{\mathbf{y}}, \tag{2.3-2}$$

and so we can readily extract equations for x and y by taking scalar products. In particular, the scalar product with $\hat{\mathbf{x}}$ yields

$$m\ddot{x} = -kx. \tag{2.3-3}$$

The equation for y is identical to this, and so we can write the general solution to equation (2.3-1) as

$$x = A \cos(\omega_0 t + \varphi_A), \tag{2.3-4a}$$

$$y = B \cos(\omega_0 t + \varphi_B), \tag{2.3-4b}$$

where the constants A, B, φ_A, and φ_B depend on the initial conditions. Both trigonometric functions repeat themselves after a time $2\pi/\omega_0$, and so the motion in the xy plane must be *periodic*. But what is the *shape* of that periodic motion?

As a first step, we can get an equation for y as a function of x by eliminating t between equations (2.3-4a and b). Let us use the phase difference

$$\varphi \equiv \varphi_B - \varphi_A$$

to write

$$y = B \cos(\omega_0 t + \varphi_A + \varphi)$$
$$= B \cos(\omega_0 t + \varphi_A) \cos \varphi - B \sin(\omega_0 t + \varphi_A) \sin \varphi.$$

Then equation (2.3-4a) enables us to eliminate the time:

$$y = \frac{Bx}{A} \cos \varphi \pm B\left(1 - \frac{x^2}{A^2}\right)^{1/2} \sin \varphi.$$

To remove the sign ambiguity, we can transpose the term in $\cos \varphi$ to the left-hand side and then square the equation:

$$A^2 y^2 - 2ABxy \cos \varphi + B^2 x^2 \cos^2 \varphi = B^2 A^2 \sin^2 \varphi - B^2 x^2 \sin^2 \varphi,$$

from which we get

$$B^2 x^2 - 2AB \cos \varphi \, xy + A^2 y^2 = A^2 B^2 \sin^2 \varphi. \tag{2.3-5}$$

The equation for the shape contains x and y in the forms x^2, y^2, and xy only, that is, in "quadratic" expressions only. Therefore the shape must be one of the classic conic sections: a circle, a hyperbola, a parabola, an ellipse, or a straight line, the limit of a narrow ellipse. Under which conditions on the quantities A, B, and φ do we get which curves? We can proceed by a process of elimination.

1. *Hyperbola.* The arms of a hyperbola extend to infinite distance, to infinite values of x or y or of both. Equations (2.3-4a and b) imply that x and y always remain finite. That eliminates the hyperbola.
2. *Parabola.* Similar reasoning eliminates the parabola.
3. *Circle.* If equation (2.3-5) is to represent a circle, we must have $A = B$ and the cross term must vanish. To get $\cos \varphi = 0$, we need $\varphi = \pm \pi/2$.

Table 2.3-1 Orbit shapes and their conditions of occurrence

Shape	φ	$\dfrac{A}{B}$
Circle	$\pm\dfrac{\pi}{2}$	1
Line	$0, \pm\pi$	any ratio
Ellipse	all other combinations	

4. *Straight line.* If the shape is to be a straight line, the ratio of dy to dx, which would be the slope of the line, must be constant. From the solutions as a function of time, we get

$$\frac{\Delta y}{\Delta x} = \frac{\dot{y}\,\Delta t}{\dot{x}\,\Delta t} = \frac{B\,\sin(\omega_0 t + \varphi_B)}{A\,\sin(\omega_0 t + \varphi_A)}.$$

The right-hand side is constant if $\varphi_B = \varphi_A$ or if $\varphi_B = \varphi_A \pm \pi$. And only if one of these conditions holds will there be constancy. Thus a straight line arises when $\varphi = 0$ or $\varphi = \pm\pi$. There is no restriction on A or B.

5. *Ellipse.* By elimination, all other combinations of φ and the ratio A/B produce an ellipse. An example would be $\varphi = \pi/2$, $A = \frac{1}{3}$, and $B = 1$:

$$\frac{x^2}{3^2} + y^2 = \frac{1}{3^2},$$

an ellipse with $x_{max} = 1$ and $y_{max} = \frac{1}{3}$.

In table 2.3-1 we have all the possibilities—circle, straight line, and ellipse—together with the conditions of amplitude and phase difference under which they arise. In a study as systematic as this one, there is something satisfying. Moreover, the results are useful well beyond the motion of a mass harmonically bound. If a monochromatic electromagnetic wave propagates along the $\hat{z}$ direction, the components of the electric field vector along $\hat{x}$ and $\hat{y}$ will vary with time as the right-hand sides of equations (2.3-4a and b). The tip of the field vector periodically will trace out a curve in space. Our analysis tells us what that curve may be, and so tells us all the possibilities for the polarization of such a wave.

2.4 SINUSOIDALLY DRIVEN OSCILLATOR

Of the driving forces that are applied to an oscillator, the most common is a force that varies sinusoidally in time, in the general sense of the word

2.4 SINUSOIDALLY DRIVEN OSCILLATOR

sinusoidal. The typical equation of motion has the structure

$$m\ddot{x} = -kx - \gamma\dot{x} + F_0 \cos \omega t, \qquad (2.4\text{-}1)\bigstar$$

which we can write suggestively as

$$\ddot{x} = -\omega_0^2 x - \frac{1}{\tau}\dot{x} + \frac{F_0}{m} \cos \omega t. \qquad (2.4\text{-}2)$$

The angular frequency ω of the driving force is distinct from the natural frequency ω_0 of the oscillator. The response of the oscillator may be complicated, but a portion of that response is surely at the driving frequency ω. The maxima associated with that portion may not coincide in time with the maxima of the driving force; there may be a phase difference. As a form of solution to try, prudence suggests

$$x = A \cos(\omega t + \varphi), \qquad (2.4\text{-}3)$$

where the amplitude A and phase difference φ are to be chosen to satisfy equation (2.4-1). If we substitute this form into equation (2.4-2) and expand the trigonometric functions by the identities for the sum of two angles, namely $\omega t + \varphi$, we find

$$\left\{ A\left[(\omega^2 - \omega_0^2) \cos \varphi + \frac{\omega}{\tau} \sin \varphi\right] + \frac{F_0}{m}\right\} \cos \omega t$$

$$+ A\left[(\omega^2 - \omega_0^2) \sin \varphi - \frac{\omega}{\tau} \cos \varphi\right] \sin \omega t = 0. \quad (2.4\text{-}4)$$

If we evaluate these expressions at $\omega t = \pi/2$, we conclude that the coefficient of $\sin \omega t$ must be zero; from that we extract an expression for $\tan \varphi$:

$$\tan \varphi = \frac{\omega/\tau}{\omega^2 - \omega_0^2}. \qquad (2.4\text{-}5)$$

If we evaluate equation (2.4-4) at $t = 0$, we find that the coefficient of $\cos \omega t$ must be zero and that it will determine A. To find the values of $\cos \varphi$ and $\sin \varphi$, we may regard the numerator and denominator in equation (2.4-5) as proportional to the opposite and adjacent sides, respectively, of a right triangle with angle φ. Then solutions are

$$\sin \varphi = \frac{-\omega/\tau}{[(\omega^2 - \omega_0^2)^2 + (\omega/\tau)^2]^{1/2}}, \qquad (2.4\text{-}6a)$$

$$\cos \varphi = \frac{-(\omega^2 - \omega_0^2)}{[(\omega^2 - \omega_0^2)^2 + (\omega/\tau)^2]^{1/2}}. \qquad (2.4\text{-}6b)$$

Inserting these, we find

$$A = \frac{F_0}{m} \frac{1}{[(\omega^2 - \omega_0^2)^2 + (\omega/\tau)^2]^{1/2}}. \qquad (2.4\text{-}7)$$

The negative signs were adopted in equations (2.4-6a and b) so that the amplitude A would emerge as positive.

We have a solution to equation (2.4-2) but not a complete one; as yet, there is no place to incorporate initial conditions. We may, however, add to $A \cos(\omega t + \varphi)$ a solution like that in equation (2.1-9). By choosing the constants A_+ and A_- appropriately, we may incorporate the initial conditions and thus complete the solution. That addition, however, will die away to insignificance in a time of order 10τ. It is only a transient. Often only the long-time behavior of the oscillator is relevant; that behavior is fully described by $A \cos(\omega t + \varphi)$, which we call the *steady-state solution*.

Figure 2.4-1 illustrates the steady-state amplitude A as a function of the driving frequency ω.

The Lorentz Atom

When an electromagnetic wave of frequency ω passes through a gas, some of the energy is absorbed by the atoms and subsequently radiated (as scattered light). If the frequency of the incident light is altered, how does the rate of absorption change? H. A. Lorentz addressed this question in the early years of the century. Here is how his analysis went.

At the location of any specific atom, the electric field of the light beam merely oscillates sinusoidally with time. The atomic electrons are set into forced motion. If we take the x axis to lie along the field's oscillation direction, then Newton II for that direction becomes

$$m\ddot{x} = -kx - \gamma\dot{x} + qE_0 \cos \omega t. \qquad (2.4\text{-}8)$$

But this equation requires some explanation; let us start from the right-hand end. The factor qE_0 is simply the product of the electron's charge q and the magnitude E_0 of the electric field. (Because the electrons move slowly relative to the speed of light, the force that arises from the magnetic field of the wave may be ignored; that force would not point along $\hat{x}$ anyway.) The term $-\gamma\dot{x}$ represents radiative damping: when the atom radiates stored energy, some electromagnetic force of reaction must act on the motion, to balance the energy budget. The term $-kx$ describes, in Lorentz's words, "a certain elastic force by which an electron is pulled back toward its position of equilibrium after having been displaced from it."*

Lorentz was writing before Rutherford discovered the atomic nucleus (in 1911). The favorite atomic model of the day was Thomson's, a sphere of uniformly distributed positive charge with the electrons *inside* that sphere, which had a radius of atomic size, roughly 10^{-10} meter. The attractive force between an electron and the positive charge would vary exactly as the gravitational force in figure 1.5-2: radially inward and proportional to distance from the orgin. That is a linear restoring force. If the electrons took up equilibruim locations away from the center—conceivable when we include the mutual

---

*H. A. Lorentz, *The Theory of Electrons* (B. G. Teubner, Leipzig, Germany, 1909).

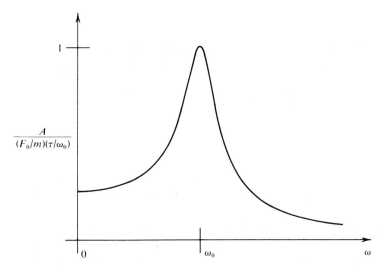

Figure 2.4-1 Amplitude response of the sinusoidally driven harmonic oscillator. The curve is drawn for $\omega_0\tau = 4$, a value that is abnormally small. However, it does keep the resonance broad, so that the behavior at low and high frequencies is visible in detail.

repulsion among the electrons—the restoring forces plausibly would be as Lorentz took them; the typical occurrence of such linear restoring forces is something we will study in section 3.1. In short, there was plenty of justification for equation (2.4-8) when Lorentz proposed it. Nonetheless, what about today? Rutherford's discovery and the rise of quantum theory certainly make the equation suspect. The absorption profile calculated from Lorentz's model does, however, coincide with the profile calculated by quantum theory. Moreover, the electron of quantum theory may be considered, in a certain sense, as bound by a linear restoring force; problem 2-14 describes the calculation. In short, we can marshall justification for the Lorentz model. While not the whole truth, it provides a way to calculate a correct absorption profile; in that spirit, let us adopt equation (2.4-8).

The light emitted by an excited atom determines numerically the natural frequency $\omega_0 = (k/m)^{1/2}$ associated with the Lorentz atom. For visible light, the angular frequency would be of order $2\pi \times 10^{15}$ radians/second. An isolated excited atom radiates spontaneously, typically taking a time of order 10^{-8} second to lose its energy; thus we should take $\tau = m/\gamma$ to be that order. The product $[\omega_0/(2\pi)]\tau$ is then of order 10^7: there are 10^7 oscillations in one decay time, implying very weak damping indeed.

To investigate absorption, we need to look at the processes of energy exchange. We take the sum of the electron's kinetic and potential energies, differentiate with respect to time, and invoke equation (2.4-8):

$$\frac{d}{dt}\left(\frac{1}{2}m\dot{x}^2 + \frac{1}{2}kx^2\right) = -\gamma\dot{x}^2 + \dot{x}qE_0\cos\omega t. \tag{2.4-9}$$

The first term on the right describes energy dissipation via radiation. The second term must describe the rate at which energy is absorbed from the light beam. Since a steady state is established in a time of order $\tau \simeq 10^{-8}$ second, the relevant quantity is the absorption in the steady state when averaged over one oscillation cycle. We can write the average as

$$\langle \text{Energy absorption rate} \rangle = \langle \dot{x} q E_0 \cos \omega t \rangle$$
$$= \langle -\omega q E_0 A \sin(\omega t + \varphi) \cos \omega t \rangle,$$

where equation (2.4-3) provides the expression for $\dot{x}$. The identity

$$\sin(\omega t + \varphi) = \sin \omega t \cos \varphi + \sin \varphi \cos \omega t,$$

together with

$$\langle \sin \omega t \cos \omega t \rangle = \langle \tfrac{1}{2} \sin 2\omega t \rangle = 0,$$

implies

$$\langle \text{Energy absorption rate} \rangle = -\omega q E_0 A \sin \varphi \langle \cos^2 \omega t \rangle$$
$$= \frac{(qE_0)^2}{2m} \frac{\omega^2/\tau}{(\omega^2 - \omega_0^2)^2 + (\omega/\tau)^2}, \quad (2.4\text{-}10)$$

after equations (2.4-6a) and (2.4-7) have been invoked.

Typically we are interested in the absorption profile when the driving frequency ω is close to the natural frequency ω_0. Then we can afford to approximate the difference in the denominator as

$$\omega^2 - \omega_0^2 = (\omega + \omega_0)(\omega - \omega_0) \simeq 2\omega_0(\omega - \omega_0).$$

Elsewhere we may simply replace ω by ω_0. Then we may factor equation (2.4-10) as

$$\langle \text{Energy absorption rate} \rangle \simeq \frac{(qE_0)^2 \tau}{2m} \times \frac{[1/(2\tau)]^2}{(\omega - \omega_0)^2 + [1/(2\tau)]^2}. \quad (2.4\text{-}11)\bigstar$$

The dimensionless factor on the far right, called a *lorentzian* function, is graphed in figure 2.4-2.

The peak of the resonance curve lies at the natural frequency ω_0. How wide is the curve? A convenient measure is the full width when the absorption is one-half its maximum value. Equation (2.4-11) tells us that the absorption drops to one-half its peak value when $|\omega - \omega_0| = 1/(2\tau)$. Thus the full width at half maximum, abbreviated FWHM, is simply this:

$$(\text{FWHM of absorption profile}) = \frac{1}{\tau}. \quad (2.4\text{-}12)$$

For an atomic system such as ours, the width is narrow relative to the resonant frequency itself:

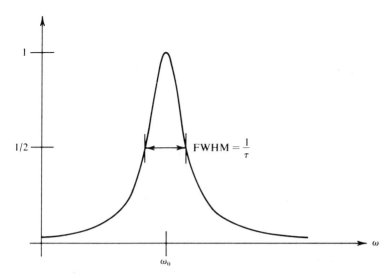

Figure 2.4-2 The lorentzian function is

$$\frac{[1/(2\tau)]^2}{(\omega - \omega_0)^2 + [1/(2\tau)]^2}.$$

The curve is drawn for $\omega_0\tau = 3$.

$$\frac{\text{FWHM}}{\text{Resonant frequency}} = \frac{1}{\omega_0\tau}$$

$$\approx \frac{1}{2\pi \times 10^{15} \times 10^{-8}} = \frac{1}{2\pi \times 10^7}.$$

2.5 QUALITY FACTOR Q

The harmonic oscillator has only two characteristic times: the natural period $2\pi/\omega_0$ and the damping time τ. Any dimensionless measure of behavior that involves time at all must be expressible in terms of these two times, together with numerical factors of order unity. We have seen three such measures already. For the oscillator *without* a driving force, we noted in passing that

$$\text{Number of oscillations in a damping time} = \frac{\tau}{\text{period}} = \frac{\omega_0\tau}{2\pi}, \quad (2.5\text{-}1)$$

and we worked out that

$$\frac{\text{Energy dissipated per period}}{\langle E \rangle} = \frac{2\pi}{\omega_0\tau}. \quad (2.5\text{-}2)$$

The third measure arose when we examined absorption by the sinusoidally driven oscillator. When we compared the width of the absorption profile to the resonance location, we found

$$\frac{\text{FWHM}}{\text{Resonant frequency}} = \frac{1}{\omega_0 \tau}. \qquad (2.5\text{-}3)$$

The ratios in equations (2.5-2) and (2.5-3) were evaluated under the specification of light damping: $\omega_0 \tau \gg 2\pi$.

The notion of "energy dissipated per period" can be applied to the driven oscillator as well as to the undriven. It is the basis for the conventional measure of performance: the *quality factor Q*, defined by

$$Q \equiv 2\pi \frac{\langle E \rangle}{\text{energy dissipated per period}}. \qquad (2.5\text{-}4)$$

When dissipation is small relative to the average energy of the oscillator, the oscillator is, of course, of good quality, and Q is large, to show that.

For the undriven oscillator, we can readily express Q in terms of ω_0 and τ. Equation (2.5-2) and the definition imply

$$Q = \omega_0 \tau \quad \text{provided } \omega_0 \tau \gg 2\pi; \qquad (2.5\text{-}5)$$

the proviso is inserted because equation (2.5-2) was computed in the limit of light damping. For the oscillator driven at resonance, the same expression holds.

Perhaps the most common use of Q is to describe the relative width of an absorption (or emission) profile. Using equation (2.5-5) in equation (2.5-3) yields

$$\frac{\text{FWHM}}{\text{Resonant frequency}} = \frac{1}{Q}. \qquad (2.5\text{-}6) \bigstar$$

High Q implies a narrow profile.

Some representative Q values are displayed in table 2.5-1.

Table 2.5-1 Some representative Q values
Where only an approximate value ($\simeq$) is cited, the quality factor will vary in individual circumstances, even by an order of magnitude or more.

System	Q
50-gram mass hanging from coil spring	$\simeq 25$
Earth, for oscillations induced by earthquake	$\simeq 200$
FM radio receiver	$\simeq 5 \times 10^3$
Tuning fork	$\simeq 10^4$
Sodium atom and yellow light	5×10^7
Iron nucleus (^{57}Fe) and gamma ray	3×10^{12}

2.6 A GUIDE TO THE MAJOR IDEAS

A chapter on the harmonic oscillator easily becomes a labyrinth. There are so many facets to the topic. Here is a guide to the major ideas.

There is a restoring force and a damping force. Both are linear forces: the restoring force is proportional to the displacement, and the damping force is proportional to the velocity. Provided the damping is not severe, the two forces lead to damped oscillations, as described by equation (2.1-10).

There is a natural angular frequency ω_0 that would apply if there were no damping. (That would be a situation with pure sinusoidal oscillations, unchanging in amplitude.) The presence of damping shifts the frequency to ω_0'. Equation (2.1-8) shows ω_0' to be smaller than ω_0, but usually the shift itself is small.

Both the amplitude and the energy of the oscillator decay exponentially. Their time scales for decay are set by the damping constant τ. Because the energy is basically quadratic in the amplitude, it decays with the shorter time scale: τ, rather than 2τ.

Often the oscillator is driven by an additional force, one that varies sinusoidally with angular frequency ω. The response consists of two parts: a transient (which damps out, by definition) and the steady-state solution, purely sinusoidal at the driving frequency ω. We can study the steady-state solution as ω is varied at fixed ω_0 and τ. Typically attention focuses on one or more of the following three quantities:

1. The amplitude A, given by equation (2.4-7).
2. The relative phase φ, defined in equation (2.4-3).
3. The energy absorption rate, given by equation (2.4-11) when ω is close to ω_0.

The quality factor Q describes the insignificance of the damping mechanism in a relative fashion, so that a dimensionless quantity emerges. A look at its definition in equation (2.5-4) will confirm this characterization. Small damping means relatively little energy dissipation; that, in turn, yields a large Q.

If the damping is small, every response of the driven oscillator will have its extreme value (as a function of ω) when ω is near ω_0, the natural frequency. The smaller the damping, the sharper the extreme. Translated into terms of Q, the sentence says, the larger the Q, the sharper the extreme. Equation (2.5-6) is a particularly useful instance of this principle.

WORKED PROBLEMS

WP2-1 A small mass is suspended vertically from a spring as in figure WP2-1. At some instant the mass is released from rest a distance 10 centimeters above its equilibrium position. The mass has made five full oscillations after $t_5 = 10$ seconds. It is again instantaneously at rest, but only at a distance 7 centimeters above the equilibrium position.

64 THE HARMONIC OSCILLATOR

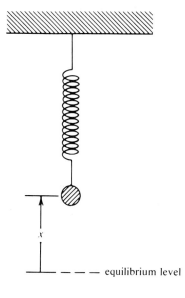

Figure WP2-1 The mass and its displacement x from the equilibrium location.

Problem. For this oscillator determine the natural angular frequency ω_0, the damping constant τ, and the quality factor Q.

The oscillator's displacement is described by equation (2.1-10), but with $\dot{x}(0)$ set to zero because the motion starts from rest:

$$x = x(0)e^{-t/(2\tau)}\left(\cos \omega_0' t + \frac{1}{2\omega_0'\tau} \sin \omega_0' t\right) \tag{1}$$

with $x(0) = 10$ centimeters. The displacement changes with time because of the oscillatory trigonometric functions and because of the exponential damping.

The velocity is somewhat simpler in appearance. Differentiating equation (1) with respect to time yields

$$\frac{dx}{dt} = -x(0)\,\omega_0'\left[1 + \frac{1}{(2\omega_0'\tau)^2}\right]e^{-t/(2\tau)} \sin \omega_0' t.$$

The sine function determines the times of zero velocity. To say that the mass is instantaneously at rest, after five full oscillations, when $t = t_5$ requires

$$\omega_0' t_5 = 5(2\pi).$$

Thus

$$\omega_0' = \frac{5(2\pi)}{10} = \pi.$$

The connection between ω_0' and ω_0 is given by equation (2.1-8). Its square is

$$(\omega_0')^2 = \omega_0^2 - \frac{1}{(2\tau)^2}, \tag{2}$$

and so we need τ before we can get ω_0.

We can extract τ from equation (1) evaluated at time t_5:

$$x(t_5) = x(0) \exp \frac{-t_5}{2\tau}.$$

To solve for τ, we take natural logarithms:

$$\ln \frac{x(t_5)}{x(0)} = \frac{-t_5}{2\tau},$$

whence

$$\tau = -\tfrac{1}{2}t_5/\ln \tfrac{7}{10} = +14 \text{ seconds}.$$

Now we can return to equation (2) and solve for ω_0:

$$\omega_0 = \omega_0'\left[1 + \frac{1}{(2\omega_0'\tau)^2}\right]^{1/2}$$

$$= \omega_0'(1 + 6.5 \times 10^{-5})$$

$$= 3.141796 \text{ radians/second}.$$

Thus ω_0 is quite close to ω_0'.

The strong inequality

$$\omega_0 \tau = 44 \gg 2\pi$$

codifies what we already know directly from the data: we have many oscillations in the time that it takes the damping to change the amplitude much. We can evaluate Q from equation (2.5-5):

$$Q = \omega_0 \tau = 44.$$

WP2-2 Suppose the oscillator that we analyzed in WP2-1 is driven sinusoidally at angular frequency ω. A steady state will develop, but it will require a continuous input of energy (because the damping continuously dissipates energy).

Problem. Is the energy absorption at $\omega = 1.1\omega_0$ much different from the rate at resonance, when $\omega = \omega_0$?

In replying, we can first ask ourselves, Is the shift from ω_0 to $1.1\omega_0$ large? Large relative to what? A natural comparison is with one-half of the full width at half maximum of the absorption profile. Equation (2.5-6) gives us

$$\text{FWHM} = \frac{1}{Q}\omega_0,$$

and so

$$\tfrac{1}{2}\text{FWHM} = \tfrac{1}{2}(\tfrac{1}{44})\omega_0 = 0.011\omega_0.$$

66 THE HARMONIC OSCILLATOR

Thus a shift of $0.1\omega_0$ is relatively large. We should expect a substantial change in energy absorption.

To be more quantitative, we can use equation (2.4-11) and its Lorentzian factor. When $\omega = \omega_0$, the lorentzian factor is unity; when $\omega = 1.1\omega_0$, it is

$$\frac{[1/(2\tau)]^2}{(0.1\omega_0)^2 + [1/(2\tau)]^2} = \frac{[1/(2\omega_0\tau)]^2}{(0.1)^2 + [1/(2\omega_0\tau)]^2}$$

$$= 0.013.$$

Thus the energy absorption rate drops by a factor of 100.

PROBLEMS

2-1 Two masses m_1 and m_2 are connected by a spring and move in frictionless one-dimensional motion. The spring resists both extension and compression relative to its natural length ℓ_0; the spring constant is k. If the masses have an initial separation not equal to ℓ_0, what is the frequency of the ensuing oscillatory motion? (You may want to write down Newton II for each mass separately and then combine the equations to get a differential equation for the relative separation.)

2-2 A cylinder of height h and density ρ floats vertically in a liquid of density ρ_{liquid}. What forces act on the cylinder? Suppose the cylinder is poked, so that it begins to bob up and down. Determine the frequency of oscillation. Can you turn this into an experimental method for determining the unknown density of a fluid? What difficulties might you encounter?

2-3 An object is tethered by a linear restoring force; there is no damping. Determine the time that the object spends in the interval between x_0 and $x_0 + \Delta x$, where Δx is perhaps large, but the endpoints of the interval do, of course, fall within the object's oscillation range: $\pm A$. Express this time as a fraction of the period. Examine the limiting situation $\Delta x \ll A$, and sketch the fraction as a function of x_0 for fixed Δx. Can you explain the trend qualitatively? (If you have studied the harmonic oscillator in quantum mechanics, does your sketch resemble anything you saw there?)

2-4 A laser works by stimulated emission from an excited atomic state. Perhaps the simplest equations that describe the essentials of laser operation are these:

$$\frac{dn}{dt} = +\frac{nN}{\tau} - \frac{n}{\tau_{out}}, \tag{1}$$

$$\frac{dN}{dt} = R - \frac{nN}{\tau} - \frac{N}{\tau_s}. \tag{2}$$

Here n = number of photons in the laser, N = number of excited atoms, nN/τ = stimulated emission rate, n/τ_{out} = laser output (plus internal loss rate),

R = rate at which atoms are pumped (by external means) to the excited state, and N/τ_s = spontaneous emission rate.

(a) Determine the values of n and N when the laser is operating in the steady state, that is, when the time derivatives are zero.

(b) Denote the results in (a) by n_0 and N_0. How do small deviations from them change with time? Write $n = n_0 + \Delta n$ and $N = N_0 + \Delta N$; substitute in equations (1) and (2); discard terms in $\Delta n \times \Delta N$; eliminate n_0 and N_0 in terms of R, etc. Can you derive an equation for ΔN alone? [What if you differentiate your new, "linearized" version of equation (2) and use the analogous version of equation (1)?] Is the structure familiar? Describe the solution for ΔN and then go on to Δn. How would you expect the laser output to vary with time?

An experimental curve, a discussion, and further references can be found in A. Yariv, *Quantum Electronics*, 2d ed. (Wiley, New York, 1975), pp. 273–275.

T**2-5** *Heavy damping.* In the context of section 2.1, specify heavy damping, so that $1/(2\tau) > \omega_0$, and find a general solution that incorporates initial conditions $x(0)$ and $\dot{x}(0)$. Under which initial conditions would you expect $|x(t)|$ to exceed $|x(0)|$ during some interval near the start? Does your solution reproduce this behavior? Sketch your solution for one set of initial conditions. If the mass is headed inward initially, can it—under any choice of initial speed and location—overshoot the origin? (If yes, can you determine them?)

T**2-6** *Critical damping.* The context is section 2.1. We never addressed the situation when $1/(2\tau) = \omega_0$. Determine the solution when that equality holds by taking the limit $\omega_0' \to 0$ in equation (2.1-10). Sketch the behavior of the solution for two specific instances: (a) $x(0) > 0$, $\dot{x}(0) = 0$ and (b) $x(0) = 0$, $\dot{x}(0) > 0$. Next, under which initial conditions will the mass reach the origin in a finite time? And what happens afterward?

Heavy damping, characterized by $\gamma/(2m) \equiv 1/(2\tau) > \omega_0$, produces motion without oscillations. The limit $\omega_0' \to 0$, as you just discovered, takes an oscillatory solution into a solution without oscillations. Thus, at fixed ω_0, the value $\gamma/(2m) = \omega_0$ specifies the least damping that will prevent oscillation.

Of all damping coefficients that prevent oscillation, we may expect $\gamma/(2m) = \omega_0$ to provide the fastest return toward zero from some specified offset value: $x(0) = x_0$ and $\dot{x}(0) = 0$. The restoring force will pull the mass toward zero, and the damping will least resist the motion. The expectation is indeed correct. Instrument design has long incorporated this optimum damping, called *critical damping*, because it stands on the border between oscillatory and nonoscillatory motion.

2-7 A potential energy $U(x) = [k/(4\ell_0^2)]x^4$ will appear later in a realistic physical context. The constant ℓ_0 is a characteristic length, and k has the units of the usual spring constant. Sketch the potential energy. If mass m has the initial conditions $x(0) = x_0$ and $\dot{x}(0) = 0$, what does the trajectory in phase space look like? What is the maximum value of $|\dot{x}(t)|$? What can dimensional analysis reveal about the period of the motion?

T2-8 The context is section 2.3. Compute the angular momentum $\mathbf{r} \times m\mathbf{v}$ of mass m. Why must it be a constant vector? What connections can you make among the phase difference φ, the angular momentum, and the shape of the orbit?

2-9 The context is section 2.3. Suppose

$$\varphi_A = \frac{\pi}{7}, \qquad \varphi_B = \frac{8\pi}{7},$$
$$A = 15, \qquad B = 15\sqrt{3}.$$

Determine the shape of the motion and sketch it.

T2-10 *Lissajous figures.* For an atom vibrating in a typical crystal, the spring constant of the restoring force will differ from one direction to another, orthogonal direction. Suppose that

$$\mathbf{F} = -k_x x \hat{\mathbf{x}} - k_y y \hat{\mathbf{y}},$$

where the constants k_x and k_y are unequal. Solve for x and y as functions of time. If the two-dimensional motion is to be periodic, that is, if the point in the xy plane is to retrace one path periodically, what condition must the ratio of frequencies ω_y/ω_x satisfy? If you have access to a computer or programmable hand calculator, choose such a ratio and plot the path.

T2-11 The context is section 2.4. Form the general solution to equation (2.4-2) by adding to $A \cos(\omega t + \varphi)$ the right-hand side of equation (2.1-9). Why is this sum indeed a solution? Choose $A_+ \pm A_-$ so that initial conditions $x(0)$ and $\dot{x}(0)$ are incorporated. As a check, does your result reduce to equation (2.1-10) if $F_0 = 0$?

2-12 You are given a spring with spring constant $k = 6$ newtons/meter and are asked to construct a lightly damped oscillator with frequency $\simeq 2$ cycles/second and $Q \simeq 30$. What values of mass m and damping constant γ should you choose (to achieve these goals within a few percent)?

2-13 Return to the harmonic oscillator of WP2-1, specified to have mass $m = 50$ grams. Suppose you drive the mass sinusoidally at angular frequency $\omega = 10\omega_0$. To maintain a steady-state oscillation amplitude of 2 centimeters, what force amplitude is required? How much smaller a force would maintain the same amplitude at resonance, $\omega = \omega_0$?

Can you work out a general expression—simple, though only approximate —for the ratio of the forces required, F_0 when $\omega = \omega_0$ relative to F_0 when $\omega \gg \omega_0$?

T2-14 *The restoring force in the Lorentz atom: A quantum version.* Quantum theory takes away the definite orbit traveled by the electron of classical physics. In its place, it puts a probability distribution, something to tell us how likely we are to find an electron here or there. There is a certain probability of finding the (entire) negative charge here or there, and in that sense there is the *equivalent* of a diffuse cloud of negative charge. Let us focus on the simplest atom, hydrogen, and on the ground state. The charge cloud associated with the electron is spherically symmetric and centered on the positive nucleus, a single proton. Now allow an electric field, upwardly directed, to grow slowly from zero magnitude. The negative charge cloud is pulled downward; the massive nucleus moves hardly at all. With what force does the nucleus pull back on the electron cloud? It is easier to ask the opposite, How much does the cloud pull on the nucleus?, and then to use Newton's third law. If the cloud is not greatly disturbed, the situation is like that sketched in figure P2-14. Why is only the charge *inside* the sphere of radius r effective in exerting a force on the nucleus? For a typical external electric field, the displacement r is small relative to the size of the atom. If the negative charge density varies little over the region near the cloud's center, how does the net force change with r? When you turn things around with Newton III, do you find a restoring force on the charge cloud that is proportional to r? And aligned along **E**? Could the location of the cloud's center replace the classical position variable x in equation (2.4-8)?

The picture described here provides a good approximation to quantum reality when **E** changes slowly. When **E** oscillates at a frequency comparable to those at which an excited atom would radiate, doubt must set in about this description, too.

2-15 The context is section 2.4. In discussing the profile of the energy absorption rate, we might want to use equation (2.4-10), rather than the lorentzian approximation in equation (2.4-11). Where does the peak of the exact expression lie (as a function of ω)? If we evaluate the exact expression at $\omega_0 \pm 1/(2\tau)$, the value is not precisely one-half the maximum value (as it is for the lorentzian approximation). Will the value, nonetheless, be within 1 percent of $\frac{1}{2}$ if $\omega_0\tau \geq 100$, say (which might be considered a threshold for weak damping)? Is the lorentzian approximation a good one?

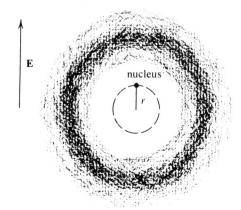

Figure P2-14

2-16 A torsional oscillator, built to study the viscosity of liquid helium, obeys the equation

$$I\ddot{\theta} = -k'\theta - \gamma'\dot{\theta}$$

for small angular displacements θ from an equilibrium orientation. The parameters have the following values: moment of inertia $I = 4.4 \times 10^{-8}$ kilogram · meter²/radian, restoring coefficient $k' = 0.1$ newton · meter/radian, damping coefficient $\gamma' = 1.4 \times 10^{-10}$ newton · meter · second/radian. Compute the natural (angular) frequency ω_0, the characteristic damping time τ, and the quality factor Q.

Suppose a sinusoidal driving torque were applied (at frequency ω). Provide a good analytic approximation for the amplitude of oscillation as a function of frequency when ω is near ω_0. What is the full width at half maximum of the response in amplitude to the driving torque?

T**2-17** *Response phase relative to driving phase.* Return to section 2.4 and the phase difference φ. Specify $-\pi \leq \varphi \leq \pi$, and sketch φ versus ω/ω_0. Can you suggest physical reasons why φ takes on the values which it does when ω/ω_0 approaches zero or infinity?

2-18 This problem addresses the question, How substantially does damping shift the frequency of an oscillator? Suppose we want ω_0' to depart from ω_0 by no more than 0.01 percent. How large must Q be? Translate the condition into a statement such as "_____oscillations in a damping time τ will suffice."

T**2-19** Recall equation (2.4-12): when the damping is light, the FWHM of the absorption profile equals $1/\tau$, which itself equals γ/m. Why should a *width* depend on the damping constant γ?

Consider the oscillator's response as a function of driving frequency ω, either the response in amplitude or the absorption. In which frequency domain is the response limited primarily by frequency mismatch, $\omega \neq \omega_0$? (You will need to examine a denominator.) Sketch the response. Next, in which domain is the response limited primarily by energy dissipation? Now lay stress on the specification *at half maximum* in the term *FWHM*. Can you explain qualitatively why the FWHM must depend on γ?

CHAPTER
THREE

NONLINEAR OSCILLATORS

3.1 A nonlinear oscillator
3.2 Amplitude jumps and hysteresis
3.3 Van der Pol's equation: The limit cycle
3.4 More uses for the averaging method
3.5 Series expansions
3.6 About the methods

Every physicist is exposed early in his career to solvable dynamical problems, for example, the harmonic oscillator and the Kepler problem.... Quite soon, one becomes aware that not all dynamical problems are explicitly solvable.... The reason for this difficulty is the fact that dynamical problems with regular equations may have solutions which behave irregularly in time.

J.-P. Eckmann
Reviews of Modern Physics, October 1981

3.1 A NONLINEAR OSCILLATOR

The harmonic oscillator is characterized by a linear restoring force. Damping, if present, is proportional to the first power of the speed, giving another linear term in the equation of motion. The harmonic oscillator arises time and again, either as an exact description of some oscillating system or as an excellent approximation, but it does not exhaust the field. The restoring force for some os-

72 NONLINEAR OSCILLATORS

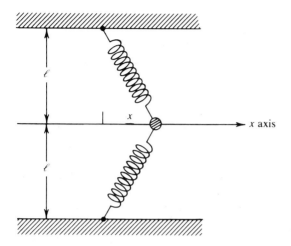

Figure 3.1-1 A mass tethered symmetrically by two springs.

cillators is inherently nonlinear, and the dynamical behavior differs sharply from that of the harmonic oscillator. For other oscillators, the force depends on the speed in a nonlinear fashion, and again the behavior is strikingly different. In this chapter we explore several aspects of nonlinear oscillators; we begin with an example that will develop some generally useful analytic tools.

The context is portrayed in figure 3.1-1. The mass m is tethered by two springs. The relaxed length of each spring is ℓ_0, which may be less than ℓ or greater. Each spring is an ideal Hooke's law spring: it exerts a force exactly k times the result of subtracting ℓ_0 from the actual length, and the force is directed along the spring. Provided we start the mass on the x axis and with initial velocity (if any) along $\hat{x}$, symmetry says that the motion will be purely along $\hat{x}$: we have a one-dimensional problem. What is the motion like?

As a first step toward answering the question, let us compute the potential energy. The actual length of each spring, denoted by s, is $s = (\ell^2 + x^2)^{1/2}$. Because the springs are ideal, the potential energy is quadratic in $s - \ell_0$, the length by which the spring is extended or compressed. We may write

$$\begin{aligned} U(x) &= 2 \cdot \tfrac{1}{2} k (s - \ell_0)^2 \\ &= k(s^2 - 2s\ell_0 + \ell_0^2) \\ &= k[\ell^2 + x^2 - 2\ell_0(\ell^2 + x^2)^{1/2} + \ell_0^2]. \end{aligned} \quad (3.1\text{-}1)$$

If the mass's displacement is relatively small, so that $x^2 \ll \ell^2$, we can profitably expand the square root by the binomial theorem. If we first factor out ℓ^2, the expansion generates

$$\ell\left(1 + \frac{x^2}{\ell^2}\right)^{1/2} = \ell\left(1 + \frac{1}{2}\frac{x^2}{\ell^2} - \frac{1}{8}\frac{x^4}{\ell^4} + \cdots\right).$$

After this has been inserted into equation (3.1-1), we can collect terms into the expression

$$U(x) = k\left[(\ell - \ell_0)^2 + \frac{\ell - \ell_0}{\ell} x^2 + \frac{\ell_0}{4\ell^3} x^4 + \cdots\right]. \quad (3.1\text{-}2)$$

The potential energy is symmetric about $x = 0$, but its behavior near the origin depends on the difference $\ell - \ell_0$. If $\ell > \ell_0$, signifying that the springs are always stretched, then $\ell - \ell_0 > 0$, and the potential energy rises immediately, as shown in figure 3.1-2. If $\ell < \ell_0$, the springs are compressed when near the origin (but may relax to natural length some distance away); then $\ell - \ell_0 < 0$, and the potential energy heads downward until the positive term in x^4 dominates the negative term in x^2 and sends the curve back up. Finally, if $\ell = \ell_0$, the springs are relaxed when at the origin. A small displacement perpendicular to that vertical orientation changes the spring's length hardly at all, and so the potential energy has no quadratic term. Growth depends on the term in x^4, which implies a comparatively broad potential well.

For each situation—ℓ relative to ℓ_0—there is at least one potential well: a location of stable equilibrium (for motions that are not too vigorous, if $\ell < \ell_0$). Let us examine the small oscillations that can occur around each such potential minimum.

Small Ocillations

In general, Newton II for our mass is

$$m \frac{d^2x}{dt^2} = -\frac{dU(x)}{dx}. \quad (3.1\text{-}3)$$

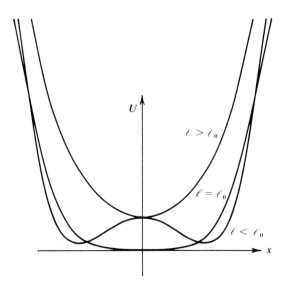

Figure 3.1-2 The potential energy $U(x)$ in approximation (3.1-2). For the sake of comparison, the difference $|\ell - \ell_0|$ is the same in the two curves where $\ell \neq \ell_0$. The exact expression for $U(x)$ would carry the curve, when $\ell < \ell_0$, all the way down to $U = 0$ before sending it up again.

74 NONLINEAR OSCILLATORS

For small displacements from an equilibrium position x_{eq}, we may expand the right-hand side in a Taylor series:

$$\frac{dU(x)}{dx} = \left.\frac{dU}{dx}\right|_{eq} + \left.\frac{d^2U}{dx^2}\right|_{eq}(x - x_{eq}) + \cdots. \tag{3.1-4}$$

By an "equilibrium position," we mean, of course, a position where the force is zero. Thus the first term, $(dU/dx)_{eq}$, is zero. The natural dependent variable is $x - x_{eq}$, the displacement from the equilibrium location. Because the derivative of a constant like x_{eq} is zero, we may use the displacement on the left-hand side of Newton II, combining equations (3.1-3) and (3.1-4), to get

$$m\frac{d^2}{dt^2}(x - x_{eq}) = -\left.\frac{d^2U}{dx^2}\right|_{eq}(x - x_{eq}) \tag{3.1-5}\bigstar$$

if we drop the higher terms in the Taylor series. The *structure* is familiar: for the displacement $x - x_{eq}$, we have just the harmonic oscillator equation. The second derivative of the potential energy (its curvature) plays the role of the spring constant. Reasoning by analogy with section 2.1, we can read off the angular frequency of the small oscillations:

$$\omega_0^2 = \frac{1}{m}\left.\frac{d^2U}{dx^2}\right|_{eq}. \tag{3.1-6}$$

Now we apply this scheme to the potential wells in figure 3.1-2.

Case 1 If $\ell > \ell_0$, the equilibrium location is $x_{eq} = 0$. Differentiating equation (3.1-2) twice and then evaluating at $x = 0$ gives

$$\left.\frac{d^2U}{dx^2}\right|_{eq} = 2k\frac{\ell - \ell_0}{\ell},$$

and so

$$\omega_0^2 = \frac{2k}{m}\frac{\ell - \ell_0}{\ell} \quad \text{if } \ell > \ell_0. \tag{3.1-7}$$

Case 2 When $\ell < \ell_0$ holds, we cannot locate the potential minimum by inspection. We can, however, differentiate U once and look for the locations that give zero slope:

$$\frac{dU}{dx} = k\left(2\frac{\ell - \ell_0}{\ell}x + \frac{4\ell_0}{4\ell^3}x^3 + \cdots\right) = 0$$

yields

$$x_{eq}^2 = \frac{2(\ell_0 - \ell)\ell^2}{\ell_0}, \tag{3.1-8}$$

if we work with just the leading terms. Taking a square root would give us the two symmetrically placed equilibrium positions. The second derivative of U produces

$$\left.\frac{d^2U}{dx^2}\right|_{eq} = k\left(2\frac{\ell-\ell_0}{\ell} + \frac{3\ell_0}{\ell^3}x_{eq}^2\right)$$

$$= 4k\frac{\ell_0-\ell}{\ell}, \qquad (3.1\text{-}9)$$

and so

$$\omega_0^2 = \frac{4k}{m}\frac{\ell_0-\ell}{\ell} \qquad \text{if } \ell < \ell_0. \qquad (3.1\text{-}10)$$

Note the factor of 4 here relative to the factor of 2 in equation (3.1-7). Even if $|\ell-\ell_0|$ were the same for the extended and compressed cases, the frequencies would differ because the curvature of $U(x)$ is different at the respective potential minima.

Case 3 The third situation is $\ell = \ell_0$. The equilibrium location is $x_{eq} = 0$, but equation (3.1-2) implies that $(d^2U/dx^2)_{eq} = 0$. Does that mean "no oscillation"? Just reasoning from a sketch of $U(x)$ tells us, "no, there is a potential minimum, and so there will be some kind of restoring force, whence oscillations are sure to occur." The zero value for the second derivative tells us merely that we need to go higher in the Taylor series, equation (3.1-4), continuing until we find a nonzero term. Indeed, the fourth derivative is nonzero. Using either the Taylor series or equation (3.1-2) directly, we find

$$m\frac{d^2x}{dt^2} = -\frac{k}{\ell_0^2}x^3, \qquad (3.1\text{-}11)$$

to lowest significant order. The right-hand side is cubic in x; this is *not* the equation for a harmonic oscillator. The solution will be periodic—our intuition tells us that—but can be expressed exactly only in terms of elliptic functions. That is not a route we want to take. Let us see what dimensional analysis can tell us about the frequency.

The differential equation has only one characteristic constant, the combination $k/(m\ell_0^2)$. Its dimensions are $1/(\text{time}^2 \cdot \text{length}^2)$. A frequency, however, has the dimensions of $1/\text{time}$, and so some other constant must enter the expression for the frequency. That constant must have the dimensions of length. The only plausible candidate is the maximum displacement during the oscillation; we will call it x_{max}. Thus we arrive at the proposition that

$$\left(\begin{array}{c}\text{Frequency}\\ \text{when } \ell=\ell_0\end{array}\right)^2 = \frac{k}{m}\left(\frac{x_{max}}{\ell_0}\right)^2\left(\begin{array}{c}\text{dimensionless constant}\\ \text{of order 1}\end{array}\right).$$

Dimensional reasoning alone cannot tell us what numerical factor must be supplied, but there is no reason for the number to depart substantially from 1; it should lie in the range $\frac{1}{10}$ to 10, say. Unlike the situation with the harmonic oscillator, which satisfies a linear equation, the frequency here depends on the amplitude of oscillation. It grows with x_{max}. We noted in passing that, when $\ell = \ell_0$, the potential well is anomalously broad. Large amplitude takes the mass out to where the walls of the well steepen, implying a large restoring force, which sends the mass hurrying back. We have a pleasantly paradoxical situation: the longer the round trip, the shorter the time it takes.

Phase Space Again

The potential energy curve with two symmetrically placed minima is the richest of the three. What kind of trajectories in phase space does it generate? Figure 3.1-3 reproduces the potential energy curve, based on approximation (3.1-2); drawn on that portion of the figure are three total-energy lines. The energy rela-

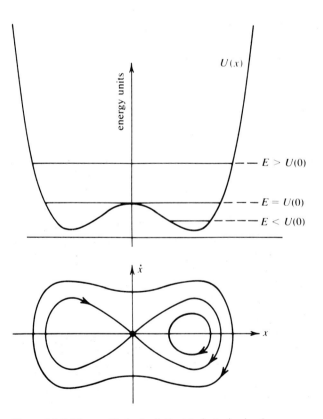

Figure 3.1-3 The qualitatively distinct trajectories in phase space.

tive to the (local) potential maximum at $x = 0$ is the crucial comparison, and all three qualitatively distinct situations are depicted. The vertical distance between E and $U(x)$ is, of course, equal to the kinetic energy. That energy is zero where the U curve intersects the total-E line. Thus the intersections are turning points of the motion along x, points where $\dot{x} = 0$ instantaneously as $\dot{x}$ changes smoothly from positive to negative or vice versa. The intersections of U and E in the energy diagram tell us where the trajectory in phase space has $\dot{x} = 0$ and hence lies on (or crosses) the x axis. We can see that in the lower portion of the figure.

When $E < U(0)$, the motion is confined to the vicinity of a single potential well. As we saw in equations (3.1-5) and (3.1-9), the restoring force is approximately linear in the displacement, at least if the displacement is small. That meant harmonic motion, and so the trajectory is, as we learned in section 2.2, an ellipse, at least to good approximation. Since we calculated the angular frequency of small oscillations in equation (3.1-10), we know the period of such motion. What about the motion for larger amplitudes of oscillation, but still confined to the well? We take up that kind of question in detail in sections 3.4 and 3.5, but already here we can derive an exact integral expression for the period. It is based on energy conservation:

$$\tfrac{1}{2} m \dot{x}^2 + U(x) = E.$$

Solving for $\dot{x}$ gives

$$\dot{x} = \pm \left(\frac{2}{m}\right)^{1/2} [E - U(x)]^{1/2}.$$

If we remember that $\dot{x} = dx/dt$, we can write

$$\text{Time to travel from one turning point to the other} = \int dt = \int \frac{dx}{\dot{x}}$$

$$= \pm \int_{\text{one point}}^{\text{the other}} \frac{dx \ (m/2)^{1/2}}{[E - U(x)]^{1/2}}. \qquad (3.1\text{-}12) \bigstar$$

The period is simply twice this one-way travel time. (We choose between plus and minus so that the result is positive.) The merit of the expression is this: We do not need to know the motion in detail, we do not need to know $x(t)$. An integral with the potential energy $U(x)$ suffices. The integral may be difficult, but it can always be evaluated numerically, to any desired accuracy. The integral expression is, of course, not limited to the present context; it applies to any one-dimensional system with turning points, one or even both of which may lie at infinity and hence not be literal turning points.

When $E > U(0)$, the mass may pass through the spatial origin, and so the trajectory in phase space extends into both half planes, $x > 0$ and $x < 0$. The kinetic energy is largest when the mass passes through the potential minima; hence the trajectory bulges out toward large $|\dot{x}|$ at those times.

The behavior when $E = U(0)$ is subtle. Suppose we place the mass at rest at the positive location x_0 where the U curve intersects the $E = U(0)$ curve. That is the positive turning point. Because $-(dU/dx)_{x_0}$ is negative, there is an inward force, and the mass starts inward, picking up speed as it goes (at least for a while). How long does it take the mass to reach the origin? Equation (3.1-12) was derived to answer this kind of question. Taking the potential energy $U(x)$ from equation (3.1-2) and remembering that $E = U(0)$ here, we can write

$$E - U(x) = -k\left(\frac{\ell - \ell_0}{\ell} x^2 + \frac{\ell_0}{4\ell^3} x^4\right)$$

$$= +k \frac{\ell_0 - \ell}{\ell} x^2 \left[1 - \frac{\ell_0}{4\ell^2(\ell_0 - \ell)} x^2\right].$$

Since $(\ell_0 - \ell) > 0$ in the present context, we may write

$$\begin{pmatrix}\text{Time to travel}\\ \text{from } x_0 \text{ to } x = 0\end{pmatrix} = -\left[\frac{m\ell}{2k(\ell_0 - \ell)}\right]^{1/2} \int_{x_0}^{0} \frac{dx}{x\left[1 - \frac{\ell_0 x^2}{4\ell^2(\ell_0 - \ell)}\right]^{1/2}}.$$

We need to examine the integrand near $x = 0$. The square root approaches 1, but the factor $1/x$ blows up badly; indeed, the integral diverges logarithmically. Formally, that means an infinite travel time. A better way to put it is this: the mass continues to approach the origin indefinitely. At any finite time, the mass is still some distance away, but creeping inward.

We can find the reason for this behavior if we examine the force that slows the particle. The force, whose magnitude is given by the slope of the potential energy curve, gets progressively weaker as the particle approaches $x = 0$. The origin in phase space corresponds to $x = 0$ and $\dot{x} = 0$, that is, to no motion. Because the retarding force gets progressively weaker, the stopping time can be indefinitely long, and the mass never reaches the origin.

The situation at a turning point such as x_0 is quite different. There the slope dU/dx is nonzero; there is a nonvanishing force; and the turning-around process is accomplished in a finite time. Although the *integrand* will diverge as $U(x)$ approaches E, the *integral* will converge to a finite time.

But back to the neighborhood of the origin (where the mass is still creeping in). The indefinitely prolonged approach saves us from an inconsistency. In section 2.2 we noted that a trajectory in phase space could be computed from a pair of coupled first-order equations. Their structure was

$$\frac{d\dot{x}}{dt} = \frac{F(x, \dot{x})}{m}, \tag{3.1-13a}$$

$$\frac{dx}{dt} = \dot{x}, \tag{3.1-13b}$$

where $F(x, \dot{x})$ denotes the force as a function of position and, possibly, of velocity. These equations enable us to extend the trajectory a step forward in time:

$$\dot{x}(t + \Delta t) = \dot{x}(t) + \frac{1}{m} F[x(t), \dot{x}(t)] \Delta t, \qquad (3.1\text{-}14a)$$

$$x(t + \Delta t) = x(t) + \dot{x}(t) \Delta t. \qquad (3.1\text{-}14b)$$

The step forward is uniquely determined by the location in phase space at time t. The same would be true of a step backward, if we took Δt to be negative. The implication of this uniqueness is that distinct trajectories can never intersect each other or merge. The proposed "impossibilities" are sketched in figure 3.1-4. The proof is by contradiction. If two trajectories were to intersect or merge, then at that location in phase space there would be two different directions for a forward step or two directions for a backward step or both. Equations (3.1-14a and b) guarantee uniqueness, however, and so the proposed intersections or mergers cannot occur.

The uniqueness does not prevent two trajectories from coming closer and closer to each other. What is required is merely that they never meet in finite time; they may approach each other asymptotically, as time goes to infinity, which is "almost meeting, but not literally." That is the situation near the origin in the phase space of figure 3.1-3. The point at the origin, $x = 0$ and $\dot{x} = 0$, is a trajectory all by itself. If the mass is placed at $x = 0$ with zero velocity, it will remain there forever, ideally at least, for that is a location of (unstable) equilibrium. The lobes on either side are two other, distinct trajectories. If the mass is started at a point on either lobe, eventually the mass will move close to the origin, but it will never get there literally. The three trajectories—two lobes and the central point—remain distinct non-intersecting trajectories for all finite times.

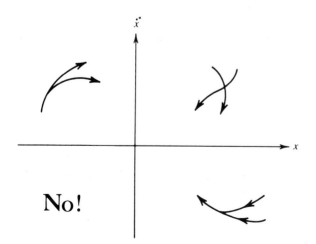

Figure 3.1-4 Intersections and mergers such as these *cannot* occur in phase space.

Instability

We just noted, in passing, that the potential hill at $x = 0$ (when $\ell < \ell_0$) is a location of unstable equilibrium. What does that imply analytically? If we insert the potential energy (3.1-2) into Newton II and retain just the leading term in the force, we have

$$m \frac{d^2x}{dt^2} = -2k \frac{\ell - \ell_0}{\ell} x$$

$$= +2k \frac{|\ell_0 - \ell|}{\ell} x. \qquad (3.1\text{-}15)$$

If the displacement x is positive, so is the force; it is anti-restoring, tending to increase the displacement.

Because equation (3.1-15) is linear and homogeneous in x, with constant coefficients, the solutions are exponentials. The exponents will be $\pm \alpha t$ where

$$\alpha = +\left[\frac{2k|\ell_0 - \ell|}{m\ell}\right]^{1/2}. \qquad (3.1\text{-}16)$$

To meet initial conditions, we can combine the two distinct solutions as

$$x(t) = x(0) \frac{e^{\alpha t} + e^{-\alpha t}}{2} + \dot{x}(0) \frac{e^{\alpha t} - e^{-\alpha t}}{2\alpha}. \qquad (3.1\text{-}17)$$

If the mass starts at rest but with $x(0) \neq 0$, the displacement grows exponentially. (We can afford to ignore the decreasing exponential relative to the growing exponential.) Similarly, if the mass is initially at the origin but $\dot{x}(0) \neq 0$, it will depart exponentially. Generic initial conditions—a random choice of $x(0)$ and $\dot{x}(0)$—lead, sooner or later, to rapid and accelerated departure. The sole exception is the special trajectory of the previous subsection, where we could head the mass inward and have it spend infinite time creeping toward the center. In real life, achieving that trajectory is a practical impossibility. A random choice of initial conditions is almost certain to see the mass make a speedy departure.

Summary of Principles

In this section we have worked our way from example to general principle and back again, indeed, more than once. Such a sequence is natural when principles are developed as the need arises, but now a summary of the principles is in order.

The first item is small oscillations about an equilibrium. We expand the potential energy (or the force itself) in a Taylor series about the equilibrium location. Then we convert Newton II into an equation for the displacement from equilibrium. If there is a linear restoring term, we have harmonic oscillations. If the linear term is absent, there may still be a restoring term farther out in the

Taylor series, but the equation will be inherently nonlinear, and so a quick answer may not be forthcoming. Finally, if there is a linear anti-restoring term, then the equilibrium is unstable, and the displacement will (typically) grow exponentially in time.

The travel time between two specific points can be calculated as the integral of $dx/\dot{x}$. This proposition is useful if all the forces can be derived from a potential energy, so that $\dot{x}$ can be expressed in terms of $U(x)$ and E.

To map out trajectories in phase space, we can convert Newton II to a pair of first-order equations. The evolution of x and $\dot{x}$ then follows by stepwise integration, one step Δt after another.

When the entire force is derivable from a potential energy U, a graph of U versus x provides a powerful aid to reasoning about the motion. We can read off equilibrium locations and can determine by inspection whether each is stable or unstable. After drawing total energy lines and recalling that $\frac{1}{2}m\dot{x}^2 = E - U$, we can see even more: we can discern the qualitative behavior of the actual motion for various initial conditions as well as the qualitative shape of the trajectory in phase space.

Trajectories in phase space can never intersect each other or merge. They may, however, approach each other asymptotically, as time goes to infinity, or they may have been close in the past. [Our proof depended on the force being a function of x and $\dot{x}$ only, so that its value was determined by the location in phase space. If the force depends on time explicitly, $F(x, \dot{x}, t)$, then new possibilities are opened. Problem 3-10 explores them.]

3.2 AMPLITUDE JUMPS AND HYSTERESIS

A hacksaw blade clamped at one end makes a fine vibrating oscillator. If we cement a ceramic magnet to the free end, we can drive the oscillator sinusoidally by running an alternating current through a coil placed near the ceramic magnet. The apparatus is sketched in figure 3.2-1. For small amplitudes of oscillation, the restoring force is proportional to the displacement, a linear restoring force. At amplitudes of a centimeter or so, the steel ceases to respond linearly. We can figure out the general form of the first correction to linear restoring. The linear force has a potential energy $\frac{1}{2}kx^2$ that is quadratic and symmetric between $+x$ and $-x$. The corrections to the potential energy also should be symmetric, and so a term in x^4 is the most likely candidate. (It would be the next nonzero term in the Taylor expansion of a symmetric potential.) To obtain the force, we need to differentiate the potential energy; the term in x^4 will produce a correction force proportional to x^3. Thus we are led to an equation of the form

$$m\ddot{x} = -kx + \frac{k}{\lambda^2} x^3 - \gamma\dot{x} + F_0 \cos \omega t. \qquad (3.2\text{-}1)$$

The length λ is a distance such that, when $|x| = \lambda$, the magnitude of the

82 NONLINEAR OSCILLATORS

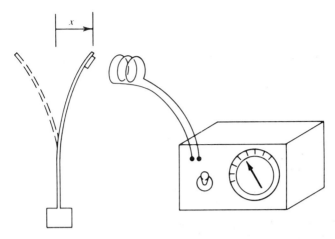

Figure 3.2-1 The saw blade oscillator. The frequency ω of the driving force can be adjusted at the voltage source.

nonlinear force equals the magnitude of the linear restoring force (or would equal it, if no higher terms were needed). The sign for the correction term is a guess; if it turns out to be wrong, we can just go back through the analysis, replacing λ^2 by $-\lambda^2$. The damping term is linear in the velocity; quadratic might be more realistic, but we have little need for the damping term, and so the issue is not significant. As it stands, equation (3.2-1) is a classic nonlinear equation, *Duffing's equation,* named after Georg Duffing, a German engineer who studied it extensively around 1915.

Duffing's Equation without Damping

For the moment, let us drop the damping term and write

$$\ddot{x} + \omega_0^2 x - \omega_0^2 \frac{x^3}{\lambda^2} = \frac{F_0}{m} \cos \omega t, \tag{3.2-2}$$

where $\omega_0 = (k/m)^{1/2}$, as before. As long as the maximum value of $|x|$ is small relative to λ, as we will suppose, the nonlinear term remains a small correction to the linear restoring term, and we can be guided—somewhat—by our experience with the driven harmonic oscillator. That oscillator, when driven at ω, has a steady state that is periodic with the period of the driving force. For equation (3.2-2), can we find an analogous periodic solution, one whose major amplitude (at least) comes in a term that oscillates at the driving frequency ω?

A trial form for the solution would start as

$$x(t) = c_1 \cos \omega t + \cdots, \tag{3.2-3}$$

where c_1 is a constant to be determined. Difficulty arises when we insert the

3.2 AMPLITUDE JUMPS AND HYSTERESIS

trial form into the nonlinear term: it generates $\cos^3 \omega t$, and that cube amounts to a linear combination of $\cos \omega t$ and $\cos 3\omega t$. A way to see this is through the identity

$$\cos a = \frac{e^{ia} + e^{-ia}}{2},$$

whose cube yields

$$\cos^3 a = \tfrac{1}{8}(e^{i3a} + 3e^{ia} + 3e^{-ia} + e^{-i3a})$$
$$= \tfrac{1}{4}(\cos 3a + 3 \cos a). \tag{3.2-4}$$

Since we will generate $\cos 3\omega t$ via the differential equation, let us include such a term in the trial solution:

$$x(t) = c_1 \cos \omega t + c_3 \cos 3\omega t + \cdots \tag{3.2-5}$$

We hope to find $|c_3/c_1| \ll 1$, or at least we hope to be able to choose parameter values so that the strong inequality holds.

Now we can substitute the augmented trial form into equation (3.2-2) and collect all terms in each frequency:

$$\left[(\omega_0^2 - \omega^2)c_1 - \frac{3}{4}\omega_0^2 \frac{c_1^3}{\lambda^2}\left(1 + \frac{c_3}{c_1} + 2\frac{c_3^2}{c_1^2}\right) - \frac{F_0}{m}\right]\cos \omega t$$

$$+ \left[(\omega_0^2 - 9\omega^2)c_3 - \frac{\omega_0^2}{4}\frac{c_1^3}{\lambda^2}\left(1 + 6\frac{c_3}{c_1} + 3\frac{c_3^2}{c_1^2}\right)\right]\cos 3\omega t + \cdots = 0. \tag{3.2-6}$$

(both cubed)

Indicated only by the ellipsis dots are terms in $\cos 5\omega t$, $\cos 7\omega t$, and $\cos 9\omega t$ generated by the nonlinear term. We trust that if c_3 is relatively insignificant, those terms are even more so. If we evaluate equation (3.2-6) at $\omega t = \pi/6$, so that $\cos 3\omega t = 0$, we find that the coefficient of $\cos \omega t$ must be zero. Then an evaluation at $\omega t = 0$, say, implies that the coefficient of $\cos 3\omega t$ must also be zero.

Can we have $|c_3/c_1| \ll 1$ self-consistently? If we assume provisionally that the strong inequality holds, then the zero coefficient of $\cos 3\omega t$ in equation (3.2-6) implies

$$c_3 \simeq \frac{1}{4}\frac{c_1^3}{\lambda^2}\frac{\omega_0^2}{\omega_0^2 - 9\omega^2},$$

which we can write as

$$\frac{c_3}{c_1} \simeq \frac{1}{4}\frac{c_1^2}{\lambda^2}\frac{1}{1 - 9\omega^2/\omega_0^2}. \tag{3.2-7}$$

Provided $|c_1| \ll \lambda$ and provided ω is not near the subharmonic $\omega_0/3$, the assumption that $|c_3/c_1| \ll 1$ is indeed self-consistent.

An equation for c_1 now follows from the zero coefficient of $\cos \omega t$ in equation (3.2-6):

84 NONLINEAR OSCILLATORS

$$(\omega_0^2 - \omega^2)c_1 - \frac{3}{4}\omega_0^2 \frac{c_1^3}{\lambda^2} \simeq \frac{F_0}{m}. \tag{3.2-8}$$

This is a cubic equation in c_1, and it may have three real roots. To extract c_1 as a function of ω^2 at fixed F_0/m, we write the equation this way:

$$\left(\omega_0^2 - \frac{3}{4}\frac{\omega_0^2}{\lambda^2}c_1^2\right) - \frac{F_0/m}{c_1} = \omega^2. \tag{3.2-9}$$

If we think of the two terms on the left-hand side *as functions of* c_1, then the first term generates a parabola, and the second, a hyperbola. These curves are graphed in figure 3.2-2; it may help to rotate the figure 90°, so that its original left-hand side is at the bottom. We may, for the moment, think of equation

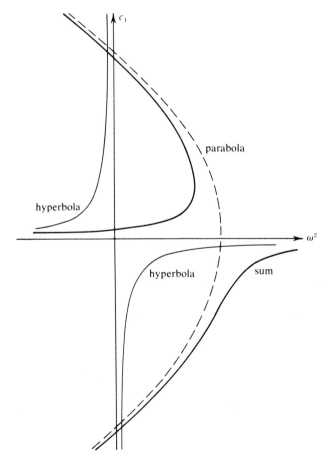

Figure 3.2-2 Solving cubic equation (3.2-9). The parabola is the dashed curve; the hyperbola consists of the two lightly drawn curves. Their sum, computed at the same value of c_1 for both terms, is drawn with the heavy line.

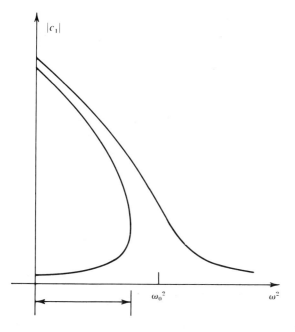

Figure 3.2-3 The amplitude c_1 is triple-valued in the interval between the arrowheads.

(3.2-9) as giving ω^2 as the sum of a parabola and a hyperbola in c_1. That sum is indicated in the figure, too; the summing is best seen with the figure rotated counterclockwise by 90°. The "sum" curve is the locus of points that satisfy equation (3.2-9). Once we have found those points by pretending to compute ω^2 as a function of c_1, we can turn things around: the sum curve gives us c_1 as a function of ω^2.

The lower sum curve describes $c_1 < 0$. The magnitude of c_1, however, is of primary interest. For a graph of $|c_1|$ versus ω^2, we need only flip over the lower sum curve, as shown in figure 3.2-3. In the actual physical problem, $\omega^2 \geq 0$. We can ignore those portions of the curves that lie in the left-hand plane; they are extraneous mathematical solutions. But in the physical half of the plane, there is an interval where the amplitude $|c_1|$ is a triple-valued function of ω^2. The ultimate cause is the cubic nonlinear term in equation (3.2-1).

Recall that we have omitted damping. We find, nevertheless, that the amplitude remains finite even if we drive the oscillator at its nominal resonant frequency ω_0. This is wholly different from the (linear) harmonic oscillator. For it, the amplitude diverges at $\omega = \omega_0$ unless damping is included. The curves in figure 3.2-3 do, however, resemble the amplitude curves of the harmonic oscillator—but twisted over. When the amplitude is large, damping will become appreciable. Just as damping rounds over the diverging curves for the harmonic oscillator, so here it will cause the two upper branches to merge smoothly at some ω^2 not far below ω_0^2. This is illustrated in figure 3.2-4; an interval of triple-valuedness is preserved.

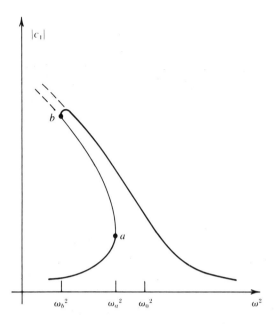

Figure 3.2-4 Amplitude c_1 when damping is included. Damping causes the two upper branches to merge smoothly and to do so at a value of ω^2 that is safely distant from $\omega_0^2/9$, a point that self-consistency—in equation (3.2-7)—dictates must be avoided. The stable portions of the single, continuous curve are drawn more heavily. The amplitude jumps occur at points a and b: upward at a, downward at b.

An experiment can suggest what the triple-valuedness implies. Figure 3.2-5 exhibits data taken with the saw blade apparatus sketched in figure 3.2-1. As the driving frequency was increased in small steps from an initial low value, the amplitude of oscillation increased slowly—and then jumped upward by a factor of 2. Further increase in ω produced a moderate, continuous decrease in amplitude. Then the frequency was decreased in small steps. At first the amplitude retraced its former path; then it continued smoothly past the location of the jump, growing ever larger. The amplitude reached a peak, descended a bit—and then plummeted to a low value. From there on, it retraced its original path.

With this information in hand, we can return to the theory in figure 3.2-4. If the driving frequency is less than ω_b, the oscillator has no options: it must have a small amplitude. As ω is increased, the amplitude increases smoothly until $\omega = \omega_a$; then it jumps discontinuously to the upper portion of the curve, which it follows as ω is increased further. If ω is now decreased, the amplitude follows the upper curve until ω reaches ω_b; then the amplitude drops discontinuously and retraces its original path. Remarkable!

Where the amplitude is triple-valued, only the upper and lower values are stable. (The middle section, the portion between points a and b, is unstable to small perturbations.) But which of the stable values the oscillator adopts depends on the oscillator's history. Here is an instance of *hysteresis:* the system's past influences its present behavior. Already Duffing's equation, with nothing more than a cubic nonlinearity, exhibits the richness of nonlinear systems.

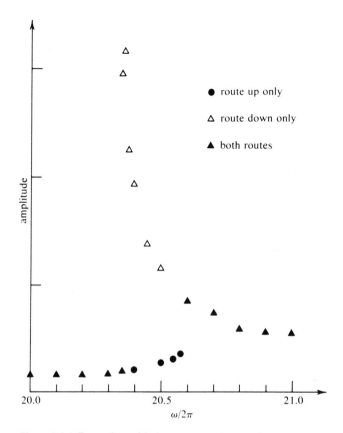

Figure 3.2-5 Data taken with the apparatus sketched in figure 3.2-1. The dots denote amplitude values observed as the frequency was *increased* from 20 cycles/second in small steps; the open triangles, as the frequency was *decreased* from 21 cycles/second. Where the values coincide, the dot and open triangle appear as a filled triangle.

If the driving force F_0 is small, the motion should be of small amplitude, in particular, much less than the characteristic length λ. Then the nonlinear term in equation (3.2-1) should be negligible relative to the linear restoring force, and the system should reduce to the driven, damped harmonic oscillator. Figure 3.2-6 confiirms this. As F_0 is decreased, the amplitude curve evolves from triple-valued to "single-valued but tipped over" and then to the almost-symmetric shape of the harmonic oscillator.

The hysteresis and amplitude jumps of Duffing's equation can be demonstrated in many ways, far more ways than just the saw blade oscillator. Several are described in four articles in the *American Journal of Physics:* D. P. Stockard, T. L. Johnson, and F. W. Sears, **35,** 961 (1967); J. N. Fox and J. J. Arlotto, **36,** 326 (1968); J. A. Warden, **38,** 773 (1070); and T. W. Arnold and W. Case, **50,** 220 (1982).

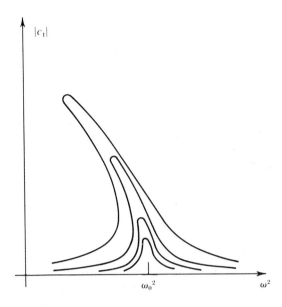

Figure 3.2-6 When the driving force F_0 is decreased in magnitude, the shape of the amplitude-versus-ω^2 curve reduces continuously to the single-valued, almost-symmetric shape of the (damped) harmonic oscillator.

3.3 VAN DER POL'S EQUATION: THE LIMIT CYCLE

An old-fashioned electric circuit with vacuum tubes and iron-core inductors led Balthasar van der Pol (around 1920) to the equation that now bears his name:

$$\ddot{x} = -\omega_0^2 x - \mu\,(x^2 - \ell_0^2)\dot{x}, \tag{3.3-1}$$

where μ is a positive constant. It is a nonlinear equation whose structure arises surprisingly often; for example, the structure reappeared in the 1960s in a theory of the laser. We will talk about van der Pol's equation as it might arise in a mechanical context.

If $|x|$ is greater than ℓ_0, which is the characteristic length in the problem, the velocity-dependent term describes some kind of damping. If, however, $|x|$ is less than ℓ_0, then the velocity-dependent force is in the direction of the motion, not opposed, and so energy is fed into the motion. This is sometimes called *anti-damping*. Let's see what qualitative reasoning in phase space has to suggest about the motion.

Suppose we start the system point at point a in figure 3.3-1. If μ were zero, the trajectory would be an ellipse with $|x| < \ell_0$ always, and so in the actual situation, energy will go *into* the motion: the trajectory will spiral outward, at least until some portion crosses the lines $|x| = \ell_0$. (The first crossing comes where $|\dot{x}|$ is small and so has little effect; the spiral should continue outward for some substantial distance.)

Now switch to point b. If μ were equal to zero, the system point would move from b in a large elliptical trajectory. Most of such a trajectory would lie in the damping region, and so—on the actual trajectory—a lot of energy is extracted from the system. To be sure, energy is fed in when $|x| < \ell_0$ and $|\dot{x}|$ is large there, but if b is very large relative to ℓ_0, then damping will surely dominate. The trajectory should spiral inward.

From point a, the trajectory would move outward. From b, it would move inward. Could they converge on a common trajectory that would be maintained indefinitely? And what shape would it have? Let us see.

Van der Pol's equation is an instance of the general form

$$\ddot{x} = -\omega_0^2 x + f(x, \dot{x}),$$

where $f(x, \dot{x})$ incorporates all the complicated terms. Appendix C outlines a method, the averaging method, for solving such equations. The form of solution is

$$x = \mathcal{A}(t) \cos[\omega_0 t + \varphi(t)],$$

where $\mathcal{A}$ is a slowly varying amplitude and φ is a time-dependent phase angle. Equations for $\mathcal{A}$ and φ are derived in appendix C.

In the context of van der Pol's equation,

$$f = -\mu(x^2 - \ell_0^2)\dot{x}. \tag{3.3-2}$$

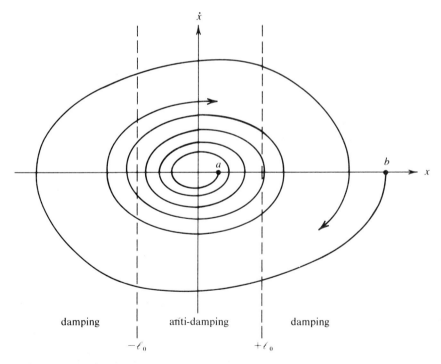

Figure 3.3-1 Two trajectories for van der Pol's equation. Damping occurs outside the vertical dashed lines; anti-damping, between them.

Then equation (C.8a) implies

$$\dot{\mathcal{A}} = -\frac{1}{\omega_0}\langle -\mu(\mathcal{A}^2\cos^2[\cdots] - \ell_0^2)(-\omega_0 \mathcal{A}\sin[\cdots])\sin[\cdots]\rangle$$

$$= -\mu\mathcal{A}\{\mathcal{A}^2\langle\cos^2[\cdots]\sin^2[\cdots]\rangle - \ell_0^2\langle\sin^2[\cdots]\rangle\}$$

$$= -\mu\mathcal{A}\{\tfrac{1}{8}\mathcal{A}^2 - \tfrac{1}{2}\ell_0^2\} \qquad (3.3\text{-}3)$$

by

$$\langle\cos^2[\cdots]\sin^2[\cdots]\rangle = \langle(\tfrac{1}{2}\sin 2[\cdots])^2\rangle$$

$$= \tfrac{1}{4}(\tfrac{1}{2}) = \tfrac{1}{8}. \qquad (3.3\text{-}4)$$

For $\dot{\varphi}$, we find

$$\dot{\varphi} = -\frac{1}{\omega_0}\left\langle -\frac{\mu}{\mathcal{A}}(\mathcal{A}^2\cos^2[\cdots] - \ell_0^2)(-\omega_0\mathcal{A}\sin[\cdots])\cos[\cdots]\right\rangle$$

$$= 0 \qquad (3.3\text{-}5)$$

because the lone factor of $\sin[\cdots]$ allows us to integrate all the cosines and get zero by cancellation at the two limits. The phase φ does not change, and we may let it drop from sight.

Here is a convenient way to integrate the equation for $\mathcal{A}$. First multiply both sides by $\mathcal{A}$ and rearrange as

$$\frac{1}{2}\frac{d\mathcal{A}^2}{dt} = -\frac{\mu}{8}\mathcal{A}^2(\mathcal{A}^2 - 4\ell_0^2),$$

so that the right-hand side is only quadratic in the "new" variable $\mathcal{A}^2$. Transfer all $\mathcal{A}$'s to the left-hand side and expand in partial fractions:

$$\left(\frac{1}{\mathcal{A}^2 - 4\ell_0^2} - \frac{1}{\mathcal{A}^2}\right)d(\mathcal{A}^2) = -\mu\ell_0^2\,dt.$$

Integration produces logarithms, which may be combined as

$$\ln\frac{\mathcal{A}^2 - 4\ell_0^2}{\mathcal{A}^2} = -\mu\ell_0^2 t + \ln\frac{\mathcal{A}^2(0) - 4\ell_0^2}{\mathcal{A}^2(0)},$$

where the constant of integration is chosen to ensure agreement at $t = 0$. Taking anti-logarithms and solving for $\mathcal{A}^2$ yields

$$\mathcal{A}^2 = \frac{4\ell_0^2}{1 + \left[\dfrac{4\ell_0^2 - \mathcal{A}^2(0)}{\mathcal{A}^2(0)}\right]\exp(-\mu\ell_0^2 t)}. \qquad (3.3\text{-}6)$$

The difference $4\ell_0^2 - \mathcal{A}^2(0)$ is central to how $\mathcal{A}$ evolves. Let us examine the possibilities.

1. $\mathcal{A}(0) < 2\ell_0$. The entire denominator in equation (3.3-6) has the structure

$$1 + (\text{a positive quantity that goes to zero as } t \to \infty),$$

and so α will increase toward $2\ell_0$.
2. $\alpha(0) > 2\ell_0$. The denominator now looks like

$$1 - \text{(a positive quantity less than 1 that goes to zero)};$$

so α will now decrease toward $2\ell_0$.
3. $\alpha(0) = 2\ell_0$. The factor multiplying the exponential vanishes, the denominator is constant, and α remains equal to $2\ell_0$.

If α was not equal to $2\ell_0$ to begin with, it approaches that value asymptotically, both if it was smaller and if it was larger. The trajectories spiral toward a single trajectory that is maintained indefinitely. Figure 3.3-2 illustrates this.

An aside. The origin in phase space is an exception. If $x = 0$ and $\dot{x} = 0$, then van der Pol's equation, equation (3.3-1), says the system point stays there for all time. Our approximation has to be applied judiciously, because of zero divisors, but the limit $\alpha(0) \to 0$ in equation (3.3-6) at fixed t does give $1/\infty$, and hence $\alpha = 0$ for all times. Since every point in the neighborhood of the origin spirals off toward $2\ell_0$, the origin is an unstable point and merits no further attention.

We have worked with the averaging method, which presumes that the influence of f is small during any single time interval of order $2\pi/\omega_0$. In the context of our solution, that means the exponent $\mu \ell_0^2 t$ must not change much in such

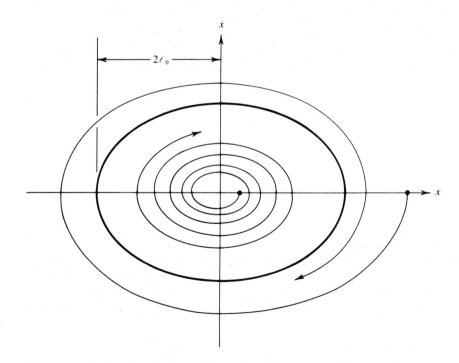

Figure 3.3-2 Trajectories spiral toward the ellipse with $x = 2\ell_0 \cos[\omega_0 t + \varphi(0)]$.

an interval. Our solution and figure 3.3-2 are quantitatively correct provided

$$\frac{2\pi\mu\ell_0^2}{\omega_0} \ll 1.$$

Our qualitative conclusions, however, are correct regardless of how small or large $2\pi\mu\ell_0^2/\omega_0$ may be. Specifically, van der Pol's equation always has an *isolated closed* trajectory among its solutions. If the system point starts on that trajectory, it returns to the starting point in a finite time and repeats the trip again and again. That is the meaning of *closed*. The term *isolated* means that there are no other closed trajectories nearby in phase space. Every nearby trajectory either spirals toward the special trajectory (which is the van der Pol situation) or spirals away, having been arbitrarily close in the distant past. (Recall that whereas a trajectory may repeat itself, it may not cross itself or any other trajectory.)

The harmonic oscillator, if not damped, has closed trajectories—they are the familiar ellipses—but there are always other such trajectories arbitrarily near any given closed trajectory: the trajectories are not isolated.

Following Poincaré, we call an isolated closed trajectory a *limit cycle*. The van der Pol equation has only a single limit cycle, and it is a stable one: all nearby trajectories spiral toward the limit cycle. Indeed, all trajectories, no matter where they start (with the sole exception of the fixed point at the origin), spiral toward it. Examples are known where all trajectories spiral away and also where those inside the limit cycle spiral toward it but those outside spiral away. Furthermore, there may be several distinct limit cycles or even infinitely

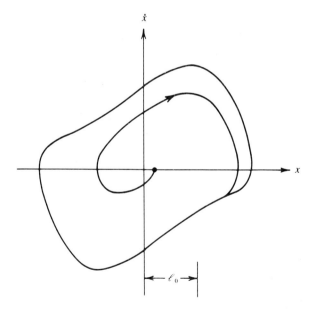

Figure 3.3-3 The limit cycle that arises when $2\pi\mu\ell_0^2/\omega_0 = 2\pi$. The outer, somewhat rectangular curve is the actual limit cycle. After the system point has gone around the origin a bit more than once, the trajectory has spiraled out so close to the limit cycle that the trajectory and the limit cycle are indistinguishable (though the trajectory never literally meets the limit cycle). If $2\pi\mu\ell_0^2/\omega_0$ is made still larger, the limit cycle becomes even more deformed relative to the ellipse that we found at small values.

many. (Some of these possibilities are explored in problems 3-17, 3-20, and 3-22.)

This section concludes with figure 3.3-3, the limit cycle when the dimensionless combination $2\pi\mu\ell_0^2/\omega_0$ is equal to 2π. Such a value is much too large for the averaging method to be applicable. The figure shows that $\mathcal{A}$ would change by an order of magnitude in one trip around the origin. A computer, however, cares little about whether $2\pi\mu\ell_0^2/\omega_0$ is small or large relative to unity. Once van der Pol's equation has been cast into first-order form—coupled equations in x and $\dot{x}$—the computer will meticulously map the trajectory and reveal the limit cycle.

3.4 MORE USES FOR THE AVERAGING METHOD

With the averaging method in hand, we can improve on the small-oscillations analysis of section 3.1. Let us take the situation $\ell > \ell_0$, so that the potential minimum falls at $x = 0$. If we use the potential energy given in equation (3.1-2), Newton II reads

$$m\ddot{x} = -2k\frac{\ell - \ell_0}{\ell}x - k\frac{\ell_0}{\ell^3}x^3. \qquad (3.4\text{-}1)$$

If we divide through by m, we can write this equation as

$$\ddot{x} = -\omega_0^2 x - \omega_0^2 \frac{x^3}{\lambda^2}, \qquad (3.4\text{-}2)$$

where

$$\omega_0^2 = \frac{2k}{m}\frac{\ell - \ell_0}{\ell}, \qquad (3.4\text{-}3)$$

$$\lambda^2 = 2\ell^2 \frac{\ell - \ell_0}{\ell_0}. \qquad (3.4\text{-}4)$$

The frequency ω_0 is the small-oscillations frequency calculated earlier. The parameter λ is the characteristic length of the nonlinear term; when the condition $|x| \ll \lambda$ ceases to hold, the nonlinear term becomes significant. If we compare equation (3.4-2) with equation (3.2-2), we see that the structure is Duffing's equation without damping or driving.

The averaging method provides us with equations for the $\mathcal{A}$ and φ that appear in the solution form

$$x = \mathcal{A}(t)\cos[\omega_0 t + \varphi(t)]. \qquad (3.4\text{-}5)$$

From equation (C.8a) we have

$$\dot{\mathcal{A}} = -\frac{1}{\omega_0}\left\langle -\frac{\omega_0^2}{\lambda^2} x^3 \sin[\cdots] \right\rangle$$

$$= \frac{\omega_0}{\lambda^2} \, \mathcal{Q}^3 \Big\langle \cos^3 [\cdots] \sin [\cdots] \Big\rangle = 0, \tag{3.4-6}$$

because the sine enables us to integrate the cosine factor and get cancellation at the two limits of integration. Next, from equation (C.8b),

$$\dot{\varphi} = -\frac{1}{\omega_0} \left\langle -\frac{\omega_0^2}{\lambda^2} x^3 \, \frac{\cos [\cdots]}{\mathcal{Q}} \right\rangle$$

$$= \frac{\omega_0}{\lambda^2} \, \mathcal{Q}^2 \langle \cos^4 [\cdots] \rangle = \frac{3}{8} \omega_0 \frac{\mathcal{Q}^2}{\lambda^2}, \tag{3.4-7}$$

by

$$\langle \cos^4 [\cdots] \rangle = \langle \cos^2 [\cdots] (1 - \sin^2 [\cdots]) \rangle$$

$$= \tfrac{1}{2} - \tfrac{1}{8} = \tfrac{3}{8},$$

when equation (3.3-4) is invoked for the step to the last line. Because $\dot{\mathcal{Q}} = 0$, the right-hand side of equation (3.4-7) is a constant, and we can immediately integrate:

$$\varphi(t) = \frac{3}{8} \omega_0 \frac{\mathcal{Q}^2(0)}{\lambda^2} t + \varphi(0). \tag{3.4-8}$$

If we insert this into equation (3.4-5) and collect terms suggestively, we have

$$x = \mathcal{Q}(0) \cos \left[\omega_0 \left\{ 1 + \frac{3}{8} \frac{\mathcal{Q}^2(0)}{\lambda^2} \right\} t + \varphi(0) \right]. \tag{3.4-9}$$

Since $\mathcal{Q}(0) = x_{\max}$, the maximum displacement, we can extract an "effective oscillation frequency" of the form

$$\omega_{\text{effective}} = \omega_0 \left(1 + \frac{3}{8} \frac{x_{\max}^2}{\lambda^2} \right). \tag{3.4-10}$$

Thus oscillations of moderate amplitude occur at a frequency somewhat higher than those of small amplitude, for which the correction term is negligible.

For a moment, let us go back to φ as given by equation (3.4-8). There we see a phase that increases uniformly with time. In an expression like (3.4-5), that uniform increase brings the maxima, say, of the cosine function sooner than the would occur otherwise. That is equivalent to a change in frequency, a conclusion which is, of course, much easier to extract from equation (3.4-9).

The increase in frequency makes sense physically, too. The x^4 term in $U(x)$ has a positive coefficient, implying that the potential energy rises faster with x than it would if only the x^2 term were present. This is sketched in figure 3.4-1. Imagine the mass starting from rest at $x_{\max}$. The steeper slope of the actual U curve, relative to just the x^2 term, means that the mass picks up kinetic energy faster and gets more of it. As the sketch shows, the kinetic energy is larger at each value of x. Since the mass moves faster, it makes the round trip in less time, and that means a higher frequency.

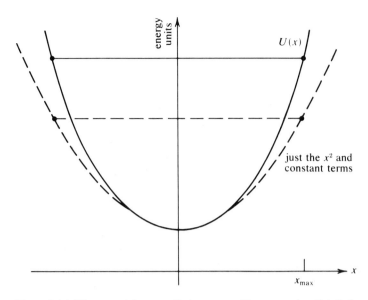

Figure 3.4-1 The potential energy $U(x)$, computed from equation (3.1-2), is compared with the potential energy when the x^4 term is omitted. Shown also are the total-energy lines associated with motion whose maximum excursion is $x_{\max}$.

Newton's Law of Damping

We can go on now to another application. A mass of any substantial size, oscillating rapidly in air, suffers a resistive drag that is quadratic in the speed, to good approximation. For the mass whose motion we have been analyzing, let us augment Newton II with such a drag, but, for simplicity's sake, let us drop the nonlinear part of the restoring force:

$$m\ddot{x} = -2k\frac{\ell - \ell_0}{\ell}x - b|\dot{x}|\dot{x}. \tag{3.4-11}$$

Here b is some positive constant; the quadratic drag is sometimes called *Newton's law of damping*. To put the equation into the form best suited to the averaging method, we divide through by m and write

$$\ddot{x} = -\omega_0^2 x - \frac{b}{m}|\dot{x}|\dot{x}. \tag{3.4-12}$$

What do we expect for changes in α and φ? Since there is a resistive drag, we certainly expect damping, which should show up as a decrease in α. Whether the phase or effective frequency will change is not obvious.

Invoking equation (C.8a), we have

$$\dot{\alpha} = -\frac{b}{m}\omega_0 \alpha^2 \langle|\sin[\cdots]|\sin^2[\cdots]\rangle.$$

The right-hand side is certainly nonzero and negative, guaranteeing us the decrease in α that we anticipated. The absolute value sign makes the averaging a little awkward to perform. The product, however, repeats itself already after half a period: $\frac{1}{2}(2\pi/\omega_0)$. Thus we may write

$$\langle |\sin[\cdots]|\sin^2[\cdots]\rangle = \frac{1}{\pi/\omega_0} \int_0^{\pi/\omega_0} \sin^3[\omega_0 t + \varphi]\, dt$$

$$= \frac{1}{\pi}\left\{\frac{1}{3\cdot 4}\cos 3[\cdots] - \frac{3}{4}\cos[\cdots]\right\}\Big|_0^{\pi/\omega_0}$$

$$= \frac{4}{3\pi},$$

and so

$$\dot{\alpha} = -\frac{4b}{3\pi m}\omega_0 \alpha^2. \tag{3.4-13}$$

The equation for $\dot\varphi$ is

$$\dot{\varphi} = -\frac{b}{m}\omega_0 \alpha \langle |\sin[\cdots]|\sin[\cdots]\cos[\cdots]\rangle$$

$$= -\frac{b}{m}\omega_0 \alpha \left\langle \frac{1}{2}\sin 2[\cdots]|\sin[\cdots]|\right\rangle = 0. \tag{3.4-14}$$

The first factor in the average oscillates in sign; a sketch will confirm complete cancellation during the average over a period.

The equation for α is readily integrated. Write

$$\frac{d\alpha}{\alpha^2} = -\frac{4b}{3\pi m}\omega_0\, dt,$$

integrate to get

$$-\frac{1}{\alpha} + \frac{1}{\alpha(0)} = -\frac{4b}{3\pi m}\omega_0 t,$$

and solve algebraically for α:

$$\alpha = \frac{\alpha(0)}{1 + \dfrac{4b\omega_0\, \alpha(0)}{3\pi m} t}. \tag{3.4-15}$$

If the mass is started from rest at $x = x_{\max}$, the full solution is

$$x = \frac{x_{\max}}{1 + \dfrac{4b\omega_0 x_{\max}}{3\pi m} t}\cos\omega_0 t. \tag{3.4-16}$$

The solution itself shows us what the characteristic damping time is: $3\pi m/(4b\omega_0 x_{\max})$. When t is large relative to that time, the t term in the denominator dominates, and the amplitude decreases as $1/t$. This is much slower than the exponential decay that linear damping $(-\gamma \dot{x})$ produces. The reason is not hard to find. When the damping is quadratic $(-b|\dot{x}|\dot{x})$, the damping itself diminishes faster as the motion decays and as $|\dot{x}|$, averaged over a cycle, becomes smaller.

3.5 SERIES EXPANSIONS

At low temperatures, around 4 kelvin, say, helium atoms stick very nicely to a solid surface such as graphite. The potential energy curve is shown in figure 3.5-1. To the right of the minimum, the force on the atom is attractive, toward the surface. To the left of the minimum, a strong repulsion sets in as the atom squashes against the surface. This makes the potential energy curve asymmetric. What is the motion like for an atom bound by such a potential well?

If the amplitude of oscillation is not too large, we can represent the potential energy by its Taylor expansion around the minimum. For convenience, we adjust the x coordinate so that $x = 0$ coincides with the potential minimum.

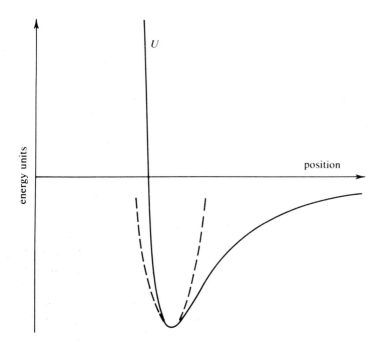

Figure 3.5-1 The potential energy U for a helium atom interacting with a solid surface. The surface is on the left. The dashed curve is the quadratic approximation to the actual potential energy, based on an expansion about the minimum.

Then the Taylor expansion is

$$U(x) = U(0) + 0 + \frac{1}{2} \left.\frac{d^2 U}{dx^2}\right|_0 x^2 + \frac{1}{2 \cdot 3} \left.\frac{d^3 U}{dx^3}\right|_0 x^3 + \cdots.$$

The second term is zero because we expand about the minimum, and the third is positive for the same reason. The dashed curve in figure 3.5-1 shows what we would get for an approximation if we used just $U(0)$ and the quadratic term. The cubic term must have a negative coefficient, to depress $U(x)$ a bit to the right of the minimum and to elevate it on the other side.

If we substitute the expansion into Newton II, we can write the equation as

$$\ddot{x} = -\omega_0^2 x + \omega_0^2 \frac{x^2}{\lambda}, \tag{3.5-1}$$

where

$$\omega_0^2 = \frac{1}{m} \left.\frac{d^2 U}{dx^2}\right|_0, \tag{3.5-2}$$

$$\lambda = -2 \left.\frac{(d^2 U/dx^2)}{(d^3 U/dx^3)}\right|_0 > 0. \tag{3.5-3}$$

The parameter λ is the characteristic length of the nonlinear part of the force. As long as the displacement is small relative to λ, one may investigate the motion with the averaging method. But one soon finds that both $\dot{\hat{a}}$ and $\dot{\varphi}$ are zero. There is no effect that accumulates from one nominal period, of duration $2\pi/\omega_0$, to another. (At least that is so at the level of approximation considered here.) Nonetheless, the asymmetry in $U(x)$ must have some effect.

We will look for a series expansion in the dimensionless parameter

$$\epsilon = \frac{x_{\max}}{\lambda},$$

where $x_{\max}$ is the right-hand maximum displacement (if the left is different). This means

$$x = x_0(t) + \epsilon x_1(t) + \epsilon^2 x_2(t) + \cdots, \tag{3.5-4}$$

an expansion that should converge well if ϵ is small, for example, if $\epsilon = 0.1$. In fact, let us be satisfied with just the first two terms. We need to substitute them into the basic equation, which we can write as

$$\ddot{x} + \omega_0^2 x - \epsilon \frac{\omega_0^2}{x_{\max}} x^2 = 0, \tag{3.5-5}$$

to display the dependence on ϵ. The substitution generates

$$(\ddot{x}_0 + \epsilon \ddot{x}_1) + \omega_0^2 (x_0 + \epsilon x_1) - \epsilon \frac{\omega_0^2}{x_{\max}} (x_0^2 + 2\epsilon x_0 x_1 + \epsilon^2 x_1^2) \simeq 0.$$

Now we collect terms with like powers of ϵ:

$$(\ddot{x}_0 + \omega_0^2 x_0) + (\ddot{x}_1 + \omega_0^2 x_1 - \frac{\omega_0^2}{x_{max}} x_0^2)\epsilon + \mathcal{O}(\epsilon^2) \simeq 0. \quad (3.5\text{-}6)$$

The symbol $\mathcal{O}(\epsilon^2)$ denotes the terms of order ϵ^2 and higher. Those terms are to be ignored, just as we ignored the corresponding term $\epsilon^2 x_2$ in the expansion. We may look at equation (3.5-6) as a problem in mathematics, where we are free to change λ and hence ϵ. The functions x_0 and x_1 are independent of ϵ. To maintain a zero value on the left if we change ϵ, the full coefficient of each power of ϵ must be zero. This generates the equations we need:

$$\ddot{x}_0 + \omega_0^2 x_0 = 0, \quad (3.5\text{-}7)$$

$$\ddot{x}_1 + \omega_0^2 x_1 - \frac{\omega_0^2}{x_{max}} x_0^2 = 0. \quad (3.5\text{-}8)$$

We can solve them sequentially.
A solution to equation (3.5-7) is

$$x_0 = A \cos \omega_0 t. \quad (3.5\text{-}9)$$

We can substitute this into equation (3.5-8) to get

$$\ddot{x}_1 = -\omega_0^2 x_1 + \frac{\omega_0^2}{x_{max}} A^2 \cos^2 \omega_0 t$$

$$= -\omega_0^2 x_1 + \frac{\omega_0^2 A^2}{x_{max}}\left(\frac{1}{2} + \frac{1}{2}\cos 2\omega_0 t\right).$$

Expressing the square of the cosine in terms of $\cos 2\omega_0 t$ and a constant makes it easier to guess a form for the solution. Indeed, a good guess for x_1 would be a term in $\cos 2\omega_0 t$ plus a constant. It works, and we find

$$x_1 = \frac{A^2}{2x_{max}} - \frac{A^2}{6x_{max}} \cos 2\omega_0 t. \quad (3.5\text{-}10)$$

Now we can go back to the basic expansion, equation (3.5-4), and write

$$x = A \cos \omega_0 t + \epsilon\left(\frac{A^2}{2x_{max}} - \frac{A^2}{6x_{max}} \cos 2\omega_0 t\right), \quad (3.5\text{-}11a)$$

$$= A \cos \omega_0 t + \frac{A^2}{2\lambda} - \frac{A^2}{6\lambda} \cos 2\omega_0 t. \quad (3.5\text{-}11b)$$

The constant term gives an offset to the right of the potential minimum. A look back at figure 3.5-1 makes that understandable: the steep repulsive rise on the left can readily push the average position $\langle x \rangle$ to the right of the minimum. The oscillatory motion is still periodic with period $2\pi/\omega_0$. A harmonic, $2\omega_0$, has appeared, however, and more would arise if we carried the expansion to higher powers of ϵ.

[In fairness, one must acknowledge that the averaging method would have produced the offset if we had pursued that route. The average of f alone is nonzero: $\langle f \rangle = \omega_0^2 \langle x^2 \rangle / \lambda > 0$. According the the last paragraph of appendix C, the nonzero average would require a constant term in the trial form for x, namely $\langle f \rangle / \omega_0^2$, which here would be $\langle x^2 \rangle / \lambda = \alpha^2/(2\lambda)$, in agreement with equation (3.5-11b).]

Sometimes we need to meet specific initial conditions, $x(0)$ and $\dot{x}(0)$. If the form in equation (3.5-11b) will not suffice, we can go back and replace $\omega_0 t$ everywhere with $\omega_0 t + \varphi_0$, where φ_0 is a constant phase term. Then the two free constants, A and φ_0, will always enable us to meet the initial conditions.

Secular Terms

When a series expansion works, it provides more detail than the averaging method. Let us apply the series method to the first context in section 3.4, the symmetric potential with an x^4 term. (That will also offer another comparison with the averaging method.) The differential equation is

$$\ddot{x} = -\omega_0^2 x - \omega_0^2 \frac{x^3}{\lambda^2}. \tag{3.5-12}$$

The characteristic length λ enters squared, and we might as well use it that way in the dimensionless expansion parameter, writing

$$\epsilon = \frac{x_{\max}^2}{\lambda^2}.$$

If we expand $x(t)$ as

$$x = x_0(t) + \epsilon x_1(t) + \cdots, \tag{3.5-13}$$

substitute into equation (3.5-12), with $1/\lambda^2$ there expressed in terms of ϵ, collect terms of like powers of ϵ, and set the full coefficients to zero, then we find the two equations we need:

$$\ddot{x}_0 + \omega_0^2 x_0 = 0, \tag{3.5-14}$$

$$\ddot{x}_1 + \omega_0^2 x_1 + \omega_0^2 \frac{x_0^3}{x_{\max}^2} = 0. \tag{3.5-15}$$

A solution to equation (3.5-14) is

$$x_0 = A \cos \omega_0 t, \tag{3.5-16}$$

and so equation (3.5-15) becomes

$$\ddot{x}_1 = -\omega_0^2 x_1 - \frac{\omega_0^2 A^3}{x_{\max}^2} \cos^3 \omega_0 t$$

$$= -\omega_0^2 x_1 - \frac{\omega_0^2 A^3}{4 x_{\max}^2} (\cos 3\omega_0 t + 3 \cos \omega_0 t), \tag{3.5-17}$$

after we use identity (3.2-4). For a solution x_1, we can guess that we will need a term in $\cos 3\omega_0 t$, but to match the term in $\cos \omega_0 t$, we need something special: a term in $t \sin \omega_0 t$. Indeed, a solution is

$$x_1 = \frac{A^3}{4x_{\max}^2} \left(\frac{1}{8} \cos 3\omega_0 t - \frac{3}{2} \omega_0 t \sin \omega_0 t \right). \tag{3.5-18}$$

[This is, moreover, the only possible solution, except for an additive term, a solution to the part of equation (3.5-17) that is homogeneous in x_1; that additive term would have the form $\cos(\omega_0 t + \text{const})$. The special term in the solution arises, mathematically, because the inhomogeneous term $3 \cos \omega_0 t$ in equation (3.5-17) oscillates at the natural frequency ω_0 of the homogeneous portion.] The term in $\omega_0 t \sin \omega_0 t$ grows with time. It threatens to destroy the utility of the expansion in ϵ; no matter how small ϵ is, the product $\epsilon \omega_0 t$ will grow to unity and beyond. Moreover, the term is not periodic. It has a long-term trend, which leads to its name, a *secular* term. Yet motion in a potential well, in the absence of damping, must be periodic. What has gone wrong?

We get a clue by looking at the solution provided by the averaging method. Section 3.4 gave us

$$x = \mathcal{A}(0) \cos[\omega_0 t + \varphi(t)]. \tag{3.5-19}$$

where

$$\varphi(t) = \frac{3}{8} \omega_0 \frac{\mathcal{A}^2(0)}{\lambda^2} t + \varphi(0). \tag{3.5-20}$$

If we suppose that $|\varphi(t)| \ll 1$, we can expand equation (3.5-19) in a Taylor series about the argument $\omega_0 t$:

$$x = \mathcal{A}(0) \{\cos \omega_0 t - \varphi(t) \sin \omega_0 t + \cdots\}. \tag{3.4-21}$$

The product $\varphi(t) \sin \omega_0 t$ reproduces the special term $\omega_0 t \sin \omega_0 t$ of the series solution. There is nothing literally wrong with our series solution, but it is a useful solution only for a short time. Our series is based on the frequency ω_0 that appears in equation (3.5-14) and then in the solution $x_0 = A \cos \omega_0 t$. Section 3.4 told us, however, that the motion actually proceeds with a different frequency, what we called the *effective* frequency in equation (3.4-10):

$$\omega_{\text{effective}} = \omega_0 \left(1 + \frac{3}{8} \frac{x_{\max}^2}{\lambda^2} \right).$$

No wonder the series has trouble keeping up with the actual motion.

With this insight—that the fundamental frequency was poorly chosen—we can go back to the series expansion and adjust the fundamental frequency so that secular terms will not arise. We write the square of the new fundamental as

$$\omega^2 = \omega_0^2 + \epsilon a_1 + \epsilon^2 a_2 + \cdots, \tag{3.5-22}$$

with constants $a_1, a_2, \ldots$ to be chosen later, as the need arises. How do we actually insert ω^2 into the mathematics? We replace the term $\omega_0^2 x$ in equation (3.5-12) with $(\omega^2 - \epsilon a_1 - \cdots)x$, because that term establishes the fundamental frequency in the expansion. If we work through first order in ϵ only, equation (3.5-12) becomes

$$\ddot{x}_0 + \epsilon \ddot{x}_1 = -(\omega^2 - \epsilon a_1)(x_0 + \epsilon x_1) - \frac{\epsilon \omega_0^2}{x_{\max}^2}(x_0^3 + \cdots).$$

Collecting terms with like powers of ϵ and equating the full coefficients to zero yields

$$\ddot{x}_0 = -\omega^2 x_0, \quad \text{should be sqd} \quad (3.5\text{-}23)$$

$$\ddot{x}_1 = -\omega^2 x_1 + a_1 x_0 - \frac{\omega_0^3}{x_{\max}^2} x_0^3. \quad (3.5\text{-}24)$$

Equation (3.5-23) has the solution

$$x_0 = A \cos \omega t. \quad (3.5\text{-}25)$$

and so equation (3.5-24) becomes

$$\ddot{x}_1 = -\omega^2 x_1 + \left(a_1 - \frac{3\omega_0^2 A^2}{4 x_{\max}^2}\right) A \cos \omega t - \frac{\omega_0^2 A^3}{4 x_{\max}^2} \cos 3\omega t.$$

To avoid a secular term in the solution, we need only choose a_1 so that the coefficient of $\cos \omega t$ vanishes. Thus

$$a_1 = \frac{3\omega_0^2 A^2}{4 x_{\max}^2}. \quad (3.5\text{-}26)$$

The solution for x_1 is then merely

$$x_1 = \frac{\omega_0^2}{\omega^2} \frac{A^3}{32 x_{\max}^2} \cos 3\omega t$$

$$\simeq \frac{A^3}{32 x_{\max}^2} \cos 3\omega t \quad (3.5\text{-}27)$$

because $\omega_0^2/\omega^2 = 1$ plus order ϵ. Putting together x_0 and x_1 and using the definition of ϵ, we get

$$x = A \cos \omega t + \frac{A^3}{32 \lambda^2} \cos 3\omega t. \quad (3.5\text{-}28)$$

The new fundamental frequency is

$$\omega = (\omega_0^2 + \epsilon a_1)^{1/2}$$

$$= \omega_0 \left(1 + \frac{\epsilon a_1}{\omega_0^2}\right)^{1/2} \simeq \omega_0 \left(1 + \frac{1}{2} \frac{\epsilon a_1}{\omega_0^2}\right)$$

$$\simeq \omega_0 \left(1 + \frac{3}{8} \frac{A^2}{\lambda^2}\right), \quad (3.5\text{-}29)$$

in perfect agreement with the effective frequency found by the averaging method.

A Note about Appearances

There is something awkward and unaesthetic about the way in which x_{max} has appeared in this section. The nonlinearity of equation (3.5-1) makes the ratio x_{max}/λ a vital parameter. That ratio determines how significant the nonlinear term can be—in the course of the motion—relative to the linear term. Thus we should expect x_{max} to play a role in determining the solution and to do so via the ratio $\epsilon = x_{max}/\lambda$. But the explicit appearance of x_{max} in equation (3.5-5) is awkward. The tidy way to formulate the mathematical problem is with *two* dimensionless ratios:

$$\epsilon = \frac{x_{max}}{\lambda}$$

and

$$y = \frac{x}{x_{max}}.$$

In terms of these quantities, equation (3.5-1) takes the neat form

$$\ddot{y} = -\omega_0^2 y + \epsilon \omega_0^2 y^2.$$

Since $|y| \leq 1$ by construction, we immediately see ϵ as measuring the relative significance of the nonlinear term. After solving the problem in terms of the variable y and the parameter ϵ, we can revert to x and λ. This two-step procedure improves appearances and may even make the mathematics easier to follow. There is a cost: introducing yet another variable, here $y = x/x_{max}$. For that reason, the section was written in terms of x.

Synopsis

A series expansion will work with an equation of the general form

$$\ddot{x} = -\omega_0^2 x + f(x, \dot{x})$$

provided the (unknown) exact solution is truly periodic in time. The basic computational need is to get equations for the successive terms in the expansion. Here, in outline form, are the steps to take (as inferred from the preceding examples).

1. Construct a dimensionless expansion parameter ϵ that characterizes the size of the nonlinear term f relative to the linear term $\omega_0^2 x$. Often we can take ϵ to be the ratio $f/(\omega_0^2 x)$, evaluated at $x = x_{max}$:

$$\epsilon = \frac{f \text{ at } x_{max}}{\omega_0^2 x_{max}}.$$

2. Rewrite the differential equation so that ϵ appears explicitly. Thus the equation may take the form
$$\ddot{x} = -\omega_0^2 x + \epsilon \omega_0^2 x_{max} \frac{f}{f \text{ at } x_{max}}.$$
3. Express the solution as a series in ϵ:
$$x = x_0(t) + \epsilon x_1(t) + \epsilon^2 x_2(t) + \cdots.$$
4. Insert the series into the differential equation, and collect terms of like order in ϵ: ϵ^0, ϵ^1, ϵ^2, ...
5. Equate to zero each set of terms of like order.
6. Solve the equations sequentially, starting with the equation of zeroth order in ϵ.

If secular terms arise or if we anticipate that the true period differs from $2\pi/\omega_0$ by a significant amount, then we must adjust the fundamental frequency in the series. Here are the additional steps.

7. Expand the true angular frequency ω in powers of ϵ:
$$\omega^2 = \omega_0^2 + \epsilon a_1 + \epsilon^2 a_2 + \cdots.$$
8. Replace ω_0^2 by $\omega^2 - \epsilon a_1 - \cdots$ in the differential equation, as part of step 2. (This inserts ω into the mathematics and ensures that the fundamental frequency in the series will be ω, not ω_0.)
9. In step 6, choose the a_i's to preclude secular terms.

3.6 ABOUT THE METHODS

Three methods for solving nonlinear equations have appeared in this chapter: the averaging method, series expansion, and the method of judicious guessing. The third was not formally announced, but it is a fair description of how we solved Duffing's equation in section 3.2. The method often consists of guessing some trial functional form and then adjusting constants to get equality, at all times, between the two sides of the equation. Sometimes it works exceedingly well, but one cannot count on it.

The first and second methods are systematic. If some modest conditions hold, they are pretty sure to work. What are the conditions?

The averaging method arose in the context
$$\ddot{x} = -\omega_0^2 x + f(x, \dot{x}).$$
As long as the function f has only a small influence on the motion during any single time interval of order $2\pi/\omega_0$, we may average as we did. (More specifically, the changes in $\mathcal{A}$ and φ must be small during each interval of length $2\pi/\omega_0$, so that $\mathcal{A}$ and φ may be legitimately taken as constant in the

averaging operation. Thus the changes must satisfy the following strong inequalities: $\Delta \mathcal{C} \ll \mathcal{C}$ and $\Delta \varphi \ll 2\pi$.) There is no reason why f could not depend explicitly on time, too: $f(x, \dot{x}, t)$. The averaging method is powerful and quite general, but it sacrifices some detail; things that happen on a time scale shorter than $2\pi/\omega_0$ are averaged out and lost from sight.

The series method can be expected to work well if the motion is truly periodic. To be sure, we may need to adjust the fundamental frequency, but, once recognized, that is no great difficulty. A damped system, however, cannot be handled well. The long-term decay of the oscillation cannot be represented well—for long times—by a sum of ostensibly periodic terms.

Always, of course, the convergence of the series is a real question, typically difficult to answer. And even if the series converges, it may do so slowly. If the first few terms do not give an approximation of the desired accuracy, the series is of little use. The limit cycle of van der Pol's equation provides an example. The motion is certainly periodic. When the dimensionless parameter has the value $\mu \ell_0^2/\omega_0 = 0.1$, a small value, we would expect a series expansion to describe the limit cycle nicely. Indeed, the trajectory is close to the ellipse that we would obtain if μ were zero. If we carry the series expansion through second order in the small parameter, the calculated trajectory does depart from the ellipse, heading toward the shape of the limit cycle as determined by digitial computer. At a typical specified value of x, the values of $\dot{x}$ agree within a few tenths of a percent. That is about what we would expect; the neglected terms in the series commence with terms of order $(\mu \ell_0^2/\omega_0)^3$, which are fractionally of order 10^{-3}, or a tenth of a percent. Often the first correction term in a series does remarkably well, and the first and second, taken together, satisfy all practical needs.

The first section of this chapter developed small-oscillations theory, a general way to describe motion near an equilibrium point. What were the objectives of the other sections? One was to develop methods of approximation; we have just reviewed them. The other objective was to show a few of the surprises that nonlinear systems generate. Amplitude jumps, hysteresis, limit cycles—these are marvelous modes of behavior, and so different from what a harmonic oscillator can show. Theories of turbulence and of biochemical self-organization build on these properties of nonlinear systems—and on others that are even more curious.

WORKED PROBLEMS

WP3-1 A small mass m starts from rest at a great distance from the sun. At some later time it passes the earth's orbital distance $r_\oplus$ and finally hits the solar surface (of radius $R_\odot$).

Problem. How long is the time interval between those two events?
We can adapt the analysis that led to equation (3.1-12), writing

$$\text{Time interval} = -\int_{r_\oplus}^{R_\odot} dr \frac{(m/2)^{1/2}}{[E - U(r)]^{1/2}}. \tag{1}$$

The minus sign applies because $dr < 0$ as r decreases from $r_\oplus$ to $R_\odot$. The potential energy is $U(r) = -GM_\odot m/r$, where $M_\odot$ denotes the sun's mass. Since $U \simeq 0$ "at a great distance from the sun," a mass that starts from rest there has $E \simeq 0$. Thus equation (1) becomes

$$\begin{aligned}\text{Time interval} &= -\int_{r_\oplus}^{R_\odot} dr \frac{r^{1/2}}{(2GM_\odot)^{1/2}} \\ &= \frac{2}{3} \frac{1}{(2GM_\odot)^{1/2}} (r_\oplus^{3/2} - R_\odot^{3/2}).\end{aligned}$$

The ratio $R_\odot/r_\oplus$ is about 5×10^{-3}, and so the term in $R_\odot$ is negligible. Numerical evaluation, using data from appendix A, yields an interval of about 2 weeks.

WP3-2 For a mass m constrained to one-dimensional motion, the potential energy function is

$$U = \left[-\left(\frac{x}{\lambda}\right)^3 + 12\left(\frac{x}{\lambda}\right)^2 - 45\frac{x}{\lambda}\right] U_0,$$

where λ and U_0 are positive constants with the units of length and energy, respectively.

Problem. (*a*) Determine the points at which the mass could be placed at rest and would remain at rest, i.e., the equilibrium positions.

(*b*) Compute the frequency of small oscillations about the point(s) of stable equilibrium.

A good first step is to work out the shape of the potential energy curve. The curve has a maximum or minimum where its derivative is zero:

$$\frac{dU}{dx} = \left[-3\left(\frac{x}{\lambda}\right)^2 + 24\frac{x}{\lambda} - 45\right]\frac{U_0}{\lambda} = 0,$$

with solutions $x = 3\lambda$ and $x = 5\lambda$. To see whether we have a maximum or minimum, we can evaluate the second derivative:

$$\left.\frac{d^2U}{dx^2}\right|_{x=3\lambda} = +\frac{6U_0}{\lambda^2};$$

$$\left.\frac{d^2U}{dx^2}\right|_{x=5\lambda} = -\frac{6U_0}{\lambda^2}.$$

Thus $x = 3\lambda$ is a minimum, while $x = 5\lambda$ is a local maximum. Since $x = 0$ implies $U = 0$, the curve must have a shape like that sketched in figure WP3-2.

An equilibrium position is characterized by a zero value of the force F. Since $F(x) = -dU/dx$, the points $x = 3\lambda$ and $x = 5\lambda$ are equilibrium points (and the only ones).

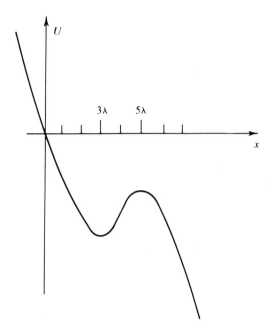

Figure WP3-2 A qualitative rendition of the potential energy curve.

The potential minimum at $x = 3\lambda$ produces a stable equilibrium. Appeal to equation (3.1-5) gives us

$$m \frac{d^2}{dt^2}(x - 3\lambda) = -\frac{d^2 U}{dx^2}\bigg|_{x=3\lambda} (x - 3\lambda)$$

$$= -\frac{6U_0}{\lambda^2}(x - 3\lambda).$$

The displacement from equilibrium $x - 3\lambda$ is subject to a restoring force $-(6U_0/\lambda^2)(x - 3\lambda)$. The corresponding angular frequency of oscillation is $[6U_0/(m\lambda^2)]^{1/2}$.

PROBLEMS

3-1 The potential energy of a mass m is

$$U = \left(17 + \cosh \frac{x}{\lambda}\right) U_0,$$

where U_0 and λ are positive constants. Sketch U and then compute the frequency of small oscillations about the equilibrium location.

3.2 An oscillator is made by allowing a negatively charged bead (charge $= q$) to slide on a thin insulating rod. The rod is horizontal, and a positive charge of magnitude Q is located a fixed distance d from the rod, as sketched in figure

108 NONLINEAR OSCILLATORS

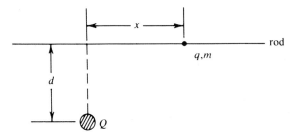

Figure P3-2

P3-2. Of course, q is attracted toward Q, but it is constrained to move on the rod and so oscillates about $x = 0$. Ignore friction. Recall that Coulomb's law for point charges gives an inverse-square force law. A note about electrical units appears at the end of appendix A.

    (a) What is the potential energy $U(x)$?
    (b) What is the restoring force $F(x)$?
    (c) Sketch a rough graph of $U(x)$ versus x and of $F(x)$.
    (d) What is the frequency of small oscillations about $x = 0$?
    (e) Now imagine a microscopic version of this situation. If $q = 1$ electron charge and $Q = |q|$, if $d = 2 \times 10^{-10}$ meters and if the mass were $m = 1$ electron mass, what would be the oscillation frequency (numerically) and where in the electromagnetic spectrum would you look for radiation?

3-3 Calculate the period of small oscillations about the equilibrium location when the potential energy is $U = U_0 \tan^2(x/\lambda)$, where U_0 and λ are positive constants. [For this potential energy one can determine the period exactly: the integral $\int dx/\dot{x}$ gives the period as $2\pi[m\lambda^2/(2U_0)]^{1/2} (1 + E/U_0)^{-1/2}$, where E is the energy of the mass m. How does your result compare with this expression?]

3-4 Refer to WP3-2 and the potential energy sketched there. What are the qualitatively distinct total energies? In a phase space with axes x and $\dot{x}$, sketch each of the qualitatively distinct trajectories. What is special about the trajectories having $E = U(5\lambda) = -50U_0$?

3-5 Suppose a mass m restricted to one-dimensional motion has a potential energy

$$U(x) = (x^4 - \lambda^2 x^2) \frac{U_0}{\lambda^4},$$

where λ is a fixed length and U_0 is a positive constant.

    (a) Sketch the potential energy. Determine the points of stable equilibrium.
    (b) Derive the frequency of small oscillations about the stable equilibrium points.
    (c) Suppose the particle has energy $E = -U_0/8$. Sketch the trajectory (or possible trajectories) in phase space. Do the same for $E = +U_0/8$.

(d) Suppose mass m is released from rest at $x = 10^{-4} \lambda$. Compute (in some approximation) the motion $x(t)$ during the interval $0 \leq t \leq 5(m\lambda^2/U_0)^{1/2}$. What is the location at the end of that interval?

3-6 The potential energy is $U(x) = -Cx^3$, let us suppose, where C is a positive constant. Sketch $U(x)$, the associated force $F(x)$, and one each of the qualitatively distinct trajectories in phase space. What is special about the trajectories having $E = 0$?

3-7 The force $F(x)$, acting on a mass m in one-dimensional motion, is specified to be discontinuous: $F(x) = +F_0$ when $x < 0$, but $F(x) = -F_0$ when $x > 0$. Here F_0 is a positive constant, and so the force is always inward. (We may take $F = 0$ at $x = 0$.) Draw, with some care, the typical trajectory in phase space. [You may find it helpful first to draw graphs of $F(x)$ and $U(x)$.] What is the period of the motion, expressed in terms of F_0, m, and x_{max}?

3-8 The context is section 3.1, with $\ell = 1.5\ell_0$. Use the integral expression (3.1-12) to calculate the period of oscillation, correct through the term of order x_{max}^2/ℓ_0^2. How does the associated frequency compare with those calculated by the averaging method or series expansion (in sections 3.4 and 3.5)?

Note that we may write

$$U(x_{max}) - U(x) = \frac{k}{3}(x_{max}^2 - x^2)\left[1 + \frac{2(x_{max}^2 + x^2)}{9\ell_0^2}\right].$$

When a root is taken, the quantity in square brackets may be expanded by the binomial theorem. Useful integrals appear in appendix A.

3-9 A positive charge (of mass m and charge q) is free to slide along a rigid plastic rod. Also present are two fixed positive charges, as indicated in figure P3-9.

(a) Sketch the potential energy of q that arises from interaction with the right-hand Q alone. A qualitative rendition will suffice. Next, do the same for the left-hand Q, and then indicate what the total potential energy might look like.

(b) What inequality must the ratio $\ell_\parallel / \ell_\perp$ satisfy if the location $x = 0$ is to be a point of *stable* equilibrium?

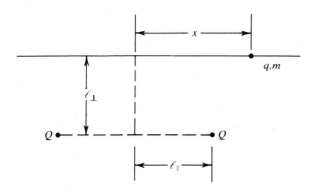

Figure P3-9

(c) Assume that your inequality in (b) is satisfied, and compute the frequency of *small* oscillations about $x = 0$.

(d) Separately from (c), assume that your inequality in (b) is satisfied, and then describe in words how you would determine the *maximum* amplitude of the oscillatory motion.

T3-10 The context is phase space, as discussed in section 3.1, and a force that depends on time explicitly: $F(x, \dot{x}, t)$. For example, suppose $F = -k(t)x$, that is, a spring constant can be changed as the experimenter wishes.

(a) Suppose the mass has been oscillating for several periods with $k = k_1$. Then when $x = 0$ but $\dot{x} > 0$, the experimenter suddenly changes k to

$$k = k_2 = 4k_1.$$

How does the trajectory continue? Describe (and sketch) it for the next few periods.

(b) Next, at an instant when both x and $\dot{x}$ are positive, the experimenter changes from k_2 to $k = k_3 = 0$. How does the trajectory continue into the future?

(c) In the context of this problem, does a trajectory ever intersect itself? Is the continuation of a trajectory uniquely determined by the current location (x and $\dot{x}$) in phase space? By the location *plus* time t?

T3-11 *Gravitational collapse.* Imagine a spherically symmetric, homogeneous cloud of dust grains, all the grains initially at rest and separated from one another by distances large relative to a grain diameter. The cloud's initial mass density and radius are ρ_0 and R_0, respectively. On the assumption that collisions and pressure may be ignored, calculate the time for gravitational collapse to zero radius. (The assumption will not be valid for the last moments of an actual collapse. Can you reason why you would still get a collapse time that is correct in order of magnitude?) The collapse time so calculated is commonly used to estimate the star formation time in gaseous clouds (or at least to provide a time scale against which to compare the effects of other processes, such as magnetic interactions).

T3-12 The context is section 3.2. If the driving force F_0 is small enough, the cubic term in Duffing's equation is negligible. How small is "small enough"? (The limitation should be couched in terms of F_0, λ, m, ω_0, and τ only.) You might try to construct a self-consistent argument, commencing with some expressions in section 2.4.

3-13 In equation (3.2-1), the nonlinear term "softens" the spring; you can see that best if the full spring force is written as $-kx(1 - x^2/\lambda^2)$. What would happen to the solution in section 3.2 if we replaced λ^2 by $-\lambda^2$, so that the nonlinearity "hardens" the spring?

T3-14 The saw blade oscillator of section 3.2 could be driven with two electromagnets simultaneously, one fed by current from a source oscillating at frequency ω_1, the other at frequency ω_2. What frequencies would you expect the blade's response to show?

T**3-15** Suppose the driving frequency in section 3.2 is fixed at $\omega = 0.9\omega_0$, but the magnitude F_0 of the driving force is changed slowly over a large range. Would you expect any discontinuous change in the saw blade's oscillation amplitude?

3-16 Suppose an oscillating mass slides on a surface whose roughness varies with position. In particular, suppose the total force is $-kx - \gamma(x^2/\ell^2)\dot{x}$, where ℓ is a fixed length and γ has the dimensions of the usual damping coefficient. Use the averaging method to analyze the motion of this nonlinear oscillator. Can you incorporate the initial conditions $x(0) = 0$ and $\dot{x}(0) = v_0$?

3-17 Suppose μ were replaced in van der Pol's equation by $-\mu'$, where $\mu' > 0$. What would the new trajectories be like?

3-18 An equation does not need to be nonlinear for the averaging method to be a useful approach. Consider, for example, a moderately damped harmonic oscillator

$$\ddot{x} = -\omega_0^2 x - \frac{\dot{x}}{\tau},$$

with $\omega_0 \tau = \mathcal{O}(10)$ and the initial conditions

$$x(0) = x_0, \qquad \dot{x}(0) = 0.$$

Treat the damping term as the "complicated function $f(x, \dot{x})$" in the averaging method, and solve approximately for $x(t)$. Compare your solution with the exact solution.

If $\omega_0 \tau = 10$ precisely, how close to the exact solution is your approximation at time $t^* = 3(2\pi/\omega_0)$, that is, after three periods, more or less? Is the failure of the approximation to generate a phase or frequency shift important? What about time $t^{**} = 300(2\pi/\omega_0)$, for both questions?

T**3-19** Consider a nonlinear oscillator for which the entire force is derivable from a potential energy: $m\ddot{x} = -dU(x)/dx$. Further, suppose the motion is investigated with the averaging method. Can you construct an argument showing why $\dot{\bar{a}} = 0$ will hold?

3-20 An acoustical problem that Lord Rayleigh tackled in 1883 led to a differential equation of the following form:

$$\ddot{x} = -\omega_0^2 x + (a - b\dot{x}^2)\dot{x},$$

where constants a and b are positive. Assume that the terms in $\dot{x}$ and $\dot{x}^3$ have a small effect on the motion during any single interval $2\pi/\omega_0$. Choose some approximation scheme and solve for $x(t)$.

Lord Rayleigh's paper appeared in the *Philosophical Magazine* **15**, 229 (1883). The anti-damping term $+a\dot{x}$ would produce unbounded motion if not compensated—at large $|\dot{x}|$—by some damping term. Rayleigh proposed to "form an idea of the state of things which then arises by adding to [his] equation a term proportional to a higher power of the velocity." He adopted the term $-b\dot{x}^3$.

3-21 This problem is a continuation of problem 3-20 on Rayleigh's nonlinear equation. Use a digital computer or a programmable hand calculator to compute numerically three trajectories in phase space. Continue your numerical integration until the asymptotic behavior is evident. The ultimate output should be in graphical form.

Dimensional analysis suggests that Rayleigh's equation has two characteristic times, $1/a$ and $2\pi/\omega_0$. Be sure to investigate both a small and a large value for their ratio $2\pi a/\omega_0$. There is also a characteristic length, $[a/(b\omega_0^2)]^{1/2}$. The qualitative behavior of a trajectory may depend on how $x(0)$ compares with that length.

Choose your three trajectories to be qualitatively distinct. Is your approximate analytical solution (in problem 3-20) supported [provided the parameters and initial values satisfy relations such as $2\pi a/\omega_0 \ll 1$ and $x(0)/[a/(b\omega_0^2)]^{1/2} \simeq 1$]?

3-22 *More about limit cycles.* Suppose Newton's second law for a nonlinear oscillator has the form

$$\ddot{x} = -\omega_0^2 x + \mu \dot{x} r (r^2 - 4r + 3),$$

where $\mu > 0$ and $r \equiv (x^2 + \dot{x}^2/\omega_0^2)^{1/2}$. Try polar coordinates in a phase plane (constructed with coordinates x and $\dot{x}/\omega_0$) as a route to determining the limit cycles for this oscillator. (Specifically, look at $\dot{r}$.) For each limit cycle that you discover, determine whether nearby trajectories spiral toward the limit cycle or away from it. Also, determine how long it takes the system point to travel around each limit cycle.

T **3-23** In discussing the harmonic oscillator in chapter 2, we worked hard to incorporate initial conditions. Typically, the solution to a dynamical equation depends on initial conditions in a significant and detailed fashion. Does that "typical" situation hold if the equation has a limit cycle? Should we make a distinction between the short-term and the long-term behavior of the solution? Can you illustrate your response by reference to figure 3.3-3? To problem 3-22?

3-24 A pendulum of length ℓ and bob mass m oscillates with moderate amplitude. The maximum angular excursion from the downward vertical is $\theta_{\max}$, specified to be less than or equal to 0.5 radian. Determine the frequency of oscillation as a function of $\theta_{\max}$, correct through effects of order $\theta_{\max}^2$.

Note: The energy expression for the pendulum may be taken as

$$\tfrac{1}{2}m(\ell\dot\theta)^2 + mg[\ell - \ell\cos\theta] = E.$$

Differentiating the expression will generate an equation in $\ddot\theta$. Trigonometric expansions are provided in appendix A.

Comparisons of theory and experiment are provided by M. K. Smith, *Am. J. Phys.*, **32**, 632 (1964) and by L. P. Fulcher and B. F. Davis, *Am. J. Phys.*, **44**, 51 (1976). An intriguing harmonic analysis is given by R. Simon and R. P. Riesz, *Am. J. Phys.*, **47**, 898 (1979) and **48**, 582 (1980).

3-25 Extend our series solution of equation (3.5-1) to second order in ϵ. Do you find that you need to adjust the frequency of the fundamental?

3-26 *The relativistic harmonic oscillator.* For a particle of rest mass m_0 subject to a linear restoring force, the relativistic generalization of Newton's second law might be

$$\frac{d}{dt}\left[\frac{m_0 \dot{x}}{(1-\dot{x}^2/c^2)^{1/2}}\right] = -kx.$$

Suppose the initial conditions are $x(0) = x_{in}$ and $\dot{x}(0) = 0$ and that $\omega_0 x_{in}/c = \mathcal{O}(\frac{1}{10})$, where $\omega_0 = (k/m_0)^{1/2}$.

(*a*) Is there an energy conservation law, perhaps of the form

$$\frac{m_0 c^2}{(1-\dot{x}^2/c^2)^{1/2}} + U(x) = \text{constant}?$$

(*b*) Determine the subsequent motion of the mass, using some approximation scheme that you justify. Your result must go beyond a purely newtonian approximation; in different words, the speed of light c must remain in your final result in a significant way.

3-27 When the electric field of a laser oscillates in only a single mode, the amplitude $E(t)$ satisfies an equation of the form

$$\frac{d^2E}{dt^2} = -\omega_0^2 E - \frac{1}{\tau}\frac{dE}{dt} + (g - \tilde{g}E^2)\frac{dE}{dt},$$

in some good approximation. Here ω_0 is the mode's natural frequency, the term in τ represents losses (including the laser output), the g term describes gain (from atomic emission), and the $\tilde{g}$ term describes saturation of such gain (as the supply of excited atoms is exhausted).

(*a*) If $\tilde{g}$ were zero, how would E vary with time? [Assume $E(0) \neq 0$ and $g > 1/\tau$.]

(*b*) How does E evolve when $g > 0$?

Note: If you examine the *structure* of the equations in (*a*) and (*b*) and compare them with structures that you have met, you can arrive at answers with very little algebraic effort.

3-28 A mass is acted on by a linear restoring force and a special velocity-dependent force, so that Newton II reads as follows:

$$m\ddot{x} = -kx - \gamma x \dot{x} \quad \text{with } \gamma > 0.$$

The second term produces damping when $x > 0$ but feeds in energy when $x < 0$. The initial conditions are $x(0) = x_{in}$ and $\dot{x}(0) = 0$. Moreover, the parameters and initial conditions have numerical values such that

$$\gamma\left(\frac{k}{m}\right)^{1/2}\frac{x_{in}^2}{kx_{in}} = \mathcal{O}\left(\frac{1}{20}\right).$$

114 NONLINEAR OSCILLATORS

Determine the subsequent motion of the particle, correct through effects of first order in the dimensionless combination cited above. (And there *is* a nonzero first-order effect.)

3-29 A mass m constrained to one-dimensional motion is acted on by two forces. The first force is purely attractive (toward the origin) and has a magnitude inversely proportional to the distance. The second force arises from an ideal spring, anchored at the origin, whose relaxed length is ℓ_0. Thus the force has the analytic form

$$F(x) = -\frac{b}{x} - k(x - \ell_0);$$

both b and k are positive constants. Only motion with $x > 0$ needs to be considered.

The parameters b, k, and ℓ_0 have numerical values such that $b/(k\ell_0^2) = O(\frac{1}{40})$; that is, the combination is of order $\frac{1}{40}$.

(a) Determine the point (or points) where the mass m could be placed at rest and would remain at rest, i.e., the equilibrium position(s).

(b) Compute the frequency of small oscillations about the point(s) of stable equilibrium.

(c) Suppose m is released from rest. Consider different locations for the release, and draw a trajectory in phase space for each (kind of) location that yields a qualitatively distinct trajectory. Take some pains to draw trajectories that are true to the qualitative features of the motion.

(d) Would the physical behavior of the mass be qualitatively different if $b/(k\ell_0^2) = 10$ were the numerical value of the combination? Why? And in which respects?

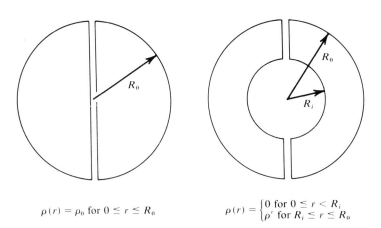

$\rho(r) = \rho_0$ for $0 \leq r \leq R_0$ $\rho(r) = \begin{cases} 0 \text{ for } 0 \leq r < R_i \\ \rho' \text{ for } R_i \leq r \leq R_0 \end{cases}$

Figure P3-30

3-30 When you land on asteroid YD8588, you discover a narrow fissure running through the center. To learn something about the mass distribution within the asteroid, you drop a small rock down the hole and note how long it takes to return. Could you then distinguish between the two mass distributions sketched in figure P3-30? The mass density is denoted by ρ. The outer radius R_0 and the total mass M are known from measurements at the surface.

On your way to a response, sketch the force on the rock (of mass m) and its potential energy as a function of r. If you can, derive an explicit expression for the round-trip time when the density is radially uniform.

CHAPTER FOUR

LAGRANGIAN FORMULATION

4.1 Fermat's principle
4.2 Calculus of variations
4.3 Newton II as an extremal principle
4.4 Lagrangians and constraints
4.5 Another instance of constrained motion
4.6 Conversion to first-order equations: Hamilton's equations
4.7 Liouville's theorem
4.8 The lagrangian and quantum mechanics
4.9 A sense of perspective

It always seems odd to me that the fundamental laws of physics, when discovered, can appear in so many different forms that are not apparently identical at first, but with a little mathematical fiddling you can show the relationship.... There is always another way to say the same thing that doesn't look at all like the way you said it before.

Richard Feynman
Nobel Prize lecture

4.1 FERMAT'S PRINCIPLE

Around 1660, the only well-established optical laws were the laws of reflection and refraction. (Today we often call the latter *Snell's law*.) *Well established* meant only experimentally; the theory was a center of dispute. Descartes had proposed a theory, but it did not meet with favor in the eyes of Pierre Fermat, the French jurist and amateur mathematician. To his great satisfaction, Fermat

4.1 FERMAT'S PRINCIPLE

found that he could express the two optical laws as a single extremal principle: When light goes from point P_1 to another point P_2, it travels by the path that takes the least time.

Stated mathematically, Fermat's principle asserts that

$$\int_{P_1}^{P_2} \frac{ds}{v(s)} \text{ is a minimum for the actual path,} \qquad (4.1\text{-}1)$$

where ds is an element of distance along the path and $v(s)$ is the speed of light at location s on the path. To see how this accounts for refraction, we can examine the passage of light from air to glass, as sketched in figure 4.1-1. The light crosses from air to glass at point P, whose y coordinate we do not yet know. Nonetheless, we can reason that no matter what y is, the least time will elapse on the trip from P_1 to P if the light travels in a straight line. The same will be true for the trip for P to P_2. If we denote the length of the straight-line sections by $s_{\text{air}}(y)$ and $s_{\text{glass}}(y)$, the integral reduces to

$$\text{Total time} = \frac{s_{\text{air}}(y)}{v_{\text{air}}} + \frac{s_{\text{glass}}(y)}{v_{\text{glass}}}. \qquad (4.1\text{-}2)$$

The two speeds of light v_{air} and v_{glass} differ. Could a single, continuous straight line from P_1 to P_2 yield the least time? No. The light could reduce its travel time by shortening the distance in the low-speed medium, even at the cost of extra

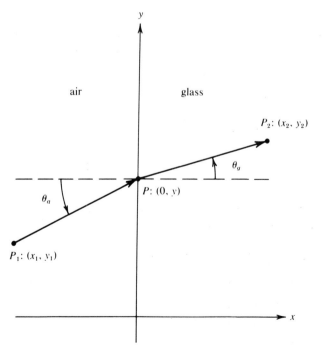

Figure 4.1-1 Refraction as derived from an extremal principle.

118 LAGRANGIAN FORMULATION

distance in the high-speed medium. There is a limit to how far this tradeoff will remain advantageous. Fermat had invented a rudimentary form of calculus (before Newton and Leibniz), and so he was able to determine the coordinate y that gave the least time. We can do it by essentially the same route.

The coordinates in figure 4.1-1 let us write

$$s_{\text{air}}(y) = [x_1^2 + (y - y_1)^2]^{1/2},$$

$$s_{\text{glass}}(y) = [x_2^2 + (y_2 - y)^2]^{1/2}.$$

These expressions go into equation (4.1-2), to give us the total time as a function of y. To find the value of y that gives the minimum, we differentiate and set the derivative to zero:

$$\frac{d}{dy}(\text{total time}) = \frac{y - y_1}{s_{\text{air}} v_{\text{air}}} - \frac{y_2 - y}{s_{\text{glass}} v_{\text{glass}}}$$

$$= \frac{\sin \theta_a}{v_{\text{air}}} - \frac{\sin \theta_g}{v_{\text{glass}}} = 0.$$

(4.1-3)

Once we express $(y - y_1)/s_{\text{air}}$ as the sine of the angle θ_a in the figure and do similarly for the glass term, we can recognize Snell's law of refraction. It works out neatly.

Fermat cast the laws of optics (as then known) into the form of an extremal principle. It is natural to ask, Can we cast Newton II into the form of an extremal principle? The answer is yes, and there are some advantages in doing so. Before we can make the transcription, however, we need to develop a general method for finding the extreme value of an integral; we do that in section 4.2. This section—apparently on optics, in a mechanics book—sets the scene historically, shows why we might try to recast Newton II.

4.2 CALCULUS OF VARIATIONS

For the calculus of variations, the central theme can be stated succinctly:

Problem: Find the function $y(x)$ that gives a certain integral, say,

$$J = \int_{x_1}^{x_2} dx\, f\left(y, \frac{dy}{dx}; x\right), \tag{4.2-1}$$

its extreme value. The function f is known; it depends on y, on the derivative dy/dx, and perhaps on x independently.

Route to solution: (*a*) Derive a differential equation that characterizes the desired function y.
(*b*) Solve the differential equation.

We plunge right in. To derive the differential equation, we compare the value J_a of the integral produced by a curve $y_a(x)$ with the value J_b produced by

4.2 CALCULUS OF VARIATIONS

a curve y_b that does not differ from y_a by much. The curves are sketched in figure 4.2-1. The difference in values is

$$J_b - J_a = \int_{x_1}^{x_2} dx \left[f\left(y_b, \frac{dy_b}{dx}; x\right) - f\left(y_a, \frac{dy_a}{dx}; x\right) \right]. \tag{4.2-2}$$

We will denote the small difference $y_b(x) - y_a(x)$ by $\delta y(x)$, so that

$$y_b(x) = y_a(x) + \delta y(x). \tag{4.2-3}$$

We can expand the first integrand in equation (4.2-2) in a Taylor series:

$$f\left(y_b, \frac{dy_b}{dx}; x\right) = f\left(y_a + \delta y, \frac{dy_a}{dx} + \frac{d(\delta y)}{dx}; x\right)$$

$$= f\left(y_a, \frac{dy_a}{dx}; x\right) + \frac{\partial f}{\partial y} \delta y + \frac{\partial f}{\partial (dy/dx)} \frac{d(\delta y)}{dx} + \mathcal{O}(\delta^2). \tag{4.2-4}$$

The parital derivatives are to be evaluated at y_a and dy_a/dx, just as the term in f is evaluated. The first partial derivative term gives the increment in f resulting from the change δy in the function's first argument. The second such term gives the increment owing to the change $d(\delta y)/dx$ in the function's second argument. (These "first order" increments are independent of each other and thus enter additively.) Terms of higher order in the Taylor series are summarized by $\mathcal{O}(\delta^2)$. When we insert the expansion into equation (4.2-2), the leading term cancels, and we are left with

$$J_b - J_a = \int_{x_1}^{x_2} dx \left[\frac{\partial f}{\partial y} \delta y + \frac{\partial f}{\partial (dy/dx)} \frac{d(\delta y)}{dx} + \mathcal{O}(\delta^2) \right]$$

$$= \int_{x_1}^{x_2} dx \left(\left\{ \frac{\partial f}{\partial y} - \frac{d}{dx} \left[\frac{\partial f}{\partial (dy/dx)} \right] \right\} \delta y + \mathcal{O}(\delta^2) \right)$$

$$+ \frac{\partial f}{\partial (dy/dx)} \delta y \bigg|_{x_1}^{x_2}. \tag{4.2-5}$$

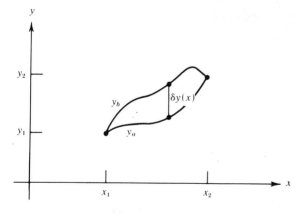

Figure 4.2-1 The two curves and their difference.

To get rid of the derivative on δy in $d(\delta y)/dx$, we integrate by parts, thus producing the second line from the first. In all contexts that we will encounter, the values of y at the endpoints are fixed. Then $y_b(x_2)$ and $y_a(x_2)$ are both equal to some fixed value, and so $\delta y(x_2) = 0$. The same is true at the endpoint x_1, and so the last expression in equation (4.2-5) vanishes.

Equation (4.2-5) holds for any pair of curves that are close together and meet the endpoint conditions. What special property would hold if we took for the curve y_a the curve that gives J its extreme value? The change in J would have to be zero (to first order in δy) for any adjacent curve y_b. (This is what *extreme value* means.) That zero change would require

$$\frac{\partial f}{\partial y} - \frac{d}{dx}\left[\frac{\partial f}{\partial (dy/dx)}\right] = 0, \qquad (4.2\text{-}6)\bigstar$$

because otherwise we could pick y_b to depart from y_a positively where $\{\partial f/\partial y - \cdots\} > 0$ and negatively where $\{\partial f/\partial y - \cdots\} < 0$, thereby ensuring that $J_b - J_a \ne 0$ in first order. In short, equation (4.2-6) holds for the function y that gives J its extreme value, and so we have found a differential equation that characterizes the desired function.

The Economical Lampshade

To see how the general theory is applied, we should work an example. The problem is sketched in figure 4.2-2. The points (x_1, y_1) and (x_2, y_2) are fixed, and a smooth curve is drawn between them. Then the curve is rotated about the y axis to generate a surface (of revolution). How should the curve be drawn to minimize the surface area?

The surface resembles a paper lampshade. The lower and upper circles are fixed in size and location, and we ask, How should we shape the surface so that we use the least paper?

What we need first is an expression for the surface area associated with any curve $y(x)$ that meets the endpoint conditions. We can decompose the surface into a sum of strips. If the strip extends horizontally a distance dx (where it intersects the plane of the sketch) and extends vertically an amount dy, then its width is

$$(dx^2 + dy^2)^{1/2} = \left[1 + \left(\frac{dy}{dx}\right)^2\right]^{1/2} dx.$$

The strip has a circumference $2\pi x$. The product of circumference and width, summed over all the strips, gives the surface area S:

$$S = \int_{x_1}^{x_2} 2\pi x \left[1 + \left(\frac{dy}{dx}\right)^2\right]^{1/2} dx. \qquad (4.2\text{-}7)$$

If we compare the present integral with equation (4.2-1), we can see what corresponds to the function f of the general theory:

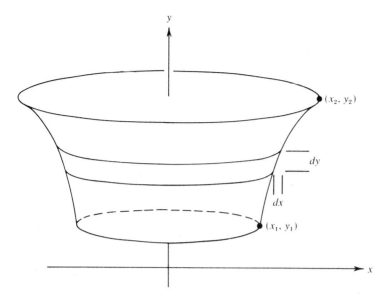

Figure 4.2-2 The lampshade: A surface of revolution. Indicated also is one of the strips into which we can decompose the entire surface.

$$f(y, \frac{dy}{dx}; x) = 2\pi x \left[1 + \left(\frac{dy}{dx}\right)^2\right]^{1/2}.$$

We need two partial derivatives,

$$\frac{\partial f}{\partial y} = 0$$

and

$$\frac{\partial f}{\partial (dy/dx)} = \frac{2\pi x (dy/dx)}{[1 + (dy/dx)^2]^{1/2}}.$$

The function $y(x)$ that gives the integral S its extreme value must satisfy equation (4.2-6); hence it must satisfy

$$0 - \frac{d}{dx}\left\{\frac{2\pi x (dy/dx)}{[1 + (dy/dx)^2]^{1/2}}\right\} = 0.$$

This is the differential equation that characterizes the most economical $y(x)$; now we must solve it.

One integration yields

$$\frac{x(dy/dx)}{[1 + (dy/dx)^2]^{1/2}} = C. \qquad (4.2\text{-}8)$$

122 LAGRANGIAN FORMULATION

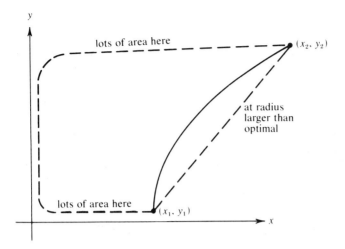

Figure 4.2-3 The extreme curve (solid) and two competitors (dashed).

Squaring and solving for dy/dx produces

$$\frac{dy}{dx} = \frac{C}{(x^2 - C^2)^{1/2}}. \tag{4.2-9}$$

If we look at equation (4.2-8), we can see that every x between x_1 and x_2 is larger than C (because the factor multiplying x is always less than 1). No zero denominator can occur in equation (4.2-9), and—more to the point—the slope dy/dx decreases in magnitude as x goes from x_1 to x_2. As illustrated in figure 4.2-3, that makes sense: A steep slope at the start keeps the curve away from the region of large radius.

Equation (4.2-9) can be integrated as an inverse hyperbolic cosine:

$$y(x) = C \cosh^{-1} \frac{x}{C} + C'; \tag{4.2-10}$$

the constants C and C' must be chosen so that y passes through the fixed endpoints. There is no need to do that here; we have seen how the calculus of variations works in an actual problem, and that was the object of the exercise. (More about the lampshade, including some cautions, can be found in problem 4-2.)

A paragraph about names is in order. In looking for the extreme value of an integral, we vary the function y in the integrand and see how much the integral changes. The notion of varying quantities leads to the name *calculus of variations*. A prescription like Fermat's for finding the path taken by light—look for the path that takes the least time—is often called a *variational principle*. Last, equation (4.2-6) is called *Euler's equation*, honoring Leonhard Euler, the remarkably prolific Swiss mathematician who derived the equation as a general

4.3 NEWTON II AS AN EXTREMAL PRINCIPLE

The success of Fermat's principle prompted us to ask, Can we cast Newton II into the form of a variational or an extremal principle? To simplify at the start, we specify a single particle in one-dimensional motion, subject to a force derivable from a potential energy $U(x)$:

$$\frac{d}{dt} m\dot{x} = -\frac{dU}{dx}. \tag{4.3-1}$$

Can we cast this equation into a variational form? In different words, can we construct an integral so that this differential equation will emerge as characterizing the function that produces the extreme value of the integral?

The first paragraph of section 4.2 stated the central theme of the calculus of variations. In the context of that description, we are trying to go backward, from part (a) of Route to Solution to the integral in Problem.

We will try to "integrate" Newton II up to a variational principle. The particle's actual position as a function of time—the literal solution to Newton II—is denoted $x(t)$. Figure 4.3-1 shows that path and another nearby path, denoted $x_b(t)$. The particle does not travel along $x_b(t)$, but—in the spirit of a

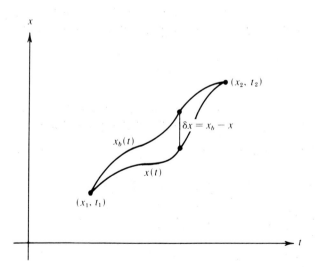

Figure 4.3-1 The actual path and a comparison path.

variational approach—we may consider that path. The difference in position at any specific time t is

$$\delta x(t) = x_b(t) - x(t).$$

Now we multiply equation (4.3-1) by δx and commence some rearrangements:

$$\left[\frac{d}{dt}(m\dot{x})\right]\delta x + \frac{dU}{dx}\delta x = 0. \tag{4.3-2}$$

The second term may be written

$$\frac{dU}{dx}\delta x \simeq U(x_b) - U(x) = \delta U.$$

We will use

δ (any quantity)   to denote   (quantity for path x_b minus quantity for path x).

In the first term of equation (4.3-2), we may pull the time derivative out in front, provided we compensate for the derivative of δx which the full derivative implies:

$$\frac{d}{dt}(m\dot{x}\,\delta x) - m\dot{x}\frac{d}{dt}(\delta x) + \delta U = 0. \tag{4.3-3}$$

The compensatory term may be written as a difference in kinetic energies:

$$m\dot{x}\frac{d}{dt}(\delta x) = m\dot{x}(\dot{x}_b - \dot{x}) \simeq m\frac{\dot{x}_b + \dot{x}}{2}(\dot{x}_b - \dot{x})$$

$$\simeq \tfrac{1}{2}m\dot{x}_b^2 - \tfrac{1}{2}m\dot{x}^2 = \delta(\tfrac{1}{2}m\dot{x}^2). \tag{4.3-4}$$

Because the compensatory term was already of first order in a small difference, being proportional to $d(\delta x)/dt$, the approximation of $\dot{x}$ by $(\dot{x}_b + \dot{x})/2$ is acceptable. The "error" is of second order, and we need to maintain accuracy only to first order in a variational calculation. If we insert equation (4.3-4) into equation (4.3-3) and integrate from t_1 to t_2, we get

$$m\dot{x}\,\delta x\,\Big|_{t_1}^{t_2} - \int_{t_1}^{t_2}\left[\delta(\tfrac{1}{2}m\dot{x}^2) - \delta U\right]dt = 0.$$

The meaning of the δ symbol permits us to make one last rearrangement:

$$m\dot{x}\,\delta x\,\Big|_{t_1}^{t_2} - \delta\int_{t_1}^{t_2}(\tfrac{1}{2}m\dot{x}^2 - U)\,dt = 0.$$

This equation holds as a consequence of Newton II, provided we vary, as indicated, away from the solution to Newton II. To get a neat variational principle, we would like to have both terms separately equal to zero. Specifying that the varied path $x_b(t)$ have the same endpoints as $x(t)$, so that $\delta x(t) = 0$ at the limits, will achieve that property.

We have worked from Newton II to an integral whose variation, relative to the actual path, is zero for nearby paths. This should dispel some of the mystery that often surrounds a variational formulation of Newton II. We may turn the route around now. To *generate* Newton II as the outcome of a variational principle, we need to start with

$$\int_{t_1}^{t_2} (\tfrac{1}{2}m\dot{x}^2 - U)\, dt, \qquad (4.3\text{-}5)\bigstar$$

and we need to say that the actual physical motion from x_1 at t_1 to location x_2 at t_2 is such that the integral has an extreme value. Then variations in path, subject to $\delta x(t_1) = 0$ and $\delta x(t_2) = 0$, lead to Newton II as the differential equation that characterizes the path which gives the extreme value.

Note the minus sign in equation (4.3-5). The integrand, whose structure is kinetic energy minus potential energy, is called the *lagrangian,* and the differential equation that results from the variational calculation is called *Lagrange's equation* (or the *Euler-Lagrange equation*), although it is simply Newton II.

The extension to three dimensions in cartesian coordinates is direct. We need only require that

$$\delta \int_{t_1}^{t_2} [\tfrac{1}{2}m(\dot{x}^2 + \dot{y}^2 + \dot{z}^2) - U(x, y, z)]\, dt = 0,$$

subject to $\delta x(t_1) = 0$, $\delta y(t_1) = 0$, etc. But here we already begin to reap benefits. The integrand is a difference of two scalars: the kinetic energy $\tfrac{1}{2}m\,\mathbf{v}\cdot\mathbf{v}$ is certainly a scalar, and the potential energy U, as constructed in section 1.4, is also a scalar. What the variational principle says is this: The time integral of the difference of these two scalars has an extremum for the actual path. Since only scalars enter, we may express the integrand in coordinates other than cartesian without any change in content or numerical value. Then the variational process will generate Newton II as written in those other coordinates. Let us see how that works.

Newton II in Cylindrical Coordinates

Figure 4.3-2 illustrates the connection between cartesian and cylindrical coordinates. For the position vector, we may write

$$\text{Position vector} = x\hat{\mathbf{x}} + y\hat{\mathbf{y}} + z\hat{\mathbf{z}}$$
$$= r\cos\theta\,\hat{\mathbf{x}} + r\sin\theta\,\hat{\mathbf{y}} + z\hat{\mathbf{z}}. \qquad (4.3\text{-}6)$$

The velocity follows upon differentiation:

$$\mathbf{v} = (\dot{r}\cos\theta - r\dot{\theta}\sin\theta)\,\hat{\mathbf{x}} + (\dot{r}\sin\theta + r\dot{\theta}\cos\theta)\,\hat{\mathbf{y}} + \dot{z}\hat{\mathbf{z}}$$
$$= \dot{r}\hat{\mathbf{r}} + r\dot{\theta}\hat{\boldsymbol{\theta}} + \dot{z}\hat{\mathbf{z}}, \qquad (4.3\text{-}7)\bigstar$$

where

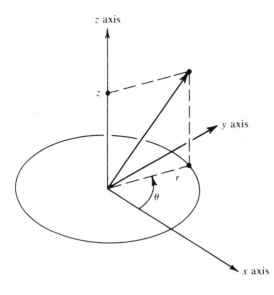

Figure 4.3-2 The connection between cartesian and cylindrical coordinates.

$$\hat{\mathbf{r}} = \cos\theta\,\hat{\mathbf{x}} + \sin\theta\,\hat{\mathbf{y}}, \qquad (4.3\text{-}8a)$$

$$\hat{\boldsymbol{\theta}} = -\sin\theta\,\hat{\mathbf{x}} + \cos\theta\,\hat{\mathbf{y}} = \hat{\mathbf{z}} \times \hat{\mathbf{r}}. \qquad (4.3\text{-}8b)$$

The unit vectors $\hat{\mathbf{r}}$ and $\hat{\boldsymbol{\theta}}$ are illustrated in figure 4.3-3. The term $r\dot\theta\hat{\boldsymbol{\theta}}$ in $\mathbf{v}$ describes the motion associated with rotation about $\hat{\mathbf{z}}$ (without requiring change in r or z). Because we have $\mathbf{v}$ in terms of orthogonal unit vectors, the square of $\mathbf{v}$ follows as

$$\mathbf{v}\cdot\mathbf{v} = \dot r^2 + r^2\dot\theta^2 + \dot z^2. \qquad (4.3\text{-}9)$$

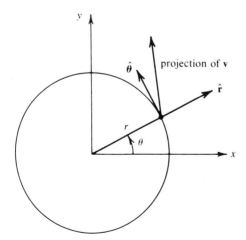

Figure 4.3-3 The motion as projected onto the plane $z = 0$.

The lagrangian is

$$L = \tfrac{1}{2}m(\dot{r}^2 + r^2\dot{\theta}^2 + \dot{z}^2) - U(r, \theta, z). \qquad (4.3\text{-}10)$$

The equations of motion follow from

$$\delta \int_{t_1}^{t_2} L\, dt = 0,$$

where we vary from the actual path $r(t), \theta(t), z(t)$ to a nearby path, one that differs by amounts $\delta r(t), \delta\theta(t),$ and $\delta z(t)$. The two paths have the same endpoints, and so $\delta r(t_1) = 0$ etc. For clarity, we should write this out in detail:

$$\begin{aligned}\delta \int_{t_1}^{t_2} L\, dt &= \int_{t_1}^{t_2} \{L(r + \delta r, \ldots) - L(r, \ldots)\}\, dt \\ &= \int_{t_1}^{t_2} \left\{\frac{\partial L}{\partial r}\delta r + \frac{\partial L}{\partial \dot{r}}\delta\dot{r} + \frac{\partial L}{\partial \theta}\delta\theta + \frac{\partial L}{\partial \dot{\theta}}\delta\dot{\theta} + \frac{\partial L}{\partial z}\delta z + \frac{\partial L}{\partial \dot{z}}\delta\dot{z}\right\} dt \\ &= \int_{t_1}^{t_2} \left\{\left[\frac{\partial L}{\partial r} - \frac{d}{dt}\left(\frac{\partial L}{\partial \dot{r}}\right)\right]\delta r + \left[\frac{\partial L}{\partial \theta} - \frac{d}{dt}\left(\frac{\partial L}{\partial \dot{\theta}}\right)\right]\delta\theta \right. \\ &\quad \left. + \left[\frac{\partial L}{\partial z} - \frac{d}{dt}\left(\frac{\partial L}{\partial \dot{z}}\right)\right]\delta z\right\} dt = 0. \qquad (4.3\text{-}11)\end{aligned}$$

The steps follow the pattern we used in the sequence (4.2-2) to (4.2-5), except that terms of order δ^2 are not even mentioned. The integration by parts costs nothing here because of the endpoint conditions, $\delta r(t_1) = 0$ and so on. We may manufacture a varied path by changing only one coordinate (such as z alone) or two or all three. The variation of the integral can be zero for all such variations only if each of the square brackets in equation (4.3-11) is individually zero. That generates three equations that the actual path must satisfy. We examine them sequentially, starting with the third.

The third equation is

$$\frac{\partial L}{\partial z} - \frac{d}{dt}\left(\frac{\partial L}{\partial \dot{z}}\right) = 0.$$

Since $\partial L/\partial \dot{z} = m\dot{z}$ and $\partial L/\partial z = -\partial U/\partial z$, the equation asserts that

$$\frac{d}{dt}(m\dot{z}) = -\frac{\partial U}{\partial z}. \qquad (4.3\text{-}12)$$

This we can recognize as just the component along $\hat{z}$ of Newton II. It is reassuring.

For the next equation, we need $\partial L/\partial \dot{\theta} = mr^2\dot{\theta}$. Since the coefficient of $\delta\theta$ in equation (4.3-11) must be zero, we deduce that

$$\frac{d}{dt}(mr^2\dot{\theta}) = -\frac{\partial U}{\partial \theta}. \qquad (4.3\text{-}13)$$

To interpret this, let us note that $mr^2\dot{\theta}$ is the component along $\hat{z}$ of the angular momentum:

$$\hat{z} \cdot [m(r\hat{r} + z\hat{z}) \times (\dot{r}\hat{r} + r\dot{\theta}\hat{\theta} + \dot{z}\hat{z}] = mr^2\dot{\theta}.$$

(The expression is cumbersome in cylindrical coordinates, where the position vector is $r\hat{r} + z\hat{z}$.) Thus equation (4.3-13) says that the angular momentum along $\hat{z}$ will be constant in time if U does not depend on θ. This corresponds to what we learned in section 1.7 from an invariance argument: Cylindrical symmetry for U implies that the angular momentum along the symmetry axis is conserved.

The final equation, from the coefficient of δr, is

$$\frac{d}{dt}\left(\frac{\partial L}{\partial \dot{r}}\right) - \frac{\partial L}{\partial r} = 0. \qquad (4.3\text{-}14)$$

The derivative $\partial L/\partial r$ produces two terms, one from the kinetic energy and the other from the potential energy. An instructive arrangement of the terms in equation (4.3-14) is this:

$$\frac{d}{dt}(m\dot{r}) - mr\dot{\theta}^2 = -\frac{\partial U}{\partial r}. \qquad (4.3\text{-}15)$$

On the right is the radial component of the force:

$$\hat{r} \cdot \mathbf{F} = -\frac{\partial U}{\partial r}.$$

If we specialize for a moment to pure circular motion, where $\dot{r} = 0$ always but $\dot{\theta}^2 > 0$, we see that a force is needed to maintain circular motion. Indeed, what is needed is a force directed radially inward. (The velocity vector must be tipped continuously to follow a circular path; a sketch will show that each little change $\Delta \mathbf{v}$ is radially inward, and so must be the force that produces the change.)

What we have found in this section for one particle can be generalized to many particles. The lagrangian, denoted L, always has the form "kinetic energy minus potential energy." The variation of $\int L\, dt$ with respect to the coordinates of each particle generates equations equivalent to Newton II.

4.4 LAGRANGIANS AND CONSTRAINTS

We have seen how Newton II for a single particle can be formulated as a variational principle. The method, we noted, can be extended to two particles (or more). An example of the extension is in order. For simplicity, we work with one-dimensional motion. Then we have

$$L = \tfrac{1}{2}m_1\dot{x}_1^2 + \tfrac{1}{2}m_2\dot{x}_2^2 - U_1(x_1) - U_2(x_2) - U_{12}(x_1, x_2). \qquad (4.4\text{-}1)$$

The mass m_1 interacts with the world outside the system via the potential energy U_1, and m_2 interacts via U_2. The two masses share a potential energy

U_{12} that describes their mutual interaction. There are now two paths, $x_1(t)$ and $x_2(t)$. In working out $\delta \int L \, dt$, we must vary both paths, but we may do so independently. Shortly we will see how that works out in detail.

At this point, we can raise another issue: In the lagrangian method, how do we describe the motion of two masses connected by a *rigid* (massless) rod? The situation is sketched in figure 4.4-1. The coordinates obey the equation

$$x_1 - x_2 = \ell_0 \qquad (4.4\text{-}2)$$

always, as x_1 and x_2 change. This is an example of a *constraint*, a limitation on the motion of the masses. To discover an answer to the question that introduced this paragraph, let us replace the rigid rod by a spring whose spring constant k we later let go to infinity. Then we are back to the context of equation (4.4-1) but with a specific mutual potential energy:

$$U_{12}(x_1, x_2) = \tfrac{1}{2} k \, (x_1 - x_2 - \ell_0)^2,$$

where the relaxed length of the spring is ℓ_0.

We could vary L to get Newton II in $\ddot{x}_1$ and $\ddot{x}_2$, but more useful for us will be the center of mass motion and the relative motion. To simplify, take equal masses: $m_1 = m_2 = m$. Then the definitions of center of mass X and relative separation x,

$$X = \frac{x_1 + x_2}{2} \qquad \text{and} \qquad x = x_1 - x_2,$$

imply

$$x_1 = X + \tfrac{1}{2}x \qquad \text{and} \qquad x_2 = X - \tfrac{1}{2}x.$$

Computing time derivatives and inserting them in equation (4.4-1) gives

$$L = \frac{1}{2}(2m)\dot{X}^2 + \frac{1}{2}\left(\frac{m}{2}\right)\dot{x}^2 - U_1\!\left(X + \frac{1}{2}x\right) - U_2\!\left(X - \frac{1}{2}x\right) - \frac{1}{2}k(x - \ell_0)^2.$$

$$(4.4\text{-}3)$$

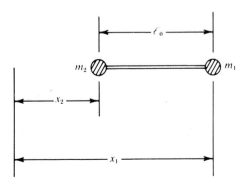

Figure 4.4-1 One-dimensional motion for two masses connected by a rigid rod of length ℓ_0.

(The cross terms in $\dot X \dot x$ cancel.) We are to imagine varied paths constructed by adding $\delta X(t)$ and $\delta x(t)$ to the actual motion. (That is fully equivalent to varying x_1 and x_2 independently.) Thus we write

$$\delta \int_{t_1}^{t_2} L(X, \dot X, x, \dot x)\, dt = \int_{t_1}^{t_2} \left[2m\dot X\, \delta \dot X - \frac{dU_1}{d(X + \tfrac{1}{2}x)} \delta X - \frac{dU_2}{d(X - \tfrac{1}{2}x)} \delta X \right.$$
$$+ \frac{m}{2} \dot x\, \delta \dot x - \frac{dU_1}{d(X + \tfrac{1}{2}x)} \left(\tfrac{1}{2} \delta x \right) - \frac{dU_2}{d(X - \tfrac{1}{2}x)} \left(-\tfrac{1}{2} \delta x \right)$$
$$\left. - k(x - \ell_0)\, \delta x \right] dt = 0. \qquad (4.4\text{-}4)$$

The potential energies U_1 and U_2 are handled with the chain rule for differentiation. The derivatives give the forces from outside the system; we may abbreviate them as

$$-\frac{dU_1}{d(X + \tfrac{1}{2}x)} = F_1(X + \tfrac{1}{2}x)$$

for the force on m_1 at location $x_1 = X + \tfrac{1}{2}x$ and similarly for m_2. Integration by parts on the derivative in $\delta \dot X = d(\delta X)/dt$ can be done without cost because $\delta X = 0$ at the endpoints, and similarly for $\delta \dot x$. Asserting that $\int L\, dt$ is an extremum with respect to independent variations δX and δx implies that the coefficients of δX and δx in equation (4.4-4) separately are zero. Hence

$$2m\ddot X = F_1(X + \tfrac{1}{2}x) + F_2(X - \tfrac{1}{2}x), \qquad (4.4\text{-}5)$$

$$\frac{m}{2} \ddot x = \tfrac{1}{2}F_1(X + \tfrac{1}{2}x) - \tfrac{1}{2}F_2(X - \tfrac{1}{2}x) - k(x - \ell_0). \qquad (4.4\text{-}6)$$

The first equation says that the total mass ($2m$) times the acceleration of the center of mass is equal to the sum of the external forces. That is right. The second equation is easier to understand if we multiply by 2:

$$\frac{d^2}{dt^2}(mx) = F_1 - F_2 - 2k(x - \ell_0). \qquad (4.4\text{-}7)$$

A glance at figure 4.4-1 helps us understand this. The force F_1, if positive, tends to increase the relative separation, while F_2 tends to decrease it. The spring acts at both ends to decrease x (if $x > \ell_0$), and that is why the 2 appears. Equation (4.4-7) is, in fact, just what we would get from Newton II for x_1 and x_2 if we formed the difference $m\ddot x_1 - m\ddot x_2$.

If the spring is very stiff and if we start the system off with $x = \ell_0$, then we know that x will remain close to ℓ_0 forever, and so in equation (4.4-5) we may say—in the limit of infinite stiffness—that

$$2m\ddot X = F_1(X + \tfrac{1}{2}\ell_0) + F_2(X - \tfrac{1}{2}\ell_0). \qquad (4.4\text{-}8)$$

Moreover, we may forget about the detailed equation for $\ddot x$ and $x(t)$.

The limit of infinite stiffness corresponds to the rigid rod. Moreover, we could get the limiting differential equation, namely equation (4.4-8), from the lagrangian (4.4-3) by setting

$$x = \ell_0 \quad \text{and} \quad \dot{x} = 0$$

to start with; that is, we impose the constraint right in the lagrangian, which started out as the lagrangian for two moderately interacting masses. This suggests a general procedure for coping with constraints.

Constraints

The constraints, let us specify, are similar to equation (4.4-2); that is, the constraints are a set of equations connecting the cartesian coordinates of the particles. If we reason by analogy with what we found for the constraint (4.4-2), we can infer the following prescription.

1. Think of the lagrangian for unconstrained but interacting masses.
2. Change to variables that will be fully free or fully fixed if the constraints are imposed. The former variables are called the *generalized coordinates*.
3. Impose the constraints by fixing the latter variables in the lagrangian and then leave them.
4. Vary the lagrangian with respect to the fully free variables, that is, with respect to the generalized coordinates.

This prescription works, as we have seen; it generates the correct combinations of Newton II. It must because, in reality, all constraints arise from physical forces.

A shortcut is often possible. By the time step 3 has been implemented, the lagrangian is a function of the generalized coordinates only. It suffices, therefore, to construct only that part of the total potential energy which changes when those coordinates change. We will see how that works in the next section.

4.5 ANOTHER INSTANCE OF CONSTRAINED MOTION

A bead is free to move frictionlessly on a circular wire that is rotating with constant angular frequency about its vertical diameter. The angular frequency is ω, say, and the radius is R. (The angular frequency is 2π times the number of revolutions per second.) The bead and wire are sketched in figure 4.5-1. The problem is to determine the equilibrium positions and their stability or instability.

The wire severely constrains the bead's location and motion. The distance from the center always must be equal to R, and the bead's azimuthal location (relative to the rotation axis) always must coincide with the wire's orientation. There is, in fact, only one fully free variable, corresponding to location along the circular loop. We may take the angle β as the generalized coordinate.

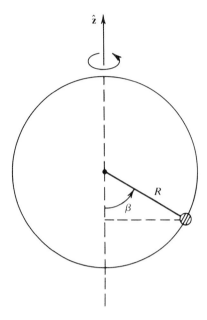

Figure 4.5-1 The circular wire rotates about the vertical line. The bead is free to slide along the wire.

The equations of constraint are

$$x^2 + y^2 + z^2 = R^2 \tag{4.5-1}$$

and

$$\frac{y}{x} = \tan \omega t, \tag{4.5-2}$$

but we can deduce the kinetic energy without using them in detail. Imagine decomposing the velocity **v** into three orthogonal pieces, with the directions based on the bead's instantaneous location: along the wire, parallel to radius R, and perpendicular to both of these. The velocity components are $R\dot\beta$, 0, and $(R \sin \beta)\omega$, respectively. Thus the kinetic energy is

$$\tfrac{1}{2}m\,[(R\dot\beta)^2 + (R \sin \beta)^2 \omega^2],$$

neatly expressed in terms of the generalized coordinate β.

Potential energy can arise from two sources: gravity and the action of the wire on the bead. Gravity contributes an amount $mg(R - R \cos \beta)$, if we take the zero to be at the bottom of the loop. The wire exerts a substantial force on the bead, ensuring that the bead never departs from the wire. We can think of the associated potential energy as zero when the bead is on the wire and infinite if the bead departs from the wire. Such an infinitely steep rise keeps the bead on the wire. When we specify that the bead is always a distance R from the center, the potential energy contributed by the wire remains constant, and we may ignore it or choose the constant to be zero, as we have done here.

The lagrangian L is the difference of the kinetic and potential energies; hence

$$L = \tfrac{1}{2}m\,[(R\dot\beta)^2 + (R\sin\beta)^2\omega^2] - mg(R - R\cos\beta). \qquad (4.5\text{-}3)$$

The equation of motion is to emerge from the assertion

$$\delta \int_{t_1}^{t_2} L(\beta, \dot\beta)\,dt = 0,$$

where the variation is from the actual "path" $\beta(t)$. The steps that we followed in section 4.2 to derive Euler's equation lead here to

$$\frac{d}{dt}\left(\frac{\partial L}{\partial \dot\beta}\right) - \frac{\partial L}{\partial \beta} = 0.$$

A look back at L shows that $\partial L/\partial \beta$ contributes two terms, so that we have

$$\frac{d}{dt}(mR^2\dot\beta) = mR^2\omega^2 \sin\beta \cos\beta - mgR \sin\beta$$

$$= mR^2\omega^2 \sin\beta \left(\cos\beta - \frac{g}{R\omega^2}\right). \qquad (4.5\text{-}4)$$

This is the Lagrange equation of motion for the generalized coordinate β.

Equilibrium

Equilibrium means β is constant, independent of time. The right-hand side of equation (4.5-4) must be zero, to preclude acceleration away from a proposed location of equilibrium:

$$\sin\beta \left(\cos\beta - \frac{g}{R\omega^2}\right) = 0. \qquad (4.5\text{-}5)$$

If $\omega^2 < g/R$, so that $g/(R\omega^2) > 1$, the expression will be zero if and only if $\sin\beta = 0$. That means $\beta = 0$ or $\beta = \pi$: the very bottom and very top of the loop are equilibrium locations.

If $\omega^2 > g/R$, then $\beta = 0$ and $\beta = \pi$ still give zeros, but another solution exists because the second factor may be zero:

$$\cos\beta = \frac{g}{R\omega^2}.$$

As ω increases, the angle β approaches $\pi/2$; the bead moves toward the horizontal periphery of the rotating wire. These possibilities are illustrated in figure 4.5-2. As an aside, let us see how they would emerge from a direct use of Newton II.

134 LAGRANGIAN FORMULATION

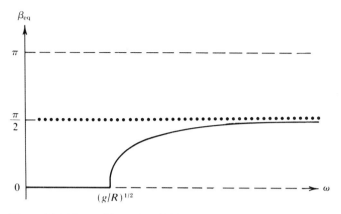

Figure 4.5-2 The positions of equilibrium β_{eq} as a function of ω. The heavy line denotes the position of stable equilibrium; the dashed lines, the positions of unstable equilibrium.

Interlude on Newton II

When in equilibrium, the bead is in uniform circular motion under the combined forces of gravity and a contact force $\mathbf{F}_c$ perpendicular to the wire. This is sketched in figure 4.5-3. No vertical acceleration implies

$$mg = F_c \cos \beta.$$

Uniform circular motion in the horizontal plane has an inward acceleration ($R \sin \beta)\omega^2$; the horizontal component of $\mathbf{F}_c$ must produce this acceleration. Hence

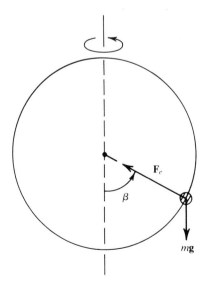

Figure 4.5-3 The newtonian analysis of equilibrium.

$$m(R \sin \beta)\omega^2 = F_c \sin \beta.$$

Solving the first equation for F_c in terms of β and inserting into the second yields

$$mR\omega^2 \sin \beta = \frac{mg \sin \beta}{\cos \beta},$$

which we may write as

$$mR^2\omega^2 \sin \beta \left(\cos \beta - \frac{g}{R\omega^2}\right) = 0.$$

This condition for equilibrium is precisely equation (4.5-5). It checks.

Stability

To test the equilibrium positions for stability, we need to study the evolution of a small displacement from equilibrium. Denote the equilibrium position by β_{eq} and the small displacement by $\Delta\beta$, the latter a function of time:

$$\beta(t) = \beta_{eq} + \Delta\beta(t). \tag{4.5-6}$$

Inserting this in the left-hand side of equation (4.5-4) gives us $d^2\Delta\beta/dt^2$. We can expand the right-hand side (RHS) about the location $\beta = \beta_{eq}$:

$$\text{RHS} = 0 + \left.\frac{\partial(\text{RHS})}{\partial \beta}\right|_{\beta_{eq}} \Delta\beta + \cdots$$

$$= mR^2\omega^2\left[\cos \beta \left(\cos \beta - \frac{g}{R\omega^2}\right) - \sin^2 \beta\right]\Bigg|_{\beta_{eq}} \Delta\beta + \cdots$$

Thus equation (4.5-4) tells us how $\Delta\beta$ evolves:

$$\frac{d^2 \Delta\beta}{dt^2} = \omega^2\left[\cos \beta_{eq}\left(\cos \beta_{eq} - \frac{g}{R\omega^2}\right) - \sin^2 \beta_{eq}\right] \Delta\beta. \tag{4.5-7}$$

Although this equation applies to each equilibrium location, its implication may differ from one to another. We should examine them individually.

If $\omega^2 > g/R$ and we look at the equilibrium given by $\cos \beta_{eq} = g/(R\omega^2)$, then equation (4.5-7) becomes

$$\frac{d^2\Delta\beta}{dt^2} = -\omega^2 \sin^2 \beta_{eq}\, \Delta\beta$$

$$= -\omega^2 \left[1 - \left(\frac{g}{R\omega^2}\right)^2\right] \Delta\beta.$$

For $\Delta\beta$, this implies harmonic oscillation at frequency $\omega\{1 - [g/(R\omega^2)]^2\}^{1/2}$ and hence stability.

If we look at the equilibrium at the bottom of the loop, where $\beta_{eq} = 0$, then equation (4.5-7) becomes

$$\frac{d^2\Delta\beta}{dt^2} = \omega^2 \left(1 - \frac{g}{R\omega^2}\right) \Delta\beta.$$

If $g/(R\omega^2) > 1$, the coefficient on the right is negative; there is a linear restoring force, and the location is stable. If, however, $g/(R\omega^2) < 1$, the coefficient is positive, and the force is anti-restoring. The solution consists of exponentials with positive and negative exponents: $\pm\omega[1 - g/(R\omega^2)]^{1/2}t$. (We met such a situation before, near the end of section 3.1.) If we fit generic initial conditions, meaning typical $\Delta\beta(0)$ and $\Delta\dot{\beta}(0)$, not special values, then the positive exponent will appear as well as the negative. The displacement is certain to become large, indeed, to grow large exponentially. In short, the situation is unstable.

The equilibrium at the top of the loop, where $\beta_{eq} = \pi$, remains to be examined. Because $\cos \pi = -1$ equation (4.5-7) becomes

$$\frac{d^2\Delta\beta}{dt^2} = +\omega^2 \left(1 + \frac{g}{R\omega^2}\right) \Delta\beta.$$

The right-side is positive for all values of ω^2, which implies that the top of the loop is always an unstable equilibrium. That hardly comes as a surprise.

The results on stability versus instability are shown in figure 4.5-2. The lower portion of the diagram merits comment or at least a verbal description. At small ω, the equilibrium curve follows the line $\beta = 0$. Then at $\omega = (g/R)^{1/2}$, the curve splits into two branches; it *bifurcates*. One branch continues along $\beta = 0$; the other rises (with infinite initial slope) and heads toward $\beta = \pi/2$. Only one branch, however, is stable. An experiment, done with care to keep friction at a truly negligible level, would find the bead staying at the bottom as the angular frequency increased from zero toward $(g/R)^{1/2}$. Once ω exceeded that critical frequency, the bead would move sharply upward on the loop. The subsequent approach to $\beta = \pi/2$ would occur only asymptotically.

4.6 CONVERSION TO FIRST-ORDER EQUATIONS: HAMILTON'S EQUATIONS

Both Newton II and the lagrangian approach provide second-order differential equations for the system's evolution in time. Take, for example, the situation in which the lagrangian depends on only a single generalized coordinate q and the associated *generalized velocity* $\dot{q}$:

$$L = L(q, \dot{q}). \tag{4.6-1}$$

The assertion that

$$\delta \int_{t_1}^{t_2} L(q, \dot{q})\, dt = 0, \tag{4.6-2}$$

subject to $\delta q(t_1) = 0$ and $\delta q(t_2) = 0$, implies that the extreme path—the actual path—is characterized by

4.6 CONVERSION TO FIRST-ORDER EQUATIONS: HAMILTON'S EQUATIONS

$$\frac{d}{dt}\left(\frac{\partial L}{\partial \dot{q}}\right) - \frac{\partial L}{\partial q} = 0. \tag{4.6-3} \star$$

Typically the kinetic energy term in the lagrangian is quadratic in $\dot{q}$ or has a portion that is quadratic. Then $\partial L/\partial \dot{q}$ is linear in $\dot{q}$, and so equation (4.6-3) will have a term in $\ddot{q}$, implying a second-order differential equation.

Examples

An example is provided by section 4.5, where β played the role of q. Equation (4.5-3) shows the quadratic appearance of $\dot{\beta}$ in L, and equation (4.5-4) shows the second-order equation. Having another example, one couched in the notation of the general theory, will be useful. Let us adapt the bead problem, but put the wire at rest so that the bead slides on a stationary wire of circular shape. Since the wire just maintains constant the bead's distance from the center, the problem is equivalent to a pendulum with bob mass m and string length R. That is perhaps the best way to think of it. We will let $q = \beta$ and take $\omega = 0$ in equation (4.5-3), so that the lagrangian is

$$L = \tfrac{1}{2} m R^2 \dot{q}^2 - mgR(1 - \cos q). \tag{4.6-4}$$

Then

$$\frac{\partial L}{\partial \dot{q}} = mR^2 \dot{q}. \tag{4.6-5}$$

The Lagrange equation of motion, from equation (4.6-3), is

$$\frac{d}{dt}(mR^2 \dot{q}) = -mgR \sin q. \tag{4.6-6}$$

Working out the derivative on the left produces a term in $\ddot{q}$, which signifies a second-order equation.

We can go back now to the general theory, though we will return to this example from time to time to illustrate the theory. If the jumping from theory to example and back is confusing, you may skip the example along the way, returning to it only after the development of theory has been completed.

Can we convert the system of equations (4.6-1) through (4.6-3) to a set of coupled first-order differential equations? That would be particularly useful for studying the evolution in a phase space. Equation (4.6-3) strongly suggests that we use $\partial L/\partial \dot{q}$ as a distinct, separate dependent variable. The dependent variables would then be

$$q \tag{4.6-7a}$$

and

$$p \equiv \frac{\partial L}{\partial \dot{q}}. \tag{4.6-7b}$$

The differential equations become

$$\frac{dq}{dt} = \dot{q} \qquad (4.6\text{-}8a)$$

and

$$\frac{dp}{dt} = \frac{\partial L(q, \dot{q})}{\partial q}. \qquad (4.6\text{-}8b)$$

This set of equations is certainly first-order, but $\dot{q}$ floats around in an awkward way. In principle, we can eliminate $\dot{q}$ from the right-hand side of equations (4.6-8a and b) by solving equation (4.6-7b) algebraically for $\dot{q}$ as a function of q and p:

$$p = \frac{\partial L(q, \dot{q})}{\partial \dot{q}} \quad \text{implies} \quad \dot{q} = \dot{q}(q, p). \qquad (4.6\text{-}9)$$

The substitution into the right-hand side of equation (4.6-8b) may be made, of course, only after $\partial L(q, \dot{q})/\partial q$ has been formed as a function of q and $\dot{q}$ (because it is the partial derivative of L with respect to q at fixed $\dot{q}$).

Example Continued

Equation (4.6-5) becomes

$$p = mR^2 \dot{q}, \qquad (4.6\text{-}10)$$

which we can solve readily for $\dot{q}$:

$$\dot{q} = \frac{p}{mR^2}. \qquad (4.6\text{-}11)$$

The derivative $\partial L/\partial q$ was worked out before and appears as the right-hand side of equation (4.6-6). Thus the analog of equations (4.6-8a and b) is

$$\frac{dq}{dt} = \frac{p}{mR^2}, \qquad (4.6\text{-}12a)$$

$$\frac{dp}{dt} = -mgR \sin q. \qquad (4.6\text{-}12b)$$

They are a closed set of equations: the time derivatives of q and p are given as functions of q and p themselves (and of them alone). Now back to the general theory.

The awkward role of $\dot{q}$ prompts the question, Can we find a function of q and p such that some derivative of it is equal to $\partial L(q, \dot{q})/\partial q$?

Let us examine the combination

$$p\dot{q} - L(q, \dot{q}), \qquad (4.6\text{-}13)$$

where the first term has been astutely chosen (by our predecessors) to provide

4.6 CONVERSION TO FIRST-ORDER EQUATIONS: HAMILTON'S EQUATIONS

a cancellation, as we shall see. Look at the combination as a function of all the variables explicitly displayed; the differential is then

$$d[p\dot{q} - L(q, \dot{q})] = p\, d\dot{q} + dp\, \dot{q} - \frac{\partial L}{\partial q}\, dq - \frac{\partial L}{\partial \dot{q}}\, d\dot{q} \quad (4.6\text{-}14)$$

$$= \dot{q}\, dp - \frac{\partial L}{\partial q}\, dq.$$

The second line follows after we impose the connection among p, $\dot{q}$, and q that the definition (4.6-7b) entails: $p = \partial L/\partial \dot{q}$. Then the first and last terms—those in $d\dot{q}$—cancel; only terms in dq and dp survive.

Now think of the combination in expression (4.6-13) as having had $\dot{q}$ eliminated, via equation (4.6-9), in terms of q and p. Denote the combination, thus conceived, by $H(q, p)$:

$$H(q, p) = [p\dot{q} - L(q, \dot{q})]_{\substack{\text{after } \dot{q} \text{ has} \\ \text{been expressed in} \\ \text{terms of } q \text{ and } p}} \quad (4.6\text{-}15)$$

The differential of H is

$$dH = \frac{\partial H}{\partial q}\, dq + \frac{\partial H}{\partial p}\, dp, \quad (4.6\text{-}16)$$

an expression that is purely formal at the moment. But now we can compare equations (4.6-14) and (4.6-16). The right-hand sides are differentials of the same quantity, and so the coefficients of dq and of dp must be equal:

$$\frac{\partial H}{\partial q} = -\frac{\partial L}{\partial q}, \quad (4.6\text{-}17a)$$

$$\frac{\partial H}{\partial p} = \dot{q}. \quad (4.6\text{-}17b)$$

These relations enable us to tidy up the right-hand side of equations (4.6-8a and b):

$$\frac{dq}{dt} = \frac{\partial H}{\partial p}, \quad (4.6\text{-}18a)\bigstar$$

$$\frac{dp}{dt} = -\frac{\partial H}{\partial q}. \quad (4.6\text{-}18b)\bigstar$$

These equations, so succinctly expressed, are called *Hamilton's equations of motion*, with $H(q, p)$ being called the *hamiltonian*. The derivative form of the right-hand side, together with the antisymmetry (for there is a sign change between the two equations), proves to be a great asset.

The Example Again

We can readily construct the hamiltonian for our example. From equations (4.6-4) and (4.6-11),

$$H = p\dot{q} - [\tfrac{1}{2}mR^2\dot{q}^2 - mgR(1-\cos q)]$$

$$= \frac{p^2}{2mR^2} + mgR(1-\cos q). \quad (4.6\text{-}19)$$

If we use this hamiltonian in equations (4.6-18a and b), we get

$$\frac{dq}{dt} = \frac{p}{mR^2}, \quad (4.6\text{-}20a)$$

$$\frac{dp}{dt} = -mgR\sin q. \quad (4.6\text{-}20b)$$

These equations are exactly the same as equations (4.6-12a and b), which we found from the Lagrange equation of motion and the elimination of $\ddot{q}$. The scheme checks out.

In section 4.7 we will see the utility of Hamilton's equations of motion. (They are, of course, just a restatement of Newton II, but a particularly fruitful one.) Here first we should note a name: the new dependent variable p is called the *generalized momentum* conjugate to the generalized coordinate q. That name is suggested by the structure of the basic Lagrange equation (4.6-3), in analogy with Newton II: $d(m\mathbf{v})/dt = \mathbf{F}$. Moreover, if the lagrangian in cartesian coordinates depends on $\dot{x}$, say, only in the kinetic energy term, as $\tfrac{1}{2}m\dot{x}^2$, then the conjugate p is just $\partial(\tfrac{1}{2}m\dot{x}^2)/\partial\dot{x} = m\dot{x}$, that is, the ordinary linear momentum. In the pendulum problem, p equals $mR^2\dot{q}$ and is the angular momentum about the support point.

Next, we can summarize the route to Hamilton's equations, the route we would take in a specific physical problem:

1. Form the lagrangian $L(q, \dot{q})$.
2. Determine the generalized momentum: $p = \partial L(q, \dot{q})/\partial \dot{q}$.
3. Express $\dot{q}$ as a function of q and p: $\dot{q} = \dot{q}(q, p)$.
4. Form the hamiltonian according to equation (4.5-15).
5. Substitute H into Hamilton's equations (4.6-18a and b).

If there are several generalized coordinates, then, of course, the algebra must be extended to encompass them all.

4.7 LIOUVILLE'S THEOREM*

Beam design for an accelerator is a subtle task. There is a bunch of particles, all protons, say, but differing slightly in location and momentum; the protons are spread over some region in phase space. How does the distribution change?

---

*This section presumes a familiarity with the divergence theorem. A good reference is E. M. Purcell, *Electricity and Magnetism* (McGraw-Hill, New York, 1965); furthermore, the theorem is developed in most texts on multivariable calculus. The section may be skipped without loss of continuity.

4.7 LIOUVILLE'S THEOREM

How should we design bending magnets and electric field regions to achieve a specific distribution in the target area? For example, we might want much higher momentum but a distribution spread out longitudinally in space (so that observing equipment will not be swamped by collisions occurring in too rapid a succession). We may want to focus the beam transversally, so that all particles see nearly the same fields and hence respond uniformly. Or we may want to focus the beam to increase the probability of collision when one bunch is sent to intersect a second bunch, nearly head on.

Hamilton's equations imply a remarkable limitation, an inevitable restriction on beam design.

It is a good idea to uncover the limitation in motion restricted to one spatial dimension before we go to the three spatial dimensions of an actual beam. Let $f(q, p, t)$ denote the number of particles per unit volume in the two-dimensional phase space. The function f is a particle density in phase space; the notion is illustrated in figure 4.7-1. To be sure, each particle is only *represented* by its own system point in phase space. Properly, we should say that f is the density of system points in phase space. That is cumbersome language, however, and so we will continue to speak of "the density of particles in phase space."

The number of particles in any fixed phase-space volume can change in only one way: by particles passing through the boundary surface of the volume. This is a way of saying that the particles are neither created nor destroyed; they just move around. If we focus on the small volume sketched in figure 4.7-1, we can express particle conservation as

$$\frac{\partial f}{\partial t} \Delta q \, \Delta p = - \text{(net flow of particles through boundary surface)}. \quad (4.7\text{-}1)$$

If the net flow is positive, say, then f should be decreasing; the minus sign incorporates this fact.

In turn, we can express the net flow of particles as the surface integral of a particle flux. The flux is the product of the particle density f and the velocity $\mathbf{V}$ of a particle in phase space:

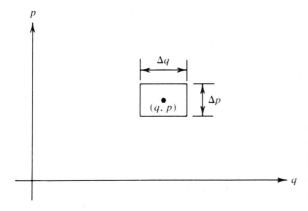

Figure 4.7-1 A small "volume" $\Delta q \, \Delta p$ around the point (q, p) in the two-dimensional phase space that is associated with motion restricted to one spatial dimension. The product $f(q, p, t) \Delta q \, \Delta p$ gives the number of particles in the volume at time t.

$$\mathbf{V} = \dot{q}\hat{\mathbf{q}} + \dot{p}\hat{\mathbf{p}}.$$

(More properly, $\mathbf{V}$ is the velocity of a particle's system point in phase space.) Thus equation (4.7-1) becomes

$$\frac{\partial f}{\partial t} \Delta q \, \Delta p = -\int (f\mathbf{V}) \cdot d\mathbf{a}.$$

With the divergence theorem, we can transform the surface integral to a volume integral:

$$\frac{\partial f}{\partial t} \Delta q \, \Delta p = -\int_{\text{volume}} \text{div} \, (f\mathbf{V}) \, dq \, dp.$$

Because the volume is tiny, the integrand is essentially constant over it; the integral reduces to the integrand times $\Delta q \, \Delta p$. Canceling the volume element on both sides, we arrive at

$$\frac{\partial f}{\partial t} = - \text{div} \, (f\mathbf{V}). \qquad (4.7\text{-}2) \bigstar$$

This is the local or differential expression of particle conservation in phase space.

Next, we can work out the divergence. As a differential operator, div affects both f and $\mathbf{V}$:

$$\text{div} \, (f\mathbf{V}) = f \, \text{div} \, \mathbf{V} + \mathbf{V} \cdot \text{grad} \, f. \qquad (4.7\text{-}3)$$

To compute div $\mathbf{V}$, we need to write $\mathbf{V}$ as a function of q and p. Hamilton's equations, derived as equations (4.6-18a and b), enter here; with them, we can write

Then
$$\mathbf{V} = \frac{\partial H}{\partial p} \hat{\mathbf{q}} + \left(-\frac{\partial H}{\partial q}\right) \hat{\mathbf{p}}.$$

$$\text{div} \, \mathbf{V} = \frac{\partial}{\partial q}\left(\frac{\partial H}{\partial p}\right) + \frac{\partial}{\partial p}\left(-\frac{\partial H}{\partial q}\right) \equiv 0.$$

Because div $\mathbf{V}$ is identically zero, only the second term in equation (4.7-3) contributes in equation (4.7-2), which reduces to

$$\frac{\partial f}{\partial t} + \mathbf{V} \cdot \text{grad} \, f = 0. \qquad (4.7\text{-}4) \bigstar$$

Equation (4.7-4) has a nice interpretation: The density f, when viewed at points along a particle's trajectory in phase space, is constant. To see this, start with figure 4.7-2 and its caption. We can express the density at the second point as

$$f(q + \dot{q}\,\Delta t, p + \dot{p}\,\Delta t, t + \Delta t) = f(q, p, t) + \frac{\partial f}{\partial t}\Delta t + \frac{\partial f}{\partial q}\dot{q}\,\Delta t + \frac{\partial f}{\partial p}\dot{p}\,\Delta t + \cdots$$

$$= f(q, p, t) + \left(\frac{\partial f}{\partial t} + \mathbf{V} \cdot \text{grad} \, f\right)\Delta t + \cdots.$$

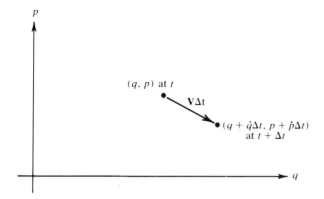

Figure 4.7-2 Imagine that we ride through phase space with a particle. Its trajectory carries us from the point (q, p) at time t to the specific nearby point $(q + \dot{q}\,\Delta t, p + \dot{p}\,\Delta t)$ at the later time $t + \Delta t$.

Equation (4.7-4) says that the coefficient of Δt is zero. Thus the densities at the two points differ by only order $(\Delta t)^2$ or higher. For the rate of change, we need the difference divided by Δt; in the limit as $\Delta t \to 0$, that will vanish. In short, equation (4.7-4) implies that if we ride with a particle, the density in our neighborhood will remain constant.

What about three spatial dimensions and a phase space with six dimensions? Our vectorial reasoning carries over without change, and so do the conclusions.

Equation (4.7-4) is often called *Liouville's theorem*. Several equivalent statements also carry that name, for example, the statement that div $\mathbf{V} = 0$ when Hamilton's equations apply in phase space.

It was beam design that introduced this section. What implication does Liouville's theorem have in that context? For example, suppose we try—by adroit choice of electric and magnetic fields—to reduce the differences in momenta, to cluster the particles tightly around some single momentum value. We can imagine riding along with some typical particle and finding more and more particles with nearly the same momentum. If that were all that was happening, the density f in our neighborhood would be increasing. Since f must remain constant, the particles must be spreading out simultaneously in the spatial directions in phase space. Perhaps we can arrange that to our advantage; perhaps not. Either way, it is inevitable.

What we have here is a lot more than just particle conservation in phase space. By introducing Hamilton's equations, we have introduced dynamics: forces. In return, we get a surprising implication: No matter how cleverly we choose the forces, we cannot compress a bunch of particles simultaneously in all directions of phase space. It is a little like squeezing a soft rubber ball or kneading dough: Push it in in one direction, and it pops out in another.

A clear picture of Liouville's theorem does require that we note the suppositions and limitations. In the way we analyzed the bunch of particles, all the

144 LAGRANGIAN FORMULATION

particles are subject to the same force fields, and those are externally determined force fields. Interactions among the particles have been ignored. If we wanted to include such interactions and do so in a rigorous way, we would need a separate six dimensions for each particle, and so we would need to use a much larger phase space, where the entire bunch of particles is described by a single system point. We could not draw the same conclusions that we can when we have many system points in a single phase space. (To be sure, there are approximate ways to include mutual interactions and to stay with a single six-dimensional phase space. That is a story all in itself, however.)

Moreover, we presumed that the forces are derivable from some kind of potential, however generalized, that fits into the hamiltonian framework. Excluded by this are dissipative forces, such as friction or the radiative reaction associated with energy loss by radiation. If such forces were allowed to act, they could collect all the particles into a tiny volume of phase space, increasing the value of f as we rode along. A collection of damped harmonic oscillators provides a nice example; all the system points spiral in to the origin in phase space, as was illustrated in figure 2.2-3.

4.8 THE LAGRANGIAN AND QUANTUM MECHANICS

There are at least three independent routes from classical physics to quantum mechanics. Heisenberg discovered the first route, via matrices, in 1925. (Incidentally, he was working with the harmonic oscillator as a mechanical model for an atomic electron.) Almost simultaneously—in 1926—Schrödinger discovered the route via wave functions. A third route, not nearly so well known, was discovered by Richard Feynman in the mid-1940s. His route is based on the lagrangian and on a sum over all conceivable ways in which something can occur. A quick tour along Feynman's route will show us why the *variation* of the lagrangian is so central to the classical description of motion.

In every form of quantum mechanics, a probability can be written as the square of a complex probability amplitude. In particular we can write

$$\text{Probability, if the particle is at location } \mathbf{x}_1 \text{ at time } t_1, \text{ that we will find the particle in tiny volume } dx\,dy\,dz \text{ around } \mathbf{x}_2 \text{ at } t_2 = |\psi(\mathbf{x}_2, t_2; \mathbf{x}_1, t_1)|^2\, dx\, dy\, dz, \quad (4.8\text{-}1)$$

where ψ is the requisite probability amplitude. The novel element in Feynman's route is the way in which the amplitude is computed:

$$\psi(\mathbf{x}_2, t_2; \mathbf{x}_1, t_1) = \sum_{\substack{\text{all conceivable} \\ \text{paths from } \mathbf{x}_1 \text{ at } t_1 \\ \text{to } \mathbf{x}_2 \text{ at } t_2}} \text{const} \times \exp\left(\frac{i}{\hbar}\int_{t_1}^{t_2} L\, dt\right). \quad (4.8\text{-}2)$$

Here $\hbar$ is Planck's constant h divided by 2π. Each path contributes to the probability amplitude with the same magnitude—that is the factor "const"—but with a phase that depends on the integral of the lagrangian along the specific path. (The constant amounts to a normalizing factor. We may think of it as being inversely proportional to the number of paths in the sum, which actually must be construed as a subtle integration over a continuum of paths. Those are difficult technical details.)

The phrase *all conceivable paths* is meant almost literally. A path may do figure eights, twist itself into knots, run back to x_1 several times before going to x_2 and *still* count in the sum. The path need satisfy no law of physics. Every path that we can scrawl from x_1 at t_1 to x_2 at t_2 enters the sum. As long as the integral of L exists, no path is too outlandish to be included.

Equations (4.8-1) and (4.8-2) apply equally to an electron and to a rock. There is a quantitative distinction, however. For a macroscopic object such as a rock and a macroscopic time interval such as $t_2 - t_1 = 1$ second, say, the integral of L is enormously larger than $\hbar$, at least typically. Here is an example: $m = 0.1$ kilogram, $v = 10$ meters/second, $U = 0$; $L = \frac{1}{2}mv^2 - 0 = 5$ kilogram $\cdot$ meter^2/second^2; if $t_2 - t_1 = 1$ second, then $\int L \, dt/\hbar = 5/(1.05 \times 10^{-34}) = 4.7 \times 10^{34}$. Let us note this disparity symbolically by writing

$$\left| \int_{t_1}^{t_2} L \, dt \right|_{\substack{\text{macroscopic system;} \\ \text{typical path}}} \gg \hbar. \tag{4.8-3}$$

When the strong inequality holds, there is violent oscillation in the contributions to the sum in equation (4.8-2). Even a macroscopically tiny change in path will (typically) change the phase by an amount large relative to π, and so contributions from neighboring paths tend to cancel, very efficiently, in fact. The paths that dominate the sum are those for which the exponent changes little as the summation goes from one path to another. The dominant paths, therefore, are those in the vicinity of the specific path for which

$$\delta \int_{t_1}^{t_2} L \, dt = 0. \tag{4.8-4}$$

When equation (4.8-4) holds, the adjacent paths have the same value of $\int L \, dt$, at least to first order, and so all those adjacent paths contribute with the same phase, that is, contribute constructively to the sum. In short, when the system is macroscopic, the dominant paths are those close to the path that satisfies Newton II.

What is the physical significance of this "dominance" in the sum? If the macroscopic object actually appears near x_2 at time t_2, we may say that the object traveled by a path close to the classical path. To be sure, we may say only *close* to the classical path. Paths that depart from the classical path so little that their $\int L \, dt$ differs by order $\hbar$ or less contribute constructively to the probability amplitude. The dominant paths form a narrow bundle of nearly parallel paths. The edges of the bundle are diffuse, of course, not precisely defined; the bundle would appear in the mind's eye like the condensation trail left by a high-flying jet airplane.

We cannot claim that the path traveled by the object is defined more sharply than the statement "the object traveled through the region swept out by paths whose $\int L \, dt$ differs by order $\hbar$ or less from the corresponding integral for the Newton II path." In this, we find an aspect of the Heisenberg uncertainty principle.

In a narrow sense, we have recovered Newton II from quantum mechanics. But, in a wider sense, there is much more. Previously, the variational principle was just another way to phrase Newton II; there was no new substance, just a difference in mathematical language. Now we can see the variational principle as playing a distinctive, significant role. All conceivable paths really are to be considered; there is physical significance to that vast continuum. The variational principle selects the dominant paths by the criterion of vanishing phase variation: constructive interference.

No route to quantum mechanics came easily. Feynman described his journey—with his customary wit—in his Nobel lecture, reprinted in *Physics Today*, August 1966. His formulation is laid out (in clear textbook fashion) in R. P. Feynman and A. R. Hibbs, *Quantum Mechanics and Path Integrals* (McGraw-Hill, New York, 1965).

4.9 A SENSE OF PERSPECTIVE

There is an intimate connection between differential equations and extremal principles. When an integral depends on some unknown function in the integrand, as in equation (4.2-1), the process of finding the extreme value of the integral leads to a differential equation for the unknown function. This is the central element in the calculus of variations.

Conversely, given a specific differential equation, often we can reverse the procedure: We construct an integral whose variation produces the differential equation. Newton II can be treated in this fashion. In section 4.3 we constructed an integral whose variation produces Newton II. The integrand—the lagrangian L—emerged as the *difference* of the kinetic and potential energies. Varying the integral produces equations of motion whose form is

$$\frac{d}{dt}\left(\frac{\partial L}{\partial \dot{q}}\right) = \frac{\partial L}{\partial q}. \tag{4.9-1}$$

One such equation is produced for each generalized coordinate q.

What advantages are there for us in the lagrangian formulation? Two have been mentioned.

First, the lagrangian is constructed from scalars, rather than from the vectors that are prominent in Newton II. What the variational principle says is this: The time integral of the difference of two scalars has an extremum for the actual path. Because only scalars enter, we may express the integrand in coordinates other than cartesian without any change in content or numerical value. Then the variational process will generate Newton II as written in those other coordinates, which may be more natural for the problem at hand.

Second, a lagrangian route can handle constraints neatly. We need not introduce explicitly the forces that hold a bead to a wire or that act at a swivel joint. It suffices to notice which coordinates remain fully free when such constraints act and to use them in constructing the lagrangian. (Another example is given in WP4-1.)

The equations that emerge from the lagrangian variational principle are second order in time. If we promote the quantity $\partial L/\partial \dot{q}$ to the status of a separate dependent variable, we can get equations that are of first order in time. We have gone two steps in a process of reformulation: from Newton II to the lagrangian and then to Hamilton's equations. The intrinsic content of the theory does not change. What does change is the facility with which we can do problems, prove theorems, and see connections with other parts of physics. For example, when quantum mechanics was developed in the first thirty years of this century, it grew out of a hamiltonian formulation of classical mechanics, not a newtonian or even a lagrangian view.

WORKED PROBLEM

WP4-1 A mass M is free to slide on a frictionless air track, as sketched in figure WP4-1. Suspended by a pivot and a light rod of length ℓ is another mass m.

Problem. Construct a lagrangian for this system and derive the equations of motion. Then determine the angular frequency when the system undergoes small oscillations.

In the context of motion in a plane, we need two cartesian coordinates for the mass M, coordinates X and Y, say, and two for m, coordinates x and y, say. There are, however, two constraints. The mass M *slides* on the track; so $Y = 0$ permanently. The *rod* fixes the distance between m and M; specifically,

$$[(x - X)^2 + y^2]^{1/2} = \ell.$$

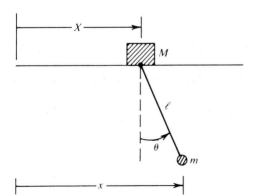

Figure WP4-1 The mass m swings freely from a pivot at the base of mass M. All motion occurs in the plane of the sketch.

148 LAGRANGIAN FORMULATION

Because of the constraints, two generalized coordinates suffice to describe the locations of both M and m. We may take them to be X and the angle θ in the sketch. The cartesian coordinates of m are then

$$x = X + \ell \sin \theta,$$
$$y = -\ell \cos \theta.$$

Differentiating these with respect to time gives the cartesian velocity components.

For the lagrangian we need the total kinetic energy,

$$\text{KE} = \tfrac{1}{2} M \dot{X}^2 + \tfrac{1}{2} m \left[(\dot{X} + \ell \dot{\theta} \cos \theta)^2 + (\ell \dot{\theta} \sin \theta)^2 \right].$$

The potential energy is the gravitational energy of m; we may write this as

$$U = mgy = -mg\ell \cos \theta,$$

taking the zero of potential energy at $y = 0$. The difference $\text{KE} - U$ provides a lagrangian.

The equations of motion now follow by differentiation. Taking first X and implementing equation (4.9-1), we find

$$\frac{d}{dt} [M\dot{X} + m(\dot{X} + \ell \dot{\theta} \cos \theta)] = 0. \tag{1}$$

This expresses conservation of momentum in the horizontal direction, along which no external forces act. Proceeding to θ, we find

$$\frac{d}{dt} (m\ell^2 \dot{\theta} + m\dot{X}\ell \cos \theta) = -mg\ell \sin \theta - m\ell \dot{X} \dot{\theta} \sin \theta. \tag{2}$$

Deriving the differential equations was easy—we never even had to discuss the contact forces that act at the swivel—but solving them is still hard work.

To determine the frequency of small oscillations, we can suppose the system was released from rest with some small nonzero angle θ. The quantity inside the square brackets in equation (1) was zero to start with and remains so. That gives us

$$\dot{X} = \frac{-m\ell \dot{\theta} \cos \theta}{m + M}.$$

After this has been used to eliminate $\dot{X}$ in equation (2) and after common constants have been canceled, we get

$$\frac{d}{dt}\left(\dot{\theta} - \frac{m}{m+M} \dot{\theta} \cos^2 \theta \right) = -\frac{g}{\ell} \sin \theta + \frac{m}{m+M} \dot{\theta}^2 \sin \theta \cos \theta.$$

When $|\theta| \ll 1$, we may approximate: $\sin \theta \simeq \theta$ and $\cos \theta \simeq 1$. The second term on the right is of third order, and so we may drop it. The equation reduces to

$$\frac{d\dot\theta}{dt} \simeq -\frac{g}{\ell}\frac{m+M}{M}\theta.$$

The angular frequency is the square root of the constant factors on the right. When $M \gg m$, the frequency approaches $(g/\ell)^{1/2}$, the value for a pendulum with a fixed support, as it should.

PROBLEMS

4-1 A beam of light is sent into a material that becomes progressively more dense, so that the speed of light changes continuously. As described by plane cartesian coordinates, the speed of light is a function of x: $v = v(x)$. The light beam leaves the origin ($x = 0$, $y = 0$) at a 45° angle ($dy/dx = 1$).

(a) Use Fermat's principle and the calculus of variations to derive a differential equation for the path taken by the beam: $y = y(x)$. (Recall that the arc length is $ds = [1 + (dy/dx)^2]^{1/2}\, dx$.) Solve as far as you can without knowing more about $v(x)$.

(b) If the speed of light is specifically $v(x) = c/(1 + x/\ell)^{1/2}$ for $x \geq 0$, where ℓ is a characteristic length, compute the path analytically and sketch it.

Fermat's principle is set lovingly into historical context by Vasco Ronchi, *The Nature of Light* (Harvard, Cambridge, Mass., 1970).

T**4-2** *The economical lampshade: Words of caution.* If the extreme lampshade curve in section 4.2 were to bend in toward the axis before heading out to point (x_2, y_2), then y would not always be a single-valued function of x. Mathematical trouble might ensue. To preclude such trouble, regard x as the dependent variable and y as the independent one. Generate the corresponding Euler equation. Does equation (4.2-10) provide a solution?

Can you specify some pairs of points (x_1, y_1) and (x_2, y_2) such that the extreme curve is sure to bend in?

If $y_2 - y_1$ is larger than both x_1 and x_2, another difficulty may arise. A curve such as the "competitor" on the left in figure 4.2-3 may be a serious competitor. Indeed, its limiting form—a disk at the bottom, a vanishingly narrow vertical column, and another disk at the top—may be the most economical surface (though hardly a good lampshade). The three pieces meet at right angles: curves with discontinuous tangents and infinite slopes. Can you see why a variational computation might have missed it?

The variational calculation searches for a "local extremum" (by comparing nearby curves). Take the symmetric situation: $x_2 = x_1$. Can you show that if $(y_2 - y_1)/2 > x_1/1.5089$, there is no cosh-like solution and hence no local extremum (aside from the limiting curve described in the preceding paragraph)?

4-3 *The brachistochrone problem.* Newton and Leibniz invented the calculus independently—and then came to quarrel over priority and competence. In 1696 Johann Bernoulli issued a challenge to all mathematicians, but especially to Newton, because Bernoulli was a Leibniz partisan. Here is Bernoulli's problem: Determine the curve connecting two given points $[(x_1, y_1)$ and $(x_2, y_2)]$, at different distances above the horizon and not in the same vertical line, along which a body passing by its own gravity will descend to the lower point in the shortest time possible. (The problem's name comes from the Greek word for "shortest time.") Over the course of 6 months only four men on the continent of Europe proposed solutions: Johann and Jacob Bernoulli, L'Hospital, and Leibniz.

(a) Approach the problem in the spirit of the calculus of variations. If the curve $y(x)$ is to give the traversal time

$$\int \frac{ds}{v} = \int_{x_1}^{x_2} \frac{ds/dx}{v} dx$$

its smallest value, what differential equation must y satisfy? Here ds is arc length along the curve, and v is the speed as a function of position, expressible via energy conservation.

(b) Specify that the object starts from rest. Can you confirm that a parametric solution is provided by

$$y_1 - y = R(1 - \cos \varphi),$$
$$x - x_1 = R(\varphi - \sin \varphi),$$
$$0 \le \varphi \le \text{some } \varphi_{max}?$$

Here φ is an auxiliary variable, and R is a constant whose size is chosen so that the curve, a cycloid, will pass through the lower point.

(c) Sketch the solution. What is the initial slope? Under which condition on the ratio $(x_2 - x_1)/(y_1 - y_2)$ will the curve dip below the level y_2 before reaching the second point?

Newton received the problem on January 29, 1697, at 4 p.m., when he arrived home from a hard day as an administrator of the British mint (during the Great Recoinage). Piqued by the challenge, he set to work—and solved the problem by 4 a.m. He published his solution anonymously in the *Philosophical Transactions*, February 1697. When Bernoulli received a copy, he recognized the author (in his words) *tanquam ex ungue leonem* (just as, from the pawprint alone, one recognizes the lion). Chagrin could not suppress his admiration.

4-4 A mass m moves in a vertical plane with velocity $\mathbf{v} = \dot{x}\hat{\mathbf{x}} + \dot{y}\hat{\mathbf{y}}$ and is subject to a gravitational potential energy $U = mgy$. Construct a lagrangian; then vary it to get the equations of motion. Do you find familiar equations?

T4-5 Starting with equations (4.3-6) through (4.3-8a and b), compute the acceleration $\mathbf{a}$ in cylindrical coordinates, expressing the result in terms of the unit vectors $\hat{\mathbf{r}}, \hat{\boldsymbol{\theta}}$, and $\hat{\mathbf{z}}$. (You will need to work out how $\hat{\mathbf{r}}$ and $\hat{\boldsymbol{\theta}}$ change with time.) Can you cast your result into the form

$$\mathbf{a} = (\ddot{r} - r\dot{\theta}^2)\hat{\mathbf{r}} + \frac{1}{r}\frac{d(r^2\dot{\theta})}{dt}\hat{\boldsymbol{\theta}} + \ddot{z}\hat{\mathbf{z}}?$$

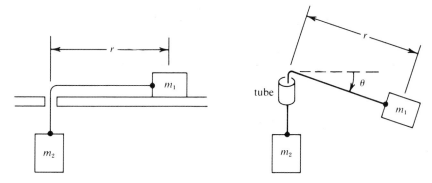

Figure P4-6

Is the Lagrange equation (4.3-15) simply the radial component of Newton II?

4-6 *Mundane physics.* On a frictionless table (realizable in practice to a high degree) a mass m_1 goes in a circle of radius r with angular frequency ω. (Thus ω is 2π times the number of revolutions per second.) The mass m_2 just hangs there on the end of the string, as in the left-hand sketch of figure P4-6. Compute the value of the mass m_2 that we must hang there to keep the system in balance. (Remember, a string transmits a pull in *both* directions. Try it.)

Now we remove the table and replace it with a small tube. Of necessity, the mass m_1 hangs at some angle θ, as shown in the right-hand sketch. The values of m_1, ω, and r are unchanged, although r is no longer the radius of the circle. Will mass m_2 with the same value as above suffice to keep the system in balance? What is the value of θ?

4-7 Here is a novel pendulum: a mass m_2 at the end of a rod of length ℓ and a mass m_1 at the midpoint, as in figure P4-7. Treat the two masses as subject to constraints, and write down a lagrangian. Then vary your lagrangian to find the equation of motion. What is the frequency of small oscillations? Does your result reduce to what Galileo knew in the limit $m_1/m_2 \to 0$?

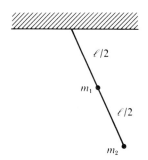

Figure P4-7

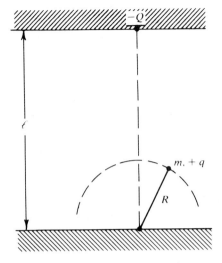

Figure P4-8

4-8 A mass m with positive charge q is attached to the end of a light rod of length R, pivoted at one end, as in figure P4-8. Directly above the pivot and at a distance ℓ is located a negative charge $-Q$, fixed in position. Because of the attractive force, the mass oscillates in the plane of the sketch (unless it is lined up in the beginning). Gravity is to be ignored; after all, if we make q and Q big enough, ...

(a) Write a lagrangian. Then vary it to find the equation of motion.

(b) Suppose that $\ell \geq 2R$ and that the amplitude of oscillation is less than or equal to $R/20$. Extend your results in part (a) to compute the frequency of such oscillations.

4-9 The context is section 4.5. Compute the cartesian velocity components $\dot{x}$, $\dot{y}$, $\dot{z}$ in terms of β, $\dot{\beta}$, and ωt, using equations (4.5-1) and (4.5-2) and the relation $x = R \sin \beta \cos \omega t$. Do your results, when used to compute the kinetic energy, confirm the expression we used?

4-10 Suppose the rotating hoop of section 4.5 is split open at the top and the wires are straightened, so that they form a rotating V, where each arm of the V makes an angle θ_0 with the vertical. Choose a generalized coordinate that will adequately describe the bead's location, and construct a lagrangian. Then repeat the analysis of section 4.5: determine the equation of motion, find the equilibrium location(s), and investigate the stability. Can you explain in simple terms why there is such a qualitative difference?

4-11 Figure P4-11 shows a circular metal loop suspended by three strings of length ℓ. The loop has radius R and mass M. If we twist the loop, it rises somewhat (because the strings are no longer vertical) but remains horizontal. When released, the loop will oscillate. Calculate the frequency of those oscillations when they are small.

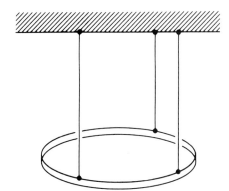

Figure P4-11

Construct a lagrangian, although you can do the problem more directly, too. You will need to establish a connection between the rotation angle and the height change; doing that in the small-angle approximation will suffice. Incidentally, the apparatus is easy to build, and you can test your prediction for the oscillation frequency.

T **4-12** *Invariance and conservation.* In section 1.7 we saw how an invariance property for the potential energy U implied a corresponding conservation law. How does the notion "an invariance implies a conservation law" appear in a lagrangian context?

Would you want to translate the phrase *an invariance* as *the lagrangian does not depend on the (specific) generalized coordinate q?* What would be the corresponding conservation law? Buttress your presentation with two examples.

4-13 An ordinary pendulum consists of an inextensible cord or rod with a large mass m at one end, swinging in a vertical plane. Suppose the cord or rod is replaced by a spring with relaxed length ℓ_0 and spring constant k.

(a) Construct a lagrangian for the two-dimensional motion of the mass; then work out the equations of motion.

(b) Suppose the mass, initially at rest at the equilibrium location, is given a small kick, so that it commences to swing and jiggle. Investigate the motion on the assumption that your generalized coordinates depart only a little from the equilibrium values, so that you may expand all terms in the differential equations through first order in the departures and may ignore contributions of higher order.

(c) When you examine special cases, do your equations in parts (a) and (b) reduce to expressions that you know are correct?

4-14 For a particle of mass m, work out the hamiltonian and Hamilton's equations in cylindrical coordinates, r, θ, and z. The potential energy is $U(r, \theta, z)$. Three generalized momenta will emerge; describe them in terms of more familiar notions, such as components of linear and angular momentum. Can you specify conditions on U such that one or more of the generalized momenta are conserved?

Suppose $U = \frac{1}{2}kr^2 + \frac{1}{2}k'z^2$, with $k' \neq k$ but both positive. Can you specify initial conditions on the generalized coordinates and momenta so that $r = r_0 > 0$ for all time but z oscillates between $\pm z_{\max} \neq 0$? Describe the motion in words.

T **4-15** *The hamiltonian and the energy.* Often the kinetic energy is a quadratic function of the generalized velocity $\dot{q}$: $\text{KE} = \tilde{f}(q)\dot{q}^2$, where $\tilde{f}$ (f tilde) is some function of the generalized coordinate q. Often, too, the potential energy does not depend on $\dot{q}$: $U = U(q, t)$. Can you show that under these circumstances the hamiltonian is equal to the energy? Can you cite an example?

T **4-16** *The hamiltonian as a constant of the motion.* In a context with one generalized coordinate and the associated generalized momentum, the hamiltonian can depend on time in three ways: $H = H[q(t), p(t), t]$. Two ways are implicit—through the temporal evolution of q and p. The third is explicit—through an explicit appearance of t in H. (For example, the change with time of an external electric field would introduce an explicit time dependence.) Work out the total time derivative of H, and then use Hamilton's equations to eliminate $\dot{q}$ and $\dot{p}$. Under which circumstances is the hamiltonian a "constant of the motion," that is, a function whose value remains constant by virtue of the equations of motion? Can you cite examples?

4-17 *Apropos section 4.7.* Start with one spatial dimension. Suppose the particles in a bunch are free except for a damping force

$$F = -\frac{m}{\tau}\dot{q} = -\frac{1}{\tau}p.$$

How does the density f evolve as you ride along? Can you solve explicitly for f in your neighborhood as a function of time?

How would f evolve if there were three spatial dimensions and the force were

$$\mathbf{F} = -k\mathbf{r} - \frac{1}{\tau}\mathbf{p}?$$

T **4-18** *Coupled pendula.* Two simple pendula, each of length ℓ and mass m, are supported from a ceiling at a horizontal separation ℓ_0. The masses are connected by a spring of relaxed length equal to ℓ_0, as shown in figure P4-18.

(a) Construct a lagrangian for this system of two coupled pendula.

(b) Anticipate small oscillations about the equilibrium: expand the trigonometric functions in the lagrangian, retaining only terms linear or quadratic in θ_1 and θ_2.

(c) From the ensuing equations of motion, construct equations for $\theta_1 + \theta_2$ (which is proportional to the displacement of the center of mass from its equilibrium location) and for $\theta_1 - \theta_2$ (which describes the relative separation). What are the corresponding angular frequencies?

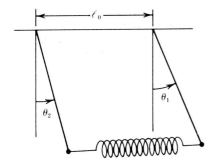

Figure P4-18

(d) What would the general solution for θ_1 as a function of time look like? What frequencies would appear? Under which circumstances would only a single frequency appear, and what would the corresponding motion of m_1 (and of m_2) be like? (The modes of motion that occur with a single frequency are called the *normal modes* of the system.)

(e) Can you confirm your theoretical results by experiment?

CHAPTER FIVE

TWO-BODY PROBLEM

5.1 Reduction to motion in a plane
5.2 Effective potential energy
5.3 Orbit shape
5.4 Orbits around the spherical sun
5.5 The oblate sun
5.6 Stability of circular orbits
5.7 The orbit in time
5.8 Compendium on central-force motion

Now many things lead me to believe that the comet of the year 1531, observed by Apian, is the same as that which, in the year 1607, was described by Kepler and Longomontanus, and which I saw and observed myself, at its return in 1682.... I may, therefore, with confidence predict its return in the year 1758. If this prediction be fulfilled, there is no reason to doubt that the other comets will return.

Edmund Halley
Philosophical Transactions, March 1705

5.1 REDUCTION TO MOTION IN A PLANE

Newton's analysis of planetary motion, Bohr's model of the atom, and the launching of an artificial earth satellite share certain features. Each was a big step in the development of our civilization. Each is an instance of the *two-body problem:* what are the motions of the two interacting bodies? The question con-

5.1 REDUCTION TO MOTION IN A PLANE

tinues to arise in research, and so—for reasons both historical and anticipatory—this chapter is devoted to the two-body problem.

Two bodies of mass m_1 and m_2 interact with one another and move about in space. For each we can write Newton II:

$$\frac{d}{dt}(m_1 \mathbf{v}_1) = \mathbf{F}_1, \tag{5.1-1a}$$

$$\frac{d}{dt}(m_2 \mathbf{v}_2) = \mathbf{F}_2. \tag{5.1-1b}$$

With these equations as the basis, what can we say about the motion?

Typically we are interested in the relative motion. So let us introduce the relative separation $\mathbf{r}$ and the center of mass $\mathbf{R}$:

$$\mathbf{r} = \mathbf{r}_1 - \mathbf{r}_2,$$

$$\mathbf{R} = \frac{m_1 \mathbf{r}_1 + m_2 \mathbf{r}_2}{m_1 + m_2}.$$

These vectors are illustrated in figure 5.5-1. Adding equations (5.1-1a and b) gives an equation for $\mathbf{R}$:

$$\frac{d}{dt}[(m_1 + m_2)\dot{\mathbf{R}}] = \mathbf{F}_1 + \mathbf{F}_2. \tag{5.1-2}$$

Dividing each of those equations by the corresponding mass and subtracting gives an equation for $\mathbf{r}$:

$$\frac{d\dot{\mathbf{r}}}{dt} = \frac{1}{m_1}\mathbf{F}_1 - \frac{1}{m_2}\mathbf{F}_2. \tag{5.1-3}$$

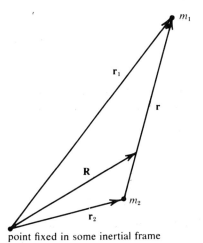

Figure 5.1-1 The relative separation and center-of-mass vectors. In this illustration, the mass ratio is $m_1/m_2 = \frac{1}{3}$, and so $\mathbf{R}$ is close to m_2.

158 TWO-BODY PROBLEM

If $\mathbf{F}_2$ equals $-\mathbf{F}_1$, that is, if the two forces satisfy Newton III, then the right-hand side of equation (5.1-2) is zero. The equation becomes $d\dot{\mathbf{R}}/dt = 0$, which implies that the center of mass moves with constant velocity. We can summarize this conclusion with the statement

$$\dot{\mathbf{R}} = \text{constant vector.}$$

Furthermore, when $\mathbf{F}_2$ equals $-\mathbf{F}_1$, we can replace $\mathbf{F}_2$ in equation (5.1-3) by $-\mathbf{F}_1$. That equation simplifies to

$$\frac{d\dot{\mathbf{r}}}{dt} = \left(\frac{1}{m_1} + \frac{1}{m_2}\right) \mathbf{F}_1 = \frac{m_1 + m_2}{m_1 m_2} \mathbf{F}_1$$

$$= \frac{1}{\mu} \mathbf{F}_1,$$

where

$$\mu \equiv \frac{m_1 m_2}{m_1 + m_2},$$

the so-called *reduced mass*. Note that

$$\mu < \text{lesser of } m_1 \text{ and } m_2$$

because we can factor the expression for μ as

$$\mu = (\text{lesser mass}) \left(\frac{\text{greater mass}}{\text{lesser + greater}}\right).$$

The inequality makes sense. There is a force acting at *each end* of $\mathbf{r}$, with both forces acting to expand $\mathbf{r}$ or to contact $\mathbf{r}$ (or to swing it around). So, of course, we would expect $\mathbf{r}$ to change more strongly than if just one force (and one mass) were present.

We will specify $\mathbf{F}_2 = -\mathbf{F}_1$ henceforth. Moreover, we will take $\dot{\mathbf{R}} = 0$, which we can do without loss of generality by choosing an inertial frame that moves with the center of mass. The essential equation is then

$$\frac{d}{dt} \mu \dot{\mathbf{r}} = \mathbf{F}_1. \tag{5.1-4}\bigstar$$

If $\mathbf{F}_1 \propto \hat{\mathbf{r}}$, that is, if $\mathbf{F}_1$ lies along the line joining the two masses, then the vector product of $\mathbf{r}$ with equation (5.1-4) yields

$$\frac{d}{dt} [\mathbf{r} \times (\mu \dot{\mathbf{r}})] = \mathbf{r} \times \mathbf{F}_1 = 0,$$

whence

$$\mathbf{r} \times (\mu \dot{\mathbf{r}}) = \text{constant vector} \equiv \mathbf{L}.$$

This is angular momentum conservation, and it has several implications. The first is that $\mathbf{r}$ and $\dot{\mathbf{r}}$ lie in a *fixed plane* perpendicular to $\mathbf{L}$. Figure 5.1-2 helps us see this. Whereas the tip of $\mathbf{r}$ might have wandered about in a three-dimensional

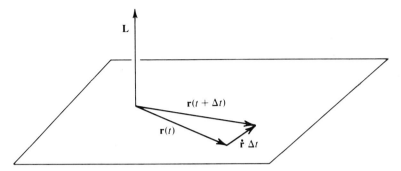

Figure 5.1-2 Conservation of angular momentum implies that **r** and **ṙ** lie, at all times, in a fixed plane perpendicular to **L**. The proof is by contradiction. If **r** began to depart from the plane, **ṙ** would have a component out of the plane. Then **r** × **ṙ** would not be along the fixed vector **L**, which would be a contradiction.

fashion, it is, in fact, restricted to moving in a fixed plane, motion that is inherently two-dimensional.

When would we *not* have $\mathbf{F}_1 \propto \hat{\mathbf{r}}$? Suppose m_2 were a small dumbbell, as in figure 5.1-3. Each ball of the dumbbell pulls gravitationally on m_1, but the unequal distances imply unequal force magnitudes. The sum of the two forces is $\mathbf{F}_1$, but that sum is not antiparallel to $\hat{\mathbf{r}}$.

We could rotate the dumbbell to generate a "mass doughnut" and get qualitatively the same effect. And then we could supplement the doughnut with a huge mass at its center. The net effect would be equivalent to the earth with an

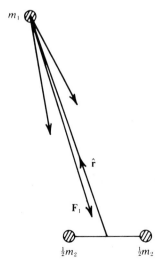

Figure 5.1-3 A situation where $\mathbf{F}_1$ would not be along $\hat{\mathbf{r}}$.

160 TWO-BODY PROBLEM

equatorial bulge. For an earth satellite, consequently, we should not expect the orbit to remain in a fixed plane.

Let us specify

$$\mathbf{F}_1 \propto \hat{\mathbf{r}} \qquad (5.1\text{-}5)$$

and

$$|\mathbf{F}_1| = \text{function of } r \text{ only.} \qquad (5.1\text{-}6)$$

The first stipulation ensures conservation of angular momentum. The two, taken together, guarantee that a potential energy $U(r)$ exists. We can solve

$$\mathbf{F}_1 = -\frac{\partial U(r)}{\partial r}\hat{\mathbf{r}}$$

for $U(r)$:

$$U(r) = -\int_{r_A}^{r} dr'\, \mathbf{F}_1(\mathbf{r}') \cdot \hat{\mathbf{r}}' + U(r_A).$$

Thus, given stipulations (5.1-5) and (5.1-6), equation (5.1-4) implies two conserved quantities:

A vector: $\qquad\qquad \mathbf{r} \times (\mu\dot{\mathbf{r}}) = \mathbf{L}, \qquad (5.1\text{-}7)$ ★

A scalar: $\qquad\qquad \tfrac{1}{2}\mu\, \dot{\mathbf{r}} \cdot \dot{\mathbf{r}} + U(r) = E. \qquad (5.1\text{-}8)$ ★

A force that meets conditions (5.1-5) and (5.1-6) is called a *central force*. (In the most common usage, both conditions must be met if a force is to be called a central force. We could be satisfied with just the first condition: any force directed along the line joining the centers of the two interacting objects would be called a central force, even if the magnitude were not a function of r only. We will, however, insist on both conditions.)

Plane Polar Coordinates

To reap the benefits of equation (5.1-7)—that the relative motion is restricted to a plane fixed in space—we should go to polar coordinates in that plane, as indicated in figure 5.1-4. Although in real life both m_1 and m_2 move, in the mathematical problem—the problem set by equation (5.1-4)—we may treat one end of $\mathbf{r}$ as fixed.

Most of the kinematic analysis has been done already, in section 4.3. Equation (4.3-7) implies

$$\dot{\mathbf{r}} = \dot{r}\hat{\mathbf{r}} + r\dot{\theta}\hat{\boldsymbol{\theta}}. \qquad (5.1\text{-}9)$$

(Note that we must distinguish between $\dot{\mathbf{r}}$ and $\dot{r}$. The former is the derivative of a vector; the latter, the derivative of a scalar.) The angular momentum is then

$$\mathbf{L} = \mu \mathbf{r} \times (\dot{r}\hat{\mathbf{r}} + r\dot{\theta}\hat{\boldsymbol{\theta}})$$
$$= \mu r^2 \dot{\theta}\, \hat{\mathbf{r}} \times \hat{\boldsymbol{\theta}}.$$

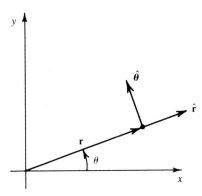

Figure 5.1-4 Polar coordinates in the plane of the motion.

The vector product $\hat{r} \times \hat{\theta}$ is a fixed unit vector perpendicular to the plane of the motion. (In figure 5.1-4, the product $\hat{r} \times \hat{\theta}$ always points out of the plane of the paper.) The constancy of the vector **L** implies the constancy of $\mu r^2 \dot{\theta}$, which we abbreviate as

$$L = \mu r^2 \dot{\theta}. \qquad (5.1\text{-}10) \bigstar$$

To cast the energy expression into polar form, we need only square the right-hand side of equation (5.1-9) and insert it:

$$E = \tfrac{1}{2}\mu(\dot{r}^2 + r^2\dot{\theta}^2) + U(r).$$

As it stands, the energy expression depends on $\dot{r}$, r, and $\dot{\theta}$. The last two variables, however, are connected by angular momentum conservation. We may use L to eliminate $\dot{\theta}$ in terms of r and L:

$$E = \tfrac{1}{2}\mu \dot{r}^2 + \frac{L^2}{2\mu r^2} + U(r). \qquad (5.1\text{-}11) \bigstar$$

The implications of this form are central to the chapter.

5.2 EFFECTIVE POTENTIAL ENERGY

The energy expression (5.1-11) contains only r and $\dot{r}$ as variables. We may regard $\mu \dot{r}^2/2$ as the kinetic energy for the radial motion of a particle of mass μ. Then it is natural to regard the other terms as an *effective potential energy* for the radial part of the motion:

$$U_{\text{eff}}(r; L) \equiv \frac{L^2}{2\mu r^2} + U(r). \qquad (5.2\text{-}1) \bigstar$$

Thus we may write, suggestively,

$$\tfrac{1}{2}\mu \dot{r}^2 + U_{\text{eff}}(r; L) = E. \qquad (5.2\text{-}2)$$

We can best see the implications of these two equations if we look at examples.

Gravitational Attraction

If masses m_1 and m_2 attract each other gravitationally and do so as spherically symmetric objects, then

$$U(r) = -\frac{Gm_1m_2}{r}$$

and

$$U_{\text{eff}}(r; L) = \frac{L^2}{2\mu r^2} + \left(-\frac{Gm_1m_2}{r}\right).$$

Which term in U_{eff} dominates? We should examine the ratio

$$\frac{L^2/(2\mu r^2)}{|U|} \propto \frac{1}{r}.$$

The ratio goes to zero as $r \to \infty$, and so U dominates at large r. When $r \to 0$, however, the ratio diverges, and so $L^2/(2\mu r^2)$ must dominate at small r. This behavior is displayed in figure 5.2-1.

For the radial motion, there is an effective potential well, as sketched in figure 5.2-2. If $E < 0$, the radial motion is constrained to lie between the two radial distances at which $U_{\text{eff}}(r; L)$ equals E. At these turning points, $\dot{r} = 0$ instantaneously, as the radial motion reverses direction.

Why do these two bounds to the motion arise?

The outer turning point is easy to understand: The attractive force prevents the relative separation from growing indefinitely.

The inner turning point is subtle. Recall that an inward force is needed even for circular motion. [We saw that explicitly in equation (4.3-15).] When $L \neq 0$, it becomes difficult—at small r—to shrink r further because most of the

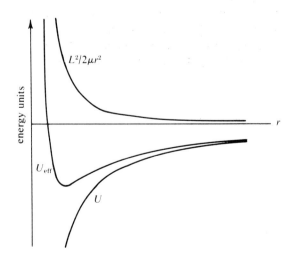

Figure 5.2-1 The effective potential energy $U_{\text{eff}}(r; L)$ when the force is gravitational attraction.

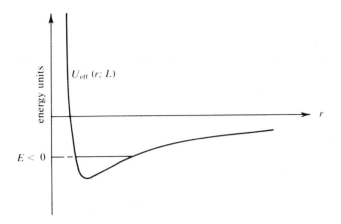

Figure 5.2-2 The effective potential well for radial motion.

attractive force is used just to *turn* the velocity $\dot{r}$ in the direction needed for almost circular motion, to keep the mass μ from flying off along the instantaneous tangent to the orbit.

The behavior at small r reminds us sharply that we are really dealing with motion in two dimensions. The conservation of angular momentum exerts a powerful influence on the radial motion. Because the dependence on L is crucial, that parameter is displayed explicitly in the symbol $U_{\text{eff}}(r; L)$. The effective potential energy for the radial motion will prove to be a powerful idea, but it will also produce puzzles unless we remember the essential two-dimensional quality of the orbital motion.

Spring Force

The force holding together the two atoms of a diatomic molecule may be represented by a spring whose relaxed length ℓ_0 is the normal separation of the atoms. This leads us to investigate a potential energy of the form

$$U(r) = \frac{k}{2}(r - \ell_0)^2,$$

where k is the "spring constant." For this system the effective potential energy is

$$U_{\text{eff}}(r; L) = \frac{L^2}{2\mu r^2} + \frac{k}{2}(r - \ell_0)^2.$$

Without question, $L^2/(2\mu r^2)$ dominates at small r, and $U(r)$ dominates at large r. Thus the curve for U_{eff} comes down from infinity at each end of the range $0 \le r \le \infty$. If we were sketching, rather than plotting numerically, how should we join those two pieces of the full curve? A single potential well seems cor-

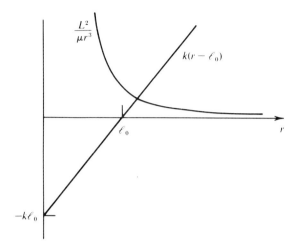

Figure 5.2-3 A graphical solution of equation (5.2-3). The left-hand side plots as a straight line; the right-hand side decreases monotonically.

rect, but we should check. We need to find the place—or places—where U_{eff} has a zero slope:

$$\frac{dU_{\text{eff}}}{dr} = -\frac{L^2}{\mu r^3} + k(r - \ell_0) = 0.$$

Thus we need the solutions of

$$k(r - \ell_0) = \frac{L^2}{\mu r^3}. \tag{5.2-3}$$

Multiplying by r^3 would give a quartic equation, not a pleasant prospect. We can learn what we need by a method that often works for complicated algebraic

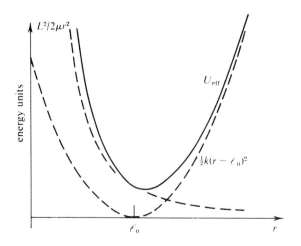

Figure 5.2-4 The effective potential energy associated with a spring of relaxed length ℓ_0.

equations: Plot both sides as a function of the dependent variable, and look for the points of intersection. This is done in figure 5.2-3. In the range $0 \le r \le \infty$, there is always one—but only one—intersection. Thus U_{eff} has a zero slope at one and only one location, which lies at $r > \ell_0$. The completed graph of U_{eff} is shown in figure 5.2-4.

As long as the mass μ has only finite energy, it will be trapped between a nonzero inner radius and a finite outer radius. The radial motion will consist of oscillation between those bounding radii. While that oscillation is occurring, however, the vector **r** will be moving azimuthally, too. The motion is inherently two-dimensional.

5.3 ORBIT SHAPE

The comments just made prompt us to look for the shape of the orbital motion. We will ignore the time dependence of the motion if we can extract the shape without it. And indeed we can, by combining the two conservation laws. Conservation of angular momentum gives us

$$\frac{d\theta}{dt} = \frac{L}{\mu r^2}, \tag{5.3-1}$$

and we can solve for $\dot{r}$ from energy conservation:

$$\frac{dr}{dt} = \left\{\frac{2}{\mu}\left[E - U_{\text{eff}}(r; L)\right]\right\}^{1/2}. \tag{5.3-2}$$

The ratio of these two equations gives us

$$\frac{d\theta}{dr} = \frac{L/(\mu r^2)}{[(2/\mu)(E - U_{\text{eff}})]^{1/2}}, \tag{5.3-3} \star$$

which we may integrate, at least in principle, as

$$\theta + \text{const} = \int \frac{[L/(\mu r^2)]\, dr}{\{(2/\mu)[E - U_{\text{eff}}(r; L)]\}^{1/2}}. \tag{5.3-4}$$

This relation is quite general, but only for a few special forms of $U(r)$ can the integral be evaluated in terms of tabulated functions. Numerical integration will always work, of course, but it gives results only for the chosen values of the parameters, such as L and E.

So that we can do the integral analytically, let us use the spring force of section 5.2 but set $\ell_0 = 0$, so that

$$U(r) = \frac{k}{2} r^2. \tag{5.3-5}$$

Equation (5.3-4) becomes

$$\theta + \text{const} = \frac{L}{(2\mu)^{1/2}} \int \frac{(1/r^2)\, dr}{[E - L^2/(2\mu r^2) - kr^2/2]^{1/2}}.$$

166 TWO-BODY PROBLEM

The variable r appears under the square root sign in terms that differ by r^2 or r^4. We may make this quartic expression look like merely a quadratic form by the substitution

$$w = \frac{1}{r^2},$$

with an accompanying change in differential:

$$dr = d(w^{-1/2}) = -\tfrac{1}{2}w^{-3/2}\, dw.$$

The integral becomes

$$\theta + \text{const} = -\frac{1}{2}\frac{L}{(2\mu)^{1/2}} \int \frac{ww^{-3/2}\, dw}{[E - wL^2/(2\mu) - k/(2w)]^{1/2}}.$$

The net factor in the numerator, $w^{-1/2}$, can be transferred to the square root as w and will clear it of the inverse power of w. The integrand is then just 1 over the square root of a quadratic expression. A table of integrals provides the integral as an inverse sine function:

$$\theta + \text{const} = \frac{1}{2}\arcsin \frac{E - L^2 w/\mu}{(E^2 - kL^2/\mu)^{1/2}}.$$

We can multiply by 2, take the sine of both sides, and solve for w:

$$\frac{1}{r^2} = \frac{\mu E}{L^2}\left\{1 - \left(1 - \frac{kL^2}{\mu E^2}\right)^{1/2}\sin[2(\theta + \text{const})]\right\}. \tag{5.3-6}$$

The basic orbit shape is determined by the values of L and E. They do that, of course, within a framework set by the specific potential energy function $U(r)$.

What determines the value of "const"? The constant is associated with how we orient our polar coordinates relative to the actual orbit. For definiteness, specify that the direction $\theta = 0$ should line up with an extreme value of r, either a minimum or a maximum. That stipulation implies

$$\frac{d}{d\theta}[\text{right-hand side of equation (5.3-6)}]\Big|_{\theta=0} = 0,$$

which reduces to

$$\cos(2\,\text{const}) = 0,$$

a solution of which is $\text{const} = \pi/4$. Then

$$\sin[2(\theta + \text{const})] = \sin\left(2\theta + 2\frac{\pi}{4}\right)$$

$$= \cos 2\theta$$

by the identity (A.2-3) for the sine of a sum. When we choose this orientation for the direction $\theta = 0$, the expression for the orbit shape is

$$\frac{1}{r^2} = \frac{\mu E}{L^2}\left[1 - \left(1 - \frac{kL^2}{\mu E^2}\right)^{1/2}\cos 2\theta\right]. \tag{5.3-7}$$

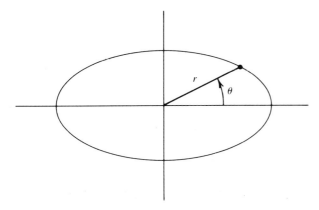

Figure 5.3-1 The shape of the orbit, as computed from equation (5.3-7), when $kL^2/(\mu E^2) = 0.6$.

The radius r goes through a full cycle when θ goes merely through an angle π.

Figure 5.3-1 displays the orbit, computed from equation (5.3-7) by solving for r as a function of θ. To the eye, the shape is quite like an ellipse. We cannot, however, trust the eye to distinguish an ellipse from a curve of similar shape but different structure in detail. For example, a graph of

$$\frac{x^2}{a^2} + \frac{y^4}{b^4} = 1$$

looks more or less elliptical, but the graph is not a genuine ellipse. The orbit in figure 5.3-1, however, is a bona fide ellipse; a proof comes readily from our analysis in section 2.3. The present potential energy, given in equation (5.3-5), implies a radially inward linear restoring force. That is precisely the same as the force in equation (2.3-1). We found that the shape of the periodic motion was an ellipse, with a circle and a straight line as special cases. Q.E.D.

The orbit given by equation (5.3-7) becomes a circle when the coefficient of $\cos 2\theta$ is zero, that is, when $E = (kL^2/\mu)^{1/2}$. A square root then gives the radius as

$$\text{Radius of circular orbit} = \left(\frac{L^2}{\mu E}\right)^{1/2} = \left(\frac{L^2}{k\mu}\right)^{1/4}.$$

Let us compare these statements with predictions based on the effective potential energy.

If we set $\ell_0 = 0$ in the latter half of section 5.2, we find, via equation (5.2-3), that the minimum in U_{eff} occurs at $r = [L^2/(k\mu)]^{1/4}$. If we want circular motion, we need to keep $\dot{r} = 0$ for all time. That means we need to reduce the energy E to its lowest possible value, so that the inner and outer turning points of the radial oscillation coalesce into a single radial location. From the general equation (5.2-2), namely,

$$\tfrac{1}{2}\mu \dot{r}^2 + U_{\text{eff}}(r; L) = E,$$

we extract the requirement on E as

$$E = U_{\text{eff}}(r; L)\bigg|_{r \text{ that gives minimum}}$$
$$= \left(\frac{kL^2}{\mu}\right)^{1/2}.$$

This energy will guarantee motion at the radius that gives the minimum is U_{eff}, namely, $r = [L^2/(k\mu)]^{1/4}$. The agreement with the detailed solution is reassuring.

Two comments are in order before we leave this section. The radial component of velocity $\dot{r}$ may be either positive or negative. Thus the square root on the right-hand side of equation (5.3-2) may be negative as well as positive. The distinction has no substantial effect on the route we took to the orbit solution in equation (5.3-7), but we should keep the plus-or-minus possibility in the back of our minds.

The second comment concerns the periodicity of the motion. Whenever there are both outer and inner turning points, the radial coordinate is periodic in time. (An analysis with U_{eff} tells us that.) The variation of r with θ—the orbit shape, independent of time—is also periodic, but the period in θ may not be 2π or $2\pi/$(integer). The linear restoring force gives a period π, and so the orbit in two dimensions begins to retrace itself whenever θ has gone through 2π. Any period of the form $2\pi/$(integer) will give a shape that retraces itself after 2π. That behavior, however, is not typical of attractive force laws. Such retracing happens to be true for the two force laws most commonly studied—linear restoring and $1/r^2$—but it is not generic. In section 5.5 we will find an orbit—indeed, a very realistic orbit—that does not retrace itself.

5.4 ORBITS AROUND THE SPHERICAL SUN

This section is both an end in itself and a prelude. Taking the sun to be perfectly spherical and calculating orbits will give us a good approximation to the actual orbits of planets and comets. Moreover, we will establish a framework for section 5.5, in which the assumption of sphericity is relaxed.

If we let m_2 denote the sun's mass and m_1, the mass of the object in orbit, then their mutual potential energy is $U = -Gm_1m_2/r$. We can construct $U_{\text{eff}}(r; L)$, and we could calculate an orbit shape via equation (5.3-3), namely, by integrating the equation

$$\frac{d\theta}{dr} = \frac{L/(\mu r^2)}{\{(2/\mu)[E - U_{\text{eff}}(r; L)]\}^{1/2}}. \tag{5.4-1}$$

While that route would go well here, it would be difficult to follow when we drop the sphericity assumption and must cope with a more complicated potential energy. Instead of integrating equation (5.4-1), we look for a differential equation in which later we can use the techniques we learned when studying nonlinear oscillators.

The reciprocal of equation (5.4-1) gives us an equation in $dr/d\theta$. We can take θ as the independent variable. In the present situation, the r dependence in U_{eff} actually appears as powers of $1/r$; that—together with experience—suggests a change in dependent variable to

$$u = \frac{1}{r}. \tag{5.4-2}$$

Then

$$\frac{du}{d\theta} = -\frac{1}{r^2}\frac{dr}{d\theta},$$

and the reciprocal of equation (5.4-1) becomes

$$\frac{du}{d\theta} = -\frac{\mu}{L}\left(\frac{2}{\mu}\right)^{1/2}(E - U_{\text{eff}})^{1/2}. \tag{5.4-3}$$

To get rid of the square root—and to get a second-order equation of the kind we studied in chapter 3—we can differentiate again:

$$\frac{d^2u}{d\theta^2} = +\frac{\mu}{L}\left(\frac{2}{\mu}\right)^{1/2}\tfrac{1}{2}(E - U_{\text{eff}})^{-1/2}\frac{dU_{\text{eff}}}{dr}\frac{dr}{d\theta}$$

$$= \frac{\mu r^2}{L^2}\frac{dU_{\text{eff}}}{dr}$$

$$= \frac{\mu r^2}{L^2}\left(-\frac{L^2}{\mu r^3} + \frac{dU}{dr}\right)$$

so that

$$\frac{d^2u}{d\theta^2} = -u + \frac{\mu r^2}{L^2}\frac{dU}{dr}. \tag{5.4-4}\bigstar$$

The step to the second line above follows when we use equation (5.4-1) to express $dr/d\theta$. The outcome, equation (5.4-4), is quite general and applies in any context in which the potential energy is a function of r only.

Substituting our potential energy into equation (5.4-4) yields

$$\frac{d^2u}{d\theta^2} = -u + \frac{\mu r^2}{L^2}\frac{Gm_1m_2}{r^2}$$

$$= -u + \frac{1}{\alpha}, \tag{5.4-5}$$

where

$$\alpha \equiv \frac{L^2}{\mu G m_1 m_2} \tag{5.4-6}$$

is a convenient abbreviation. In structure, the differential equation resembles that for a harmonic oscillator: it has a second derivative term and a linear restoring term. To make the resemblance perfect, we can combine the constant term with u to form a new dependent variable. Upon letting

$$\tilde{u} = u - \frac{1}{\alpha}, \tag{5.4-7}$$

we get

$$\frac{d^2\tilde{u}}{d\theta^2} = -\tilde{u}. \tag{5.4-8}$$

Now we can write a solution immediately:

$$\tilde{u} = A \cos(\theta + \varphi_0),$$

which implies

$$\frac{1}{r} = \frac{1}{\alpha} + A \cos(\theta + \varphi_0). \tag{5.4-9}$$

Although α is determined by the masses and L^2, we do not know immediately what value the constants A and φ_0 should have. That is because we lost the energy E in the process of going from $d\theta/dr$ to $d^2\tilde{u}/d\theta^2$. We can determine A by returning to equation (5.4-3). That equation and our solution give us two equivalent ways to express $du/d\theta$:

$$\frac{du}{d\theta} = \begin{cases} -A \sin(\theta + \varphi_0) \\ -\frac{\mu}{L}\left(\frac{2}{\mu}\right)^{1/2} [E - U_{\text{eff}}(r; L)]^{1/2}. \end{cases}$$

If we evaluate the right-hand side when $\theta + \varphi_0 = \pi/2$ and equate the equivalent expressions, we get

$$-A = -\frac{\mu}{L}\left(\frac{2}{\mu}\right)^{1/2}\left(E - \frac{L^2}{2\mu\alpha^2} + \frac{Gm_1m_2}{\alpha}\right)^{1/2}.$$

This gives us A, which can be written neatly as

$$A = \frac{\epsilon}{\alpha},$$

where ϵ denotes the eccentricity (discussed shortly):

$$\epsilon = \left[1 + \frac{E}{Gm_1m_2/(2\alpha)}\right]^{1/2}. \tag{5.4-10}$$

Finally, equation (5.4-9) becomes

$$\frac{1}{r} = \frac{1}{\alpha}[1 + \epsilon \cos(\theta + \varphi_0)]. \tag{5.4-11}\bigstar$$

5.4 ORBITS AROUND THE SPHERICAL SUN

The constant φ_0 is associated merely with the orientation of the line $\theta = 0$ relative to the actual orbit. The essential dependence of the orbit is on E (through its presence in the expression for ϵ) and on L (through the parameter α).

The Typical Bound Orbit

Figure 5.2-2 showed us that a bound orbit necessarily has $E < 0$. Then ϵ is less than 1. To learn what the shape of the typical bound orbit is like, we can tabulate r for several values of the argument $\theta + \varphi_0$ in equation (5.4-11), as is done in table 5.4-1. Interpolation between those values and many intermediate values generates a continuous curve. For $\epsilon = 0.6$, such a curve is shown in figure 5.4-1. The shape appears to be elliptical; indeed, it is the justly famous keplerian ellipse. For a proof, we convert the orbit equation from polar form (r and θ) to a cartesian expression in x and y.

We write

$$x = r \cos(\theta + \varphi_0),$$
$$r = \sqrt{x^2 + y^2}.$$

After using these relations to eliminate r and the cosine in equation (5.4-11), we get

$$\frac{1}{\sqrt{x^2 + y^2}} = \frac{1}{\alpha}\left(1 + \frac{\epsilon x}{\sqrt{x^2 + y^2}}\right).$$

Table 5.4-1 Relative separation r of a typical bound orbit as computed from equation (5.4-11)

$\theta + \varphi_0$	$\dfrac{1}{r}$	r	Comment
0	$\dfrac{1+\epsilon}{\alpha}$	$\dfrac{\alpha}{1+\epsilon}$	Smallest r
$\dfrac{\pi}{2}$	$\dfrac{1}{\alpha}$	α	
π	$\dfrac{1-\epsilon}{\alpha}$	$\dfrac{\alpha}{1-\epsilon}$	Largest r
$\dfrac{3\pi}{2}$	$\dfrac{1}{\alpha}$	α	
2π	$\dfrac{1+\epsilon}{\alpha}$	$\dfrac{\alpha}{1+\epsilon}$	Begins to repeat

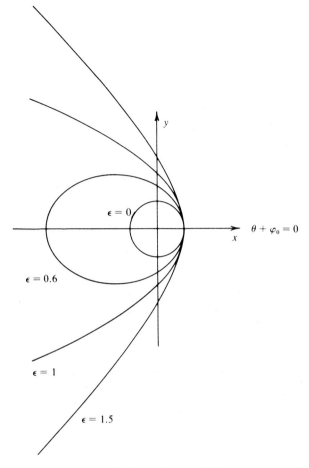

Figure 5.4-1 The qualitatively distinct orbits in an attractive $1/r^2$ force field.

To remove the square roots, first we multiply by $\alpha\sqrt{x^2 + y^2}$:

$$\alpha = \sqrt{x^2 + y^2} + \epsilon x.$$

Then we move the term in ϵx to the left-hand side and square:

$$\alpha^2 - 2\alpha\epsilon x + \epsilon^2 x^2 = x^2 + y^2.$$

We can place all terms in x^2 or x on the right-hand side and then "complete the square":

$$\alpha^2 = (1 - \epsilon^2)x^2 + 2\alpha\epsilon x + y^2;$$

$$\alpha^2 = (1 - \epsilon^2)\left(x + \frac{\epsilon\alpha}{1 - \epsilon^2}\right)^2 - (1 - \epsilon^2)\left(\frac{\epsilon\alpha}{1 - \epsilon^2}\right)^2 + y^2.$$

Last, we transfer the second term on the right-hand side to the other side of the equation and divide by $1 - \epsilon^2$:

$$\frac{\alpha^2}{(1-\epsilon^2)^2} = \left(x + \frac{\epsilon\alpha}{1-\epsilon^2}\right)^2 + \frac{y^2}{1-\epsilon^2}. \tag{5.4-12}$$

Given $\epsilon < 1$, this quadratic expression decribes an ellipse. To be sure, the center of the ellipse is displaced from the origin; it is at the point $-\epsilon\alpha/(1-\epsilon^2)$ along the x axis. The present elliptical orbit differs from the elliptical orbit that we found for the linear restoring force. In the gravitational problem, the angle θ must go though a full 2π before the radius r goes through a full cycle.

The point inside the ellipse from which we are reckoning r and θ is called the *focus*. (More precisely, the point is one of two foci, placed symmetrically with respect to the center.) The quantity $\alpha/(1-\epsilon^2)$ is one-half the major axis of the ellipse, called the *semimajor axis* and denoted a:

$$\text{Semimajor axis } a = \frac{1}{2}\left(\frac{\alpha}{1-\epsilon} + \frac{\alpha}{1+\epsilon}\right)$$

$$= \frac{\alpha}{1-\epsilon^2}.$$

Thus the eccentricity ϵ specifies the center-to-focus distance $\epsilon\alpha/(1-\epsilon^2)$ as a fraction of the semimajor axis:

$$\text{Eccentricity of ellipse} = \frac{\text{center-to-focus distance}}{\text{semimajor axis}}.$$

Classifying the Orbits

For a classification of the orbit shapes in general, there are two critical energies. The value $E = 0$ separates the bound orbits (with $E < 0$) from orbits that extend to infinity. There is, moreover, a lower limit to the energy. In an analysis with $U_{\text{eff}}(r; L)$, we determine it as the minimum value of U_{eff}; the associated orbit is circular. Equivalently, because the eccentricity ϵ must be a real quantity, the square root in equation (5.4-10) imposes a lower limit on E. The most negative energy is

$$E_* \equiv -\frac{Gm_1m_2}{2\alpha} = -\frac{(Gm_1m_2)^2\mu}{2L^2}.$$

(This energy is itself a function of the angular momentum L.)
The critical energies lead to the following classification:

Energy	Eccentricity	Orbit shape
$E = E_*$	$\epsilon = 0$	Circle
$E_* < E < 0$	$0 < \epsilon < 1$	Ellipse
$E = 0$	$\epsilon = 1$	Parabola
$E > 0$	$\epsilon > 1$	Hyperbola

To see how the orbit shapes emerge, we go back to equation (5.4-12). Setting ϵ equal to zero there gives the equation of a circle. When $\epsilon > 1$ pertains, the denominator of the y^2 term is negative. A sign difference between the two quadratic terms implies a hyperbola. Because of the denominators, the situation $\epsilon = 1$ must be handled with care. First we multiply by $1 - \epsilon^2$, expand the square containing x, and combine like terms; these steps produce

$$\alpha^2 = (1 - \epsilon^2)x^2 + 2\alpha\epsilon x + y^2.$$

Now we may safely take the limit as $\epsilon \to 1$:

$$\alpha^2 = 2\alpha x + y^2.$$

This equation tells us that x is a parabolic function of y.

Figure 5.4-1 shows all four of the qualitatively distinct orbits in an attractive $1/r^2$ force field. When the curves were generated from equation (5.4-11), the parameter α was adjusted so that the minimum r remained the same while ϵ changed.

Information about some objects in solar orbit is provided in table 5.4-2. The characteristic orbital length for planets is, by convention, one-half the major axis of the planet's elliptical orbit, the semimajor axis. For comets, however, where the trajectory may be parabolic or hyperbolic, convention settles on the perihelion distance, the distance when the object is closest to the sun. The standard length unit is the *astronomical unit* (AU), the earth's semimajor axis, equivalent to 1.496×10^{11} meters.

The cometary eccentricities are based on observations made when the comets are "close" to the sun, typically within a few astronomical units. A comet with eccentricity greater than 1 actually may be bound to the solar system. When its orbit takes it again beyond the major planets, the planetary at-

Table 5.4-2 Some objects in solar orbit

	Eccentricity	Characteristic length, AU	Period, years
		(*semimajor axis*)	
Mercury	0.206	0.387	0.24
Venus	0.007	0.723	0.62
Earth	0.017	1.000	1.00
Icarus	0.827	1.077	1.12
Mars	0.093	1.524	1.88
Ceres	0.079	2.767	4.60
Jupiter	0.048	5.203	11.86
Pluto	0.250	39.44	248.4
		(*perihelion*)	
Comet Encke	0.847	0.339	3.30
Halley's comet	0.967	0.587	76.1
Comet 1961 VIII	0.993	0.681	941.9
Comet 1914 III	1.003	3.746	
Comet 1972 IX	1.006	4.276	

tractive forces (when added to the solar force) may convert the orbit to an (approximately) elliptical one. Then again, a comet with eccentricity greater than 1 may have been periodic in the past (with $\epsilon < 1$) but have "recently" suffered a strong interaction with a large planet, gained energy, and be now unbound, destined to leave the solar system forever. A genuine one-time trip from outer space and back is possible but unlikely. More about comets may be found in the review article "The background of modern comet theory" by F. L. Whipple, *Nature*, **263**, 15 (2 September 1976).

5.5 THE OBLATE SUN

Sunspots move slowly across the face of the sun. Their motion shows a rotation of the solar surface with a period of about 25 days. Accompanying the rotation is a solar equatorial bulge, illustrated in exaggerated fashion in figure 5.5-1. What effect does the bulge have on the motion of the earth?

In section 1.6 we studied the gravitational potential energy of a mass interacting with a ring of matter. That suggests an approximation for the earth's potential energy: we regard the oblate sun as a sphere of radius $R_\odot$ plus a ring of mass M_{ring} located at the solar equator, so that

$$U \simeq U_{\text{sphere}} + U_{\text{ring}}. \tag{5.5-1}$$

We can adapt the expansion method in section 1.6 to give U_{ring} when $r \gg $ radius of ring; that just amounts to expanding in powers of $R_\odot/r$. We find

$$U_{\text{ring}}(\mathbf{r}) = -\frac{GM_{\text{ring}}m_\oplus}{r}\left\{1 + \frac{R_\odot^2}{4r^2}\left[1 - 3(\mathbf{e}_3 \cdot \hat{\mathbf{r}})^2\right] + \cdots\right\}, \tag{5.5-2}$$

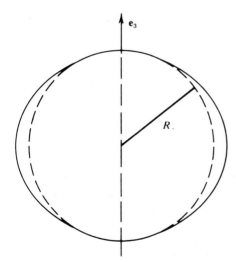

Figure 5.5-1 The sun with its equatorial bulge. The inscribed sphere has a radius $R_\odot$, to all intents and purposes.

where the subscript $\oplus$ denotes the earth, just as $\odot$ denotes the sun. The unit vector $\mathbf{e}_3$ points along the solar rotation axis; the vector $\mathbf{r}$ is the earth's position vector in the full three-dimensional space surrounding the sun.

Because U_{ring} is a function of angle (through $\mathbf{e}_3 \cdot \hat{\mathbf{r}}$) as well as of radial distance r, it is not invariant under rotation about every axis we might choose. In section 1.7 we found that rotational invariance implied conservation of angular momentum. It works in the negative sense, too: Failure of invariance implies failure of conservation. The implication here is that the angular momentum of the earth-sun relative motion may not be conserved. (In an alternative view, grad U_{ring} does not produce a force purely along $\hat{\mathbf{r}}$, at least for most locations $\mathbf{r}$.)

If the earth's orbit lies in the solar equatorial plane, however, then all is well because U_{ring} is invariant under rotations about $\mathbf{e}_3$, and so the orbital angular momentum is conserved. (To follow up the alternative view, we may reason that if $\mathbf{r}$ lies in the solar equatorial plane, then symmetry requires the gravitational attraction to lie in that plane and to be along $\hat{\mathbf{r}}$. An orbit started in that plane remains—by symmetry as well as by dynamics—in that plane, and the angular momentum is conserved.)

The planetary orbits are close to the solar equatorial plane, and so we can afford to study motion in precisely that plane. The gravitational force is then purely radial, and angular momentum is conserved.

The leading term in equation (5.5-2) varies as $1/r$; we can combine it with U_{sphere} to get the $1/r$ term produced by the sun's entire mass $M_\odot$. Thus the potential energy, when restricted to locations in the solar equatorial plane, is

$$U(r) = -\frac{GM_\odot m_\oplus}{r} - \frac{GM_{\text{ring}} m_\oplus R_\odot^2}{4r^3} \quad (5.5\text{-}3)$$

because $\mathbf{e}_3 \cdot \hat{\mathbf{r}} = 0$ for such locations. The oblateness term is smaller than the $1/r$ term by the ratio

$$\frac{1}{4} \frac{M_{\text{ring}}}{M_\odot} \frac{R_\odot^2}{r_\oplus^2}.$$

The ratio of the solar radius $R_\odot$ to the earth's orbital distance $r_\oplus$ is about 5×10^{-3}. Since $M_{\text{ring}}/M_\odot \ll 1$, the oblateness term is many orders of magnitude smaller than the leading term.

The question that introduced this section was, What effect does the solar bulge have on the earth's motion? Now that we have a potential energy and have angular momentum conservation, we can investigate the orbit shape with the differential equation that we derived as equation (5.4-4). Substituting our potential energy into the equation, we find

$$\frac{d^2 u}{d\theta^2} = -u + \frac{\mu r^2}{L^2}\left(\frac{GM_\odot m_\oplus}{r^2} + \frac{3GM_{\text{ring}} m_\oplus R_\odot^2}{4r^4}\right)$$

$$= -u + \frac{1}{\alpha} + \frac{\beta}{\alpha} u^2, \quad (5.5\text{-}4)$$

where

$$\alpha = \frac{L^2}{\mu G M'_\odot m_\oplus} \tag{5.5-5}$$

and

$$\beta = \frac{3}{4} \frac{M_{\text{ring}}}{M_\odot} R_\odot^2. \tag{5.5-6}$$

We almost have an equation with structure like

$$\frac{d^2x}{dt^2} = -\omega_0^2 x + f(x, \dot{x}),$$

where f describes some "small" influence on what would otherwise be harmonic motion. To get that structure, we can combine $1/\alpha$ with u, as before. Letting

$$\tilde{u} = u - \frac{1}{\alpha}$$

yields

$$\frac{d^2\tilde{u}}{d\theta^2} = -\tilde{u} + \frac{\beta}{\alpha}\left(\tilde{u} + \frac{1}{\alpha}\right)^2. \tag{5.5-7}$$

Earlier we noted that the bulge term in the potential energy was small relative to the $1/r$ term; we may expect the bulge to have a small effect on the motion. The averaging method, developed in appendix C, should provide a good approximate solution to equation (5.5-7).

Let us try the form

$$\tilde{u} = \mathcal{Q}(\theta) \cos[\theta + \varphi(\theta)]. \tag{5.5-8}$$

The analog of ω_0 in equation (5.5-7) is 1, and so the argument of the cosine is $1 \times \theta + \varphi(\theta)$. The analog of equation (C.8a) is

$$\frac{d\mathcal{Q}}{d\theta} = -\left\langle \frac{\beta}{\alpha}\left[\mathcal{Q}\cos[\cdots] + \frac{1}{\alpha}\right]^2 \sin[\cdots] \right\rangle$$

$$= 0.$$

The sine function enables us to integrate the squared expression; cancellation at the limits of integration, separated by 2π in angle, gives zero. The zero value can be seen also as a consequence of energy conservation. There cannot be any secular change in $\mathcal{Q}$, and so the average of $d\mathcal{Q}/d\theta$ over a cycle must be zero.

For φ, the analog of equation (C.8b) gives

$$\frac{d\varphi}{d\theta} = -\left\langle \frac{\beta}{\alpha}\left[\mathcal{Q}\cos[\cdots] + \frac{1}{\alpha}\right]^2 \frac{\cos[\cdots]}{\mathcal{Q}} \right\rangle$$

$$= -\frac{\beta}{\alpha}\left[\mathcal{Q}\langle\cos^3[\cdots]\rangle + \frac{2}{\alpha}\langle\cos^2[\cdots]\rangle + \frac{1}{\alpha^2\mathcal{Q}}\langle\cos[\cdots]\rangle\right]$$

$$= -\frac{\beta}{\alpha^2}$$

because the first and third terms average to zero. The solution for φ is

$$\varphi(\theta) = -\frac{\beta}{\alpha^2}\theta + \varphi(0),$$

and so

$$\frac{1}{r} = \frac{1}{\alpha} + \tilde{u}$$

$$= \frac{1}{\alpha} + \mathcal{Q}(0)\cos\left[\left(1 - \frac{\beta}{\alpha^2}\right)\theta + \varphi(0)\right]. \quad (5.5\text{-}9)$$

[Properly, we should add to the present solution the average of the small term in equation (5.5-7), as explained in the last paragraph of appendix C. That constant, however, would not contribute to the variation of r with angle or time, which is the goal of our study.]

The solution tells us that r does not repeat itself after an angular sweep of 2π. Rather, the sweep $\Delta\theta$ required for repetition is determined by

$$\left(1 - \frac{\beta}{\alpha^2}\right)\Delta\theta = 2\pi,$$

which yields

$$\Delta\theta = 2\pi \bigg/ \left(1 - \frac{\beta}{\alpha^2}\right)$$

$$\simeq 2\pi\left(1 + \frac{\beta}{\alpha^2}\right).$$

More than 2π is required for repetition:

$$\Delta\theta - 2\pi \simeq 2\pi\frac{\beta}{\alpha^2}, \quad (5.5\text{-}10)$$

and this implies a precession of the almost elliptical orbit, as sketched in figure 5.5-2.

Let us assess the angular shift. From the definition of β in equation (5.5-6), we have

$$\frac{\beta}{\alpha^2} = \frac{3}{4}\frac{M_{\text{ring}}}{M_\odot}\frac{R_\odot^2}{\alpha^2}.$$

When the eccentricity is small, as it certainly is for the earth, $\alpha \simeq \langle r \rangle$, the average value of the relative separation (which may be taken as either an angular or a time average, the distinction being insignificant for us). Thus

$$\Delta\theta - 2\pi \simeq (2\pi)\frac{3}{4}\frac{M_{\text{ring}}}{M_\odot}\left(\frac{R_\odot}{\langle r \rangle}\right)^2. \quad (5.5\text{-}11)$$

For the earth, $R_\odot/\langle r \rangle \simeq 4.7 \times 10^{-3}$, and so

5.5 THE OBLATE SUN

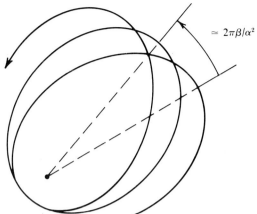

$\simeq 2\pi\beta/\alpha^2$

Figure 5.5-2 The precession induced by the solar oblateness (but much exaggerated).

$$(\Delta\theta - 2\pi)\bigg|_{\text{earth}} \simeq \frac{M_{\text{ring}}}{M_\odot} \times 10^{-4} \text{ radians/revolution.}$$

Since $M_{\text{ring}}/M_\odot$ is also small, the precession of the earth's orbit, per revolution, is tiny indeed. But the story is far from being at an end.

We change our focus from the earth to Mercury, the innermost planet, and to the year 1845. The French astronomer Leverrier had been asked to revise the calculations of Mercury's orbit and to make the comparison with observation. The perturbations by Venus and Jupiter are large and cause Mercury's orbit to precess substantially. Such effects can be calculated. Still, when Leverrier had included all the planetary perturbations, a discrepancy between theory and orbservation remained. Mercury's perihelion (the radial turning point close to the sun) precessed too rapidly by some 40 seconds of arc per century. Of course, that is a tiny amount—*too rapidly* is an exaggeration—but the discrepancy would not go away.

Certainly one could not fault Leverrier. Shortly afterward he carried out a similar analysis for the motion of Uranus. Again theory and observation did not mesh perfectly; this time Leverrier inferred a perturbation by an unknown planet. He calculated its location, and in 1846 Neptune was discovered, within one degree of where Leverrier had predicted. No, the discrepancy in Mercury's orbit could not be attributed to poor calculating.

The puzzle remained until 1915. Then Einstein, with his new geometric theory of gravitation, calculated a correction to all newtonian planetary orbits. Even when both bodies are spherically symmetric, general relativity predicts a precession. When the eccentricity is negligibly small, the precession is

$$(\Delta\theta - 2\pi)_{GR} = 2\pi \cdot 3 \cdot \frac{GM_\odot}{c^2\langle r\rangle} \quad (5.5\text{-}12)$$

in radians per revolution, where c is the speed of light. With the data for Mercury (and when the eccentricity $\epsilon = 0.2$ is included), the general relativistic prediction is 43 seconds of arc per century. The observational data available in 1915 gave a "Leverrier discrepancy" of 41.2 ± 2.1. General relativity brilliantly resolved the puzzle.

But in the early 1960s Robert Dicke* raised afresh the question, Might not a solar bulge be an important cause of precession? When the question had been asked earlier, the object had been to explain all 43 seconds of arc by a solar bulge, but the sun is too spherical for such a scheme to succeed. Dicke's question was more modest: Could the sun be responsible for, say, 10 percent of the 43 arc seconds? That would open a gap between observation and general relativity. Alternative theories of gravitation might fit the revised data better; in any event, any searching experimental test of general relativity is a welcome addition.

How much bulge do we need for a 10 percent effect? The ratio of the precession rates in equations (5.5-11) and (5.5-12) is to have the value 0.1:

$$2\pi \cdot \frac{3}{4} \cdot \frac{M_{\text{ring}}}{M_\odot} \left(\frac{R_\odot}{\langle r \rangle}\right)^2 \Big/ \left(2\pi \cdot 3 \frac{GM_\odot}{c^2 \langle r \rangle}\right) = 0.1.$$

Thus

$$\frac{M_{\text{ring}}}{M_\odot} = 0.4 \frac{GM_\odot}{c^2 \langle r \rangle} \left(\frac{\langle r \rangle}{R_\odot}\right)^2$$

$$\simeq 0.6 \times 10^{-4} \qquad (5.5\text{-}13)$$

when $\langle r \rangle$ is Mercury's distance. To convert this mass ratio to an observational bulge is not easy. If we say that the solar equatorial radius exceeds the polar radius by $\Delta R_\odot$, then M_{ring} comes from a band of mass of thickness $\Delta R_\odot$, length $2\pi R_\odot$, and width of order $R_\odot$. If the density in the band is similar to the general solar density—admittedly a questionable assumption—then the mass ratio is approximately equal to the volume ratio, and so

$$\frac{M_{\text{ring}}}{M_\odot} \simeq \frac{2\pi R_\odot^2 \Delta R}{(4\pi/3) R_\odot^3} \simeq \frac{3}{2} \frac{\Delta R_\odot}{R_\odot}.$$

Lower density in the band will require a larger $\Delta R_\odot$. When we compare this expression for the mass ratio with equation (5.5-13), we infer that an observational ratio $\Delta R_\odot / R_\odot$ of order 10^{-4} is needed to produce a solar contribution of 10 percent.

Dicke's question prompted new observations of the solar profile, both by him and by Hill and Stebbins. Reporting in the *Astrophysical Journal*, **200**, 471 (1975), Hill and Stebbins gave an intrinsic visual oblateness of

$$\frac{\Delta R_\odot}{R_\odot} = 0.96 \times 10^{-5} \times (1 \pm 0.7).$$

---

*R. H. Dicke, *The Theoretical Significance of Experimental Relativity* (Gordon and Breach, New York, 1964).

That ratio is an order of magnitude smaller than our estimate of what is needed. The authors also cite an inferred value for the mass ratio of

$$\frac{M_{\text{ring}}}{M_\odot} = (2.0 \pm 8.6) \times 10^{-6}.$$

The uncertainty is evidence for how difficult the experiment and data analysis actually are. The general inference to be drawn is this: The solar bulge is too small, by an order of magnitude, to produce even 10 percent of the observed precession. The explanation of Mercury's perihelion motion remains a triumph for Einstein's general relativity.

5.6 STABILITY OF CIRCULAR ORBITS

No orbital calculation can include all the forces that act on the orbiting object. Some are unpredictable: a burst of solar wind, a chance encounter with an asteroid. Many forces—small and of short duration—will perturb the orbit. Of necessity, there arises the question, Is the orbit stable?

In this section we investigate the stability of circular orbits. The force responsible for the basic circular orbit is, we will suppose, an attractive central force:

$$\mathbf{F} = -\frac{dU}{dr}\,\hat{\mathbf{r}}. \qquad (5.6\text{-}1)$$

With the potential energy $U(r)$ we can construct the effective potential energy

$$U_{\text{eff}}(r; L) = \frac{L^2}{2\mu r^2} + U(r) \qquad (5.6\text{-}2)$$

for the radial part of the motion. Energy conservation is expressed as

$$\tfrac{1}{2}\mu \dot{r}^2 + U_{\text{eff}}(r; L) = E.$$

Some sketches will give us an idea of what to expect in a detailed stability calculation. Two distinct potential energies,

$$-\frac{Gm_1 m_2}{r} \quad \text{and} \quad -\frac{\mathcal{K}}{r^4},$$

where $\mathcal{K}$ denotes some positive constant, are drawn in figure 5.6-1, together with the associated U_{eff}. The attractive force will give a circular orbit at some $r = r_0$ in both instances. Now perturb the orbit, for example, by giving the object a small radial kick. Adding a radial component to the instantaneous velocity will increase E but not change the angular momentum $\mathbf{L} = \mathbf{r} \times (\mu \dot{\mathbf{r}})$. In the gravitational situation, the subsequent radial motion will be bounded oscillation. Not so with the $1/r^4$ potential energy: there the object will depart from $r = r_0$, never to return. The sketches suggest that the curvature of U_{eff} at its local extremum will determine stability. We may expect the sign of $d^2 U_{\text{eff}}/dr^2$ to play a central role in the detailed analysis, to which we now turn.

182 TWO-BODY PROBLEM

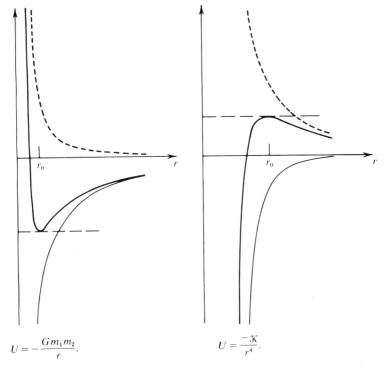

$$U = -\frac{Gm_1m_2}{r}. \qquad U = \frac{-\mathcal{K}}{r^4}.$$

Figure 5.6-1 Sketches of $U_{\text{eff}}(r; L)$ for two potential energies. The light line is $U(r)$, the dashed line is $L^2/(2\mu r^2)$, and the heavy line is U_{eff}. The horizontal dashed line denotes the total energy line that passes through the point where $dU_{\text{eff}}/dr = 0$.

How a Perturbation Evolves in Time

What happens in time if a circular orbit is nudged? To answer that, we can go back to Newton II and its time derivatives:

$$\frac{d}{dt}(\mu \dot{\mathbf{r}}) = \mathbf{F}. \qquad (5.6\text{-}3)$$

From equation (4.3-7) we have already extracted the velocity vector in plane polar coordinates:

$$\dot{\mathbf{r}} = \dot{r}\hat{\mathbf{r}} + r\dot{\theta}\hat{\boldsymbol{\theta}}.$$

To differentiate this, we need to know, from equations (4.3-8a and b), that

$$\frac{d\hat{\mathbf{r}}}{dt} = \dot{\theta}\hat{\boldsymbol{\theta}}$$

and

$$\frac{d\hat{\boldsymbol{\theta}}}{dt} = -\dot{\theta}\hat{\mathbf{r}}.$$

Then

$$\frac{d\dot{\mathbf{r}}}{dt} = \ddot{r}\hat{\mathbf{r}} + \dot{r}\dot{\theta}\hat{\boldsymbol{\theta}} + \frac{d(r\dot{\theta})}{dt}\hat{\boldsymbol{\theta}} - r\dot{\theta}^2\hat{\mathbf{r}}.$$

The terms in $\hat{\boldsymbol{\theta}}$ may be combined, so that Newton II becomes

$$\mu\left[(\ddot{r} - r\dot{\theta}^2)\hat{\mathbf{r}} + \frac{1}{r}\frac{d(r^2\dot{\theta})}{dt}\hat{\boldsymbol{\theta}}\right] = \mathbf{F}.$$

(This is the general form of Newton II in plane polar coordinates.) Taking the scalar product with $\hat{\mathbf{r}}$ gives us

$$\ddot{r} - r\dot{\theta}^2 = \frac{1}{\mu}\hat{\mathbf{r}}\cdot\mathbf{F}. \tag{5.6-4}$$

and the vector product with $\mathbf{r}$ gives

$$\frac{d(\mu r^2\dot{\theta})}{dt}\hat{\mathbf{r}}\times\hat{\boldsymbol{\theta}} = 0 \tag{5.6-5}$$

because, by equation (5.6-1), the force $\mathbf{F}$ points radially inward. The second equation is, of course, just angular momentum conservation.

Let us establish the initial circular orbit. We denote its radius by r_0 and its constant angular velocity by $\dot{\theta}_0$. The scalar product $\hat{\mathbf{r}}\cdot\mathbf{F}$ will appear often; let us abbreviate it in general by

$$\hat{\mathbf{r}}\cdot\mathbf{F} = F(r).$$

Despite its appearance, $F(r)$ is not the magnitude of $\mathbf{F}$, but, as defined, is a scalar product that will be negative because $\mathbf{F}$ is an attractive radially inward force. For the initial circular orbit, equations (5.6-4) and (5.6-5) imply

$$-r_0\dot{\theta}_0^2 = \frac{1}{\mu}F(r_0), \tag{5.6-6}$$

$$L_0 = \mu r_0^2\dot{\theta}_0; \tag{5.6-7}$$

the latter gives the initial value L_0 of the angular momentum.

Now we can imagine that the object has suffered some general perturbation; the situation at $t=0$ is sketched in figure 5.6-2. For $t \geq 0$, we may express the radial and angular variables as

$$r = r_0 + \delta r(t) \tag{5.6-8a}$$

and

$$\theta = \dot{\theta}_0 t + \delta\theta(t), \tag{5.6-8b}$$

where δr and $\delta\theta$ denote the departures from uniform circular motion. To inves-

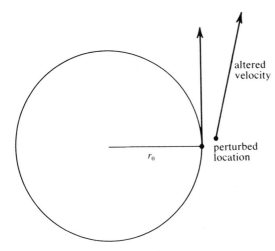

Figure 5.6-2 The object's location and velocity have been altered slightly from what they would be in uniform circular motion. (For simplicity's sake, however, all altered quantities are specified to lie in the original orbital plane.)

tigate stability is to ask, Does δr remain small as time goes on? So we turn to equation (5.6-4).

As before, angular momentum conservation enables us to eliminate $\dot{\theta}$ in terms of r. We may write

$$\mu r^2 \dot{\theta} = L_0 + \delta L; \qquad (5.6\text{-}9)$$

the change δL induced by the perturbation could be worked out from equations (5.6-8a and b), but we do not need that much detail. If we eliminate $\dot{\theta}$ in equation (5.6-4) and then substitute $r_0 + \delta r$ everywhere for r, we get

$$\frac{d^2(\delta r)}{dt^2} - \frac{(L_0 + \delta L)^2}{\mu^2 (r_0 + \delta r)^3} = \frac{1}{\mu} F(r_0 + \delta r). \qquad (5.6\text{-}10)$$

This equation is still exact, but intractable. At least during early times, the ratio $|\delta r / r_0|$ will be small—that is inherent in the notion of perturbation—and so we may expand expressions about their value when $r = r_0$. For the denominator of the second term, the binomial expansion gives

$$\frac{1}{(r_0 + \delta r)^3} = \frac{1}{r_0^3}\left(1 + \frac{\delta r}{r_0}\right)^{-3} = \frac{1}{r_0^3}\left(1 - 3\frac{\delta r}{r_0} + \cdots\right).$$

Equation (5.6-10) becomes

$$\frac{d^2(\delta r)}{dt^2} - \left(\frac{L_0^2 + 2L_0\,\delta L + \cdots}{\mu^2 r_0^3}\right)\left(1 - 3\frac{\delta r}{r_0} + \cdots\right) = \frac{1}{\mu}\left[F(r_0) + \frac{dF(r_0)}{dr}\delta r + \cdots\right].$$

Because we expand about a circular orbit, the term in $L_0^2 \times 1$ on the left equals the $F(r_0)$ term on the right; they drop out. If we then retain only the terms that are linear in δr or δL, we have

$$\frac{d^2(\delta r)}{dt^2} - \frac{2L_0\,\delta L}{\mu^2 r_0^3} + \frac{3L_0^2}{\mu^2 r_0^4}\delta r = \frac{1}{\mu}\frac{dF(r_0)}{dr}\delta r. \qquad (5.6\text{-}11)$$

5.6 STABILITY OF CIRCULAR ORBITS

The process that brought us here is an instance of a general technique: we "linearize" an exact equation, here equation (5.6-10), about the solution to the unperturbed problem, here r_0 and L_0. It is a basic stage in any "linear stability analysis."

Let us transfer the second and third terms in equation (5.6-11) to the right-hand side. Then we may combine the third term with dF/dr, which is $-d^2U/dr^2$, to arrive at the form

$$\frac{d^2(\delta r)}{dt^2} = -\frac{1}{\mu}\frac{d^2U_{\text{eff}}(r_0; L_0)}{dr^2}\delta r + \frac{2L_0\,\delta L}{\mu^2 r_0^3}. \quad (5.6\text{-}12) \star$$

Voilà! The curvature of U_{eff}, evaluated at the original circular orbit, pops out. If the perturbation is such that $\delta L = 0$, then certainly

$$\frac{d^2U_{\text{eff}}(r_0; L_0)}{dr^2} > 0 \quad (5.6\text{-}13)$$

guarantees oscillatory variation for δr. That confirms our analysis by sketch.

The more general situation is $\delta L \neq 0$. What then? The same criterion continues to hold; let's see why. The term in δL in equation (5.6-12) is just a constant term. We can incorporate its influence by writing

$$\delta r = \frac{2L_0\,\delta L}{\mu^2 r_0^3} \Big/ \left(\frac{1}{\mu}\frac{d^2U_{\text{eff}}}{dr^2}\right) + \delta\tilde{r}, \quad (5.6\text{-}14)$$

and so $\delta\tilde{r}$ will have to satsify only

$$\frac{d^2(\delta\tilde{r})}{dt^2} = -\frac{1}{\mu}\frac{d^2U_{\text{eff}}}{dr^2}\delta\tilde{r}.$$

Thus positive curvature is sufficient for stability whatever δL may be. What is novel about $\delta L \neq 0$ is this: Stable motion consists of oscillations $\delta\tilde{r}$ about a new circle, namely, about the circle given by r_0 plus the constant on the right in equation (5.6-14). [The new circle corresponds to the minimum in the new effective potential energy: $U_{\text{eff}}(r; L_0 + \delta L)$.]

If the curvature is negative, the displacement δr typically will grow with time. To assess the long-term behavior, however, we may need to go beyond the linear analysis, either by including more terms in the expansion or by looking at a graph of U_{eff}. Figure 5.6-3 shows the power of the graphical method.

The condition

$$\frac{d^2U_{\text{eff}}(r_0; L_0)}{d^2 r} > 0 \Rightarrow \text{stability} \quad (5.6\text{-}15)$$

is pictorially satisfying. The condition depends on both the force law $F(r)$ and the specific original circular orbit, through r_0 and L_0. Can we eliminate L_0 and get a statement about stability that contains only properties of the force law?

Equation (5.6-15) amounts to

$$-\frac{dF(r_0)}{dr} + \frac{3L_0^2}{\mu r_0^4} > 0.$$

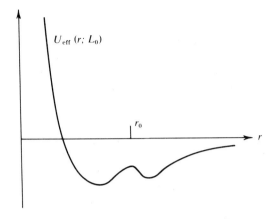

Figure 5.6-3 A circular orbit at $r = r_0$ is unstable. Nonetheless, if perturbed, the object will not escape to infinity, nor will it disappear into the origin. There will be radial oscillations of large but finite amplitude.

According to equations (5.6-6) and (5.6-7), the original circular orbit had

$$-\frac{L_0^2}{\mu^2 r_0^3} = \frac{1}{\mu} F(r_0).$$

We may use this to eliminate L_0^2:

$$-\frac{dF(r_0)}{dr} - \frac{3F(r_0)}{r_0} > 0. \qquad (5.6\text{-}16)$$

If this inequality holds for a circular orbit, the orbit will be stable.

We can apply this stability criterion to the earth's orbit in the presence of the solar bulge. In the solar equatorial plane we have

$$U(r) = -GM_\odot m_\oplus \left(\frac{1}{r} + \frac{\beta/3}{r^3} \right),$$

where

$$\beta = \frac{3}{4} \frac{M_{\text{ring}}}{M_\odot} R_\odot^2.$$

Then

$$F = -\frac{dU}{dr} = -GM_\odot m_\oplus \left(\frac{1}{r^2} + \frac{\beta}{r^4} \right).$$

The derivative dF/dr follows similarly, and we find

$$-\frac{dF}{dr} - \frac{3F}{r} = \frac{GM_\odot m_\oplus}{r^3} \left(1 - \frac{\beta}{r^2} \right).$$

If the sun were spherically symmetric, so that $\beta = 0$, then the right-hand side would be positive for all finite r, which would imply that every circular orbit was stable. The solar bulge makes $\beta > 0$, and therefore we have stability provided $r^2 > \beta$, that is, provided

$$r^2 > \left(\frac{3}{4} \frac{M_{\text{ring}}}{M_\odot}\right) R_\odot{}^2.$$

Since the mass ratio is tiny and since the earth's orbit is well outside the sun's surface, the orbit is stable. (More precisely, the orbit is guaranteed to be stable against random perturbations that leave the orbit in the solar equatorial plane, for these are the conditions that we assumed.)

Stability Criteria

In section 1.6 we noted that a minimum in the potential energy was the necessary and sufficient condition for a stable equilibrium. In this section we found that a minimum in an effective potential energy plays the decisive role. Why is there a difference? The first context envisages no motion when the mass is at the equilibrium location: a static equilibrium. The second context has motion: an angular momentum that is both nonzero and conserved. Each of the two stability requirements is correct in its context.

Yet a third context appeared in section 4.5: motion constrained by the constant angular velocity of the rotating wire hoop. When the angular velocity exceeds a critical value, the stable equilibrium is not at a potential minimum. Moreover, there is no meaningful effective potential energy (though something akin to such a function can be constructed in a hamiltonian approach). With a direct dynamical analysis, however, we did determine the stability of the off-axis location.

Stability requires that every small displacement lead to a restoring force. That is the basic criterion. How that criterion may be expressed in terms of a minimum in some function—if it may be so expressed at all—depends on the context. An impressive diversity of examples is provided by Leon Blitzer, *Am. J. Phys.*, **50**, 431 (1982).

5.7 THE ORBIT IN TIME

Thus far we have studied the shape of orbits and qualitative properties, such as boundedness and stability. For a sense of completeness, we should examine the orbit in time. Let us focus on the most important orbit, the keplerian ellipse. What we know about the shape can be summarized as follows:

$$\frac{1}{r} = \frac{1}{\alpha}(1 + \epsilon \cos \theta), \tag{5.7-1}$$

$$\alpha = \frac{L^2}{\mu G m_1 m_2}, \tag{5.7-2}$$

$$\epsilon = \left[1 + \frac{E}{G m_1 m_2/(2\alpha)}\right]^{1/2} < 1. \tag{5.7-3}$$

We derived these results in section 5.4, where the argument of the cosine was $\theta + \varphi_0$. Taking $\varphi_0 = 0$ amounts to orienting the line $\theta = 0$ along the direction for which r has its minimum value.

The first question might be, What is the temporal period of the motion? That is, how long does it take for the relative separation to go through one full cycle? To answer this question, we can adapt the travel-time scheme that led to equation (3.1-12). Here we start with equation (5.3-2), namely,

$$\frac{dr}{dt} = \left\{ \frac{2}{\mu} [E - U_{\text{eff}}(r; L)] \right\}^{1/2},$$

and we compute the period as

$$\text{Period} = 2 \int_{r_{\text{inner}}}^{r_{\text{outer}}} \frac{dr}{(2/\mu)^{1/2} [E - U_{\text{eff}}(r; L)]^{1/2}},$$

where the integration limits are the inner and outer turning points of the radial motion. After the substitution $u = 1/r$, we can look up the integral. The result is

$$\text{Period} = \left(\frac{\mu}{2}\right)^{1/2} \frac{Gm_1 m_2}{(-E)^{3/2}} \arcsin \frac{\alpha/r + \epsilon^2 - 1}{\epsilon(\alpha/r)} \Big|_{r_{\text{inner}}}^{r_{\text{outer}}}.$$

To make the argument of the inverse sine function easier to evaluate, both L^2 and $Gm_1 m_2$ have been expressed in terms of α and ϵ. Equation (5.7-1) gives $\alpha/r_{\text{inner}} = 1 + \epsilon$, and so the argument is $+1$. Similarly, $\alpha/r_{\text{outer}} = 1 - \epsilon$; the argument is -1. If we take $\arcsin 1 = \pi/2$ and $\arcsin(-1) = 3\pi/2$, to get a positive difference, which is then equal to π, we arrive at

$$\text{Period} = \frac{\pi}{2^{1/2}} \frac{\mu^{1/2} Gm_1 m_2}{(-E)^{3/2}}. \qquad (5.7\text{-}4)\bigstar$$

Although there are two conserved quantities, L and E, the period depends on only one. (This is a peculiarity of the inverse square force.)

To determine the angle θ as a function of time, we can look to angular momentum conservation:

$$\mu r^2 \dot{\theta} = L$$

implies

$$r^2 \, d\theta = \frac{L}{\mu} \, dt;$$

with equation (5.7-1), this becomes

$$(1 + \epsilon \cos \theta)^{-2} \, d\theta = \frac{L}{\mu \alpha^2} \, dt. \qquad (5.7\text{-}5)$$

At this point we should introduce the orbital period; in our context, it is far more natural than L or E. Equations (5.7-2) through (5.7-4) are an algebraic thicket from which we can extract the equality

$$\frac{L}{\mu a^2} = \frac{2\pi}{(1-\epsilon^2)^{3/2} P},$$

where P denotes the period. Inserting this into equation (5.7-5) gives

$$(1-\epsilon^2)^{3/2}(1+\epsilon\cos\theta)^{-2}\,d\theta = \frac{2\pi}{P}\,dt. \qquad (5.7\text{-}6)$$

The parameters that remain are solely the period and the eccentricity. They—and the time—determine the angular location as a function of time.

Although equation (5.7-6) can be integrated in closed form, the integration gives t as a complicated function of θ. There is no simple inversion to get θ as a function of t. Let us be satisfied with a sense of how θ varies with time: we suppose $\epsilon \ll 1$ and work to first order in ϵ. Then equation (5.7-6) becomes

$$(1 - 2\epsilon\cos\theta + \cdots)\,d\theta = \frac{2\pi}{P}\,dt,$$

which we may integrate as

$$\theta - 2\epsilon\sin\theta \simeq \frac{2\pi}{P}\,t,$$

if we take the zero of time to occur when $\theta = 0$. Look at this equation as

$$\theta \simeq \frac{2\pi t}{P} + 2\epsilon\sin\theta;$$

that is, θ equals $2\pi t/P$ plus a correction term of order ϵ. In the correction term, it suffices to use the zeroth-order form for θ, and so

$$\theta \simeq \frac{2\pi t}{P} + 2\epsilon\sin\frac{2\pi t}{P}. \qquad (5.7\text{-}7)$$

The angle does not change uniformly with time—we can see that in equation (5.7-5) or even earlier—but the angular location in the sky does have period P.

Four centuries have passed since Kepler discovered the elliptical orbit. Methods for extracting $\theta(t)$ have been invented by the score, methods much more ingenious and powerful than a series expansion. There is a touch of regret in seeing them superseded by a digital computer.

5.8 COMPENDIUM ON CENTRAL-FORCE MOTION

General results have been derived as we needed them, and so they are scattered throughout the chapter. Having them collected in one spot will be useful for future reference. Also, we will be able to survey the logical structure better.

The two bodies interact via forces that satisfy Newton III: $\mathbf{F}_2 = -\mathbf{F}_1$. Moreover, the forces act along the line joining the masses and depend, in magnitude, on only the relative separation r. The vectorial relative separation $\mathbf{r}$ then satisfies the equation

$$\frac{d}{dt}(\mu \dot{\mathbf{r}}) = -\frac{dU(r)}{dr}\hat{\mathbf{r}}, \qquad (5.8\text{-}1)$$

which characterizes central-force motion.

Equation (5.8-1) guarantees conservation of angular momentum,

$$L = \mu r^2 \dot{\theta}, \qquad (5.8\text{-}2)$$

and conservation of energy,

$$E = \tfrac{1}{2}\mu \dot{r}^2 + U_{\text{eff}}(r; L), \qquad (5.8\text{-}3)$$

where

$$U_{\text{eff}}(r; L) = \frac{L^2}{2\mu r^2} + U(r). \qquad (5.8\text{-}4)$$

The orbit lies in the plane perpendicular to **L**; polar coordinates in that plane are the natural dependent variables.

Orbit Shape

Equations (5.8-2) and (5.8-3) combine to give

$$\frac{d\theta}{dr} = \frac{L/(\mu r^2)}{\{(2/\mu)[E - U_{\text{eff}}(r; L)]\}^{1/2}}, \qquad (5.8\text{-}5)$$

which we may integrate to get $\theta = \theta(r)$, or vice versa.

Second-Order Equation for Orbit Shape

Let $u = 1/r$; take the reciprocal of equation (5.8-5) and apply $d/d\theta$ to get

$$\frac{d^2 u}{d\theta^2} = -u + \frac{\mu r^2}{L^2}\frac{dU}{dr}. \qquad (5.8\text{-}6)$$

The last element is to be expressed in terms of u after the operation d/dr has been performed.

Newton II in Polar Coordinates

$$\mu(\ddot{r} - r\dot{\theta}^2)\hat{\mathbf{r}} + \frac{1}{r}\frac{d(\mu r^2 \dot{\theta})}{dt}\hat{\boldsymbol{\theta}} = \mathbf{F}. \qquad (5.8\text{-}7)$$

Perturbation Analyses

Combine $\hat{\mathbf{r}} \cdot$ (Newton II), equation (5.8-2), and $r = r_0 + \delta r$ to get

$$\frac{d^2(\delta r)}{dt^2} = -\frac{1}{\mu}\frac{d^2 U_{\text{eff}}(r_0; L_0)}{dr^2}\delta r + \frac{2L_0\, \delta L}{\mu^2 r_0^3}, \qquad (5.8\text{-}8)$$

where r_0 is the radius of the original circular orbit and L_0 is its angular momentum.

We could also work with equation (5.8-6) to develop a perturbation analysis for the orbit shape directly. We would write $u = u_0 + \delta u$ and expand the last term about u_0 and L_0.

Orbits in an Attractive $1/r^2$ Force Field

When the attractive force is Gm_1m_2/r^2 in magnitude, the orbit is given by

$$\frac{1}{r} = \frac{\mu G m_1 m_2}{L^2}[1 + \epsilon \cos(\theta + \varphi_0)], \tag{5.8-9}$$

where

$$\epsilon = \left(1 + \frac{E}{|E_*|}\right)^{1/2} \tag{5.8-10}$$

and

$$E_* = -\frac{\mu(Gm_1m_2)^2}{2L^2}. \tag{5.8-11}$$

These expressions follow from equations (5.4-10) and (5.4-11) when the definition (5.4-6) is used to eliminate the length α. The energy E_* is the minimum energy possible at the given value of the angular momentum L; it corresponds to a circular orbit. The values of L and E determine the orbit's size and shape.

The Massive Limit

The preceding equations describe the relative separation **r**, sketched in figure 5.5-1. They hold for any value of the mass ratio m_1/m_2. A conceptual simplification occurs when $m_2 \gg m_1$. In the limit $m_1/m_2 \to 0$, the center of mass coincides with the massive object m_2. Then, in a reference frame in which the center of mass is at rest, mass m_2 is at rest. Mass m_1 is in orbit around m_2 at a distance **r**. The reduced mass $\mu \equiv m_1 m_2/(m_1 + m_2)$ becomes merely m_1. Thus in the massive limit, we may regard m_2 as being at rest and may replace μ everywhere by m_1, appropriate because "it is really a situation with m_1 in orbit about a fixed m_2."

WORKED PROBLEMS

WP5-1 A large mass M_0 holds a small mass m in *circular* gravitational orbit at radius r_0.

Problem. Use Newton II to express the kinetic energy KE_0 and the total energy E_0 in terms of the potential energy U_0.

192 TWO-BODY PROBLEM

The radial portion of equation (5.8-7) implies

$$mr_0\dot{\theta}^2 = \frac{GM_0 m}{r_0^2}; \qquad (1)$$

the reduced mass μ has been replaced by m on the left-hand side because $M_0 \gg m$. Since $v = r_0 \dot{\theta}$ in a circular orbit, we need only multiply equation (1) by $\frac{1}{2}r_0$ to get KE_0 on the left-hand side:

$$KE_0 = \tfrac{1}{2}m(r_0\dot{\theta})^2 = \frac{1}{2}\frac{GM_0 m}{r_0}$$

$$= \tfrac{1}{2}(-U_0), \qquad (2)$$

where $U_0 = -GM_0 m/r_0$.

For the total energy we have

$$E_0 = KE_0 + U_0 = +\tfrac{1}{2}U_0, \qquad (3)$$

a negative value, consistent with the circular orbit being a bound orbit.

WP5-2 Suppose the central mass in WP5-1 is suddenly increased to $M = fM_0$, where $f \simeq 2$. The inward pull on m will increase. Whereas the old pull could bend the velocity only into circular motion, the larger pull will cause the trajectory to move somewhat inward (at least for a while). We can expect the orbit to change from circular to elliptical. This suggestion is sketched in figure WP5-2.

Problem. Describe the new orbit in terms of f and parameters from the old circular orbit.

Section 5.4 provides us with three essential relations. In our context, equations (5.4-11), (5.4-6), and (5.4-10) read

$$\frac{1}{r} = \frac{1}{\alpha}[1 + \epsilon \cos(\theta + \varphi_0)], \qquad (1)$$

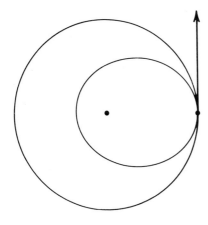

Figure WP5-2 The change from circular to elliptical orbit. The arrow represents the velocity at the instant when the central mass increases.

$$\alpha = \frac{L^2}{mGMm}, \qquad (2)$$

$$\epsilon = \left[1 + \frac{E}{GMm/(2\alpha)}\right]^{1/2}. \qquad (3)$$

A knowledge of L, the angular momentum, determines α, and then a knowldege of E, the energy, will fix ϵ. In short, the size and shape of the orbit are determined by L and E.

Because the change in central mass occurs instantaneously, neither the position of m nor its velocity changes during the transition. Consequently, the angular momentum $m\mathbf{r} \times \mathbf{v}$ does not change; for magnitudes, we have $L = L_0$. By the same reasoning, the kinetic energy does not change,

$$\mathrm{KE} = \mathrm{KE}_0,$$

but the potential energy becomes more negative:

$$U = -\frac{G(fM_0)m}{r_0} = fU_0.$$

Now we can express the new energy E in terms of the old quantities. With the aid of WP5-1, we may write

$$E = \mathrm{KE}_0 + fU_0 = -\tfrac{1}{2}U_0 + fU_0$$
$$= (2f-1)E_0. \qquad (4)$$

Next on the agenda is the eccentricity. We can combine equations (2) and (3) to eliminate α:

$$\epsilon = \left[1 + \frac{E}{(GMm)^2 m/(2L^2)}\right]^{1/2}.$$

If E and M had their original values, the ratio containing them would have to be -1, so that the eccentricity would be zero, describing a circular orbit. The actual ratio differs from -1 by the f factors in equation (4) and in $M = fM_0$. Thus

$$\epsilon = \left(1 - \frac{2f-1}{f^2}\right)^{1/2}$$
$$= \frac{f-1}{f}.$$

Because $f > 1$, the eccentricity is greater than zero but less than 1. The new orbit is indeed an ellipse.

The new value of parameter α follows from equation (2) as

$$\alpha = \frac{L_0^2}{mG(fM_0)m} = \frac{r_0}{f}$$

because the original circular orbit had $\alpha_0 = r_0$.

We can now go back to equation (1) and insert α and ϵ:

$$\frac{1}{r} = \frac{f}{r_0}\left[1 + \frac{f-1}{f}\cos(\theta + \varphi_0)\right].$$

At the instant of transition from M_0 to fM_0, the velocity was perpendicular to the radius vector $\mathbf{r}$ (because that is always true for a circular orbit). Thus the transition location is one of the extreme points on the elliptical orbit. Indeed, because we reasoned that the orbit would bend inward, it must be the location where mass m is farthest from the central mass. This fixes the orientation of the ellipse in space. If we choose our coordinates so that $\theta = 0$ at that instant, then

$$\left.\frac{dr}{d\theta}\right|_{\theta=0} = 0$$

implies $\sin\varphi_0 = 0$. Thus φ_0 equals 0 or π. Only $\varphi_0 = \pi$ will make $\theta = 0$ correspond to the largest r on the orbit, and so it is the correct choice.

This completes our description of the orbit. If $f = 2$, to pick a specific value, the eccentricity is $\epsilon = 0.5$, and the distance of closest approach is $r_0/[2(1 + 0.5)] = r_0/3$.

WP5-3 As a continuation of WP5-2, suppose mass m travels along the elliptical orbit until the polar angle θ has changed by $\pi/2$. Then the central mass drops from fM_0 to M_0, as instantaneously as before.

Problem. Describe qualitatively the subsequent motion of m.

First, let us work out the radial distance at the second transition. Since θ has increased by $\pi/2$, we have $\theta + \varphi_0 = 3\pi/2$, whose cosine is zero. Thus $r = r_0/f$ at that instant.

The radius was decreasing at that instant (because only at the two extremes of the orbit is $\dot{r} = 0$).

A graph of the effective potential energy now tells us the subsequent motion. Figure WP5-3 shows the original U_{eff}, which applies once again because the central mass is again M_0 and L has never changed. The original circular orbit lay at the minimum. The sudden shift from fM_0 back to M_0 finds m at r_0/f and with $\dot{r} < 0$. Thus the new energy line must lie above the U_{eff} curve at r_0/f, to describe the kinetic energy associated with $\dot{r}$. Mass m will continue to move inward, but only a little way. It will swing around M_0 and head outward. Because U_{eff} rises so slowly beyond r_0, the mass moves far out—many times r_0—before returning on its new elliptical orbit.

This problem and WP5-2 are not as fanciful as they may seem. When two galaxies collide, the stars in the outer portions experience a rapid increase in the gravitational attraction toward the galactic center (because the central mass is the superposition of two galactic nuclei). After the galactic nuclei have passed one another, the attraction drops quickly to its original value, more or less. The long outward swing that figure WP5-3 helped us to deduce may explain the curious appearance of "ring galaxies," where a distant ringlike concentration of stars is found. A theory based on galactic collisions is developed

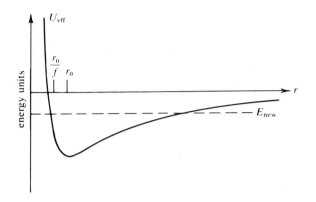

Figure WP5-3 The outward arc of the orbit will take m far beyond r_0.

in appealing fashion by Roger Lynds and Alar Toomre, *Astrophysical Journal*, **209**, 382 (1976).

PROBLEMS

⊤ **5-1** The context is two masses in interaction, as in section 5.1. We can compute the angular momentum of each mass relative to the center of mass. Is the sum of those two angular momenta equal to the angular momentum of the relative motion, $\mu \mathbf{r} \times \dot{\mathbf{r}}$? Prove your response.

Next, suppose we use a reference frame in which the center of mass is at rest. Is $\frac{1}{2}\mu\dot{\mathbf{r}} \cdot \dot{\mathbf{r}}$ then equal to the kinetic energy of the two masses?

5-2 A small particle of mass m is subject to the force

$$\mathbf{F} = -f(r)\hat{\mathbf{r}} - \gamma \mathbf{v},$$

where $-f(r)$ is an attractive central force and $-\gamma \mathbf{v}$ describes viscous damping. Compute the evolution of the angular momentum **L**.

Does the orbit lie in a single fixed plane?

If $f(r) = GMm/r^2$ (with $M \gg m$) and if the orbit is always almost circular, how do r and the speed v evolve in time (approximately)? If the speed increases, despite the viscous damping, where does the energy come from?

More about the "satellite paradox" may be found in B. D. Mills, Jr., *Am. J. Phys.* **27**, 115 (1959). The reprint collection, *Kinematics and Dynamics of Satellite Orbits* (American Institute of Physics, New York, 1963) contains Mills's paper and a fascinating selection of others, distributed over various levels of technicality.

5-3 *Proving the obvious, but learning something thereby.* A free particle of mass m starts as sketched in figure P5-3: initial velocity $\mathbf{v}_0 = -v_0 \mathbf{e}_1$ at initial location $x_0 \gg b$ and $y_0 = b$. Compute the shape of the orbit in polar coordinates by working out a differential equation of the form $d\theta/dr = f(r)$ from energy and

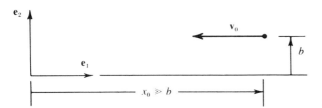

Figure P5-3

angular momentum conservation. Integrate your relationship to determine the orbit explicitly. (If all goes well, you will find the equation of a straight line. More importantly, of course, you will gain familiarity with a method and confidence in it.)

5.4 In section 5.3 we specified a potential energy of the form $U(r) = kr^2/2$ and derived the shape of the orbit under the further specification that the force was attractive: $k > 0$. Suppose, now, that the form of U persists, but the force is repulsive: $k = -|C|$.

(a) Sketch $U_{\text{eff}}(r; L)$ and determine *qualitatively* the kind of orbits that are possible.

(b) Determine the shape of the orbit $r = r(\theta)$ analytically, as a function of parameters L, E, $|C|$, and μ. Sketch the typical orbit with some attention to detail. Describe it verbally, too.

Is there any qualitative distinction between orbits for which $E < 0$ and those for which $E > 0$?

5-5 A particle of mass m is attracted to the origin by a force of magnitude $4\mathcal{K}/r^5$. The situation when the particle starts from spatial infinity is sketched in figure P5-5. Compute the particle's angular momentum (relative to the origin); then determine the shape of its orbit. Can you cast your result into the form

$$r = \frac{b}{\sqrt{2}} \coth \frac{\theta}{\sqrt{2}}?$$

After you have determined the trajectory analytically, that is, determined $r = r(\theta)$, sketch r as a function of θ with some faithfulness. Also sketch the effective potential energy U_{eff} and see whether you can understand from it—and

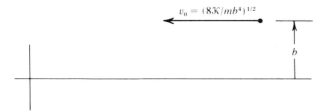

Figure P5-5

from the given data—why the orbital motion is so peculiar. (At least it strikes me as peculiar.)

5-6 As Halley's comet travels in its orbit, what is its maximum distance from the sun?

5-7 Through how large an angle (reckoned from the sun) does comet Encke move between two successive crossings of the earth's orbit (toward the sun and back out)? Data are in table 5.4-2; please answer in degrees. (You may approximate the earth's orbit by a circle.)

5-8 Suppose that $m_2 = 3m_1$ and that the relative separation has the polar description $r = r_0$ and $\theta = \dot{\theta}_0 t$, where r_0 and $\dot{\theta}_0$ are constants. The center of mass is at rest. Sketch the location of the masses, relative to the center of mass, as a function of time. Indicate specifically where the masses are at $t = \pi/(2\dot{\theta}_0)$.

What would the orbits look like if the relative separation vector **r** traced out an ellipse of eccentricity $\epsilon = 0.5$? (You may find it useful, as a first step, to solve for $\mathbf{r}_1$ and $\mathbf{r}_2$ in terms of **r** and **R**.)

5-9 In an elliptical orbit we want, let us say, a ratio

$$\mathcal{R} \equiv \frac{r_{max}}{r_{min}}$$

equal to 5. First sketch $\mathcal{R}$ as a function of the eccentricity ϵ; then determine the appropriate value of ϵ.

5-10 A mass m moves initially in a circular orbit about a fixed force center. The force is C/r^2 in magnitude and is directed radially inward. If C were suddenly reduced by a factor of 2, what would the new orbit be like? Where would the orbit lie among the qualitatively distinct trajectories shown in figure 5.4-1?

T **5-11** *Kepler's laws*. In the early 1600s Johannes Kepler announced three laws of planetary motion, quantitative descriptions extracted by his meticulous analysis of Tycho Brahe's observations. Here are the laws, together with their dates of discovery.

1. Every planet moves in an elliptical orbit with the sun at one focus. Date: c. 1605.
2. An imaginary line drawn from a planet to the sun sweeps out equal areas in equal times. Date: c. 1602.
3. The squares of the planetary periods are proportional to the cubes of the semimajor axes. Date: c. 1618.

Show, one by one, that these laws follow from our results in chapter 5 in the massive limit: $m_{planet}/M_\odot \to 0$. [For the third law, you may want to show that the semimajor axis, denoted a, equals $\alpha/(1 - \epsilon^2)$. Then equations (5.7-4) and (5.4-10) should suffice.]

5-12 A satellite orbits the earth in a circular orbit of radius r_0, traveling with (tangential) speed v_0 in this orbit. Then a rocket on the satellite fires such that the satellite acquires, *in a very short time*, a radial velocity component v_r which is equal in magnitude to v_0. Describe the subsequent motion of the satellite. Use a sketch of U_{eff} and explain what happens to the angular momentum, the potential energy, the kinetic energy, and the total energy. Does it matter whether the rocket fired radially inward or outward? (You may find one of your results counter-intuitive.)

5-13 A rocket ship of mass m coasts in a circular orbit (of radius r_0) around the (spherical) sun with speed v_0. Suddenly the rocket ship ejects a fraction f of its mass backward, opposite to its orbital motion, with relative velocity $(-\mathbf{v}_0)$. Thus the ejected mass is instantaneously at rest with respect to the sun. Assume that $m/M_\odot \ll 1$, surely a realistic assumption. The fraction f lies in the range $0 < f \le 0.1$.

(*a*) After the burst, what will be the shape of the new orbit and what will be its eccentricity?

(*b*) After the rocket has moved so that its angular position, reckoned from the sun, has changed by $\pi/2$ radians, what is its distance from the sun?

(*c*) Evaluate your expressions in parts (*a*) and (*b*) numerically for $f = 0.1$, insofar as numerical evaluation is possible.

5-14 A comet moves in a parabolic orbit in the plane of the earth's orbit. After traveling from a great distance, the comet crosses the earth's orbit at radial distance $r_\oplus$, swings around the sun at a minimum distance $fr_\oplus$, where $f < 1$, and crosses our orbit again, leaving forever. Take the earth's orbit to be exactly circular, and calculate the time the comet spends within that orbit. Express your answer as some fraction of a year. (The distance of closest approach will tell you something about the comet's angular momentum, and the earth's annual period will tell you something about $GM_\odot$, or vice versa.)

If the comet turns just outside Mercury's orbit, corresponding to $f = 0.47$, how long does the comet spend within our orbit?

Next, use your general expression to *estimate* the time that Halley's comet spends within our orbit.

5-15 The context is section 5.5, equations (5.5-1) and (5.5-2).

(*a*) Why must grad U, at point $\mathbf{r}$, lie in the plane defined by $\mathbf{r}$ and $\mathbf{e}_3$? (Recall that grad U is the *maximum* spatial rate of change of U, encompassing both direction and magnitude.)

(*b*) Compute grad U.

(*c*) Compute the torque $\mathbf{r} \times (-\text{grad } U)$, expressing the direction verbally and the magnitude algebraically.

(*d*) For which locations, relative to the oblate sun, is the torque zero?

(*e*) Would an earth orbit that starts in the solar equatorial plane (i.e., the initial location and velocity of the earth lie in that plane) remain in that plane?

5-16 When we examined the effect of the sun's equatorial bulge on planetary motion, we found the orbit equation (5.5-7).

(*a*) What combination of parameters would provide a small, *dimensionless* parameter for a series expansion for $\bar{u}$?

(*b*) Use a series expansion to solve equation (5.5-7) through first order in the small parameter. Do whatever is necessary to avoid (or to preclude) secular terms in θ, that is, terms that do not remain bounded as $\theta \to \infty$. (The "fundamental" for the series expansion, analogous to ωt, may not be $1 \times \theta$.) Express your final result for $1/r$ in terms of α, β, θ, and ϵ, where ϵ is the eccentricity of the zeroth-order orbit, i.e., the orbit if β were zero.

(c) By how much does the perihelion change (in angular location) per cycle of radial motion?

5-17 An attractive force has the form

$$\mathbf{F} \cdot \hat{\mathbf{r}} = -\frac{C}{r^2} \exp\left(-\frac{r}{\lambda}\right)$$

where C and λ are positive constants. Would a circular orbit be stable? Stability aside, can we have a circular orbit at any radius we want to choose?

5-18 The problem is to calculate the precession of an almost circular orbit in a central-force field:

$$\mathbf{F} \cdot \hat{\mathbf{r}} = -\frac{C}{r^2} \exp\left(-\frac{r}{\lambda}\right).$$

An almost circular orbit is equivalent to a circular orbit plus a perturbation. Try to find *two distinct methods* for doing the problem. The average orbit radius is specified to be less than the characteristic length λ of the force field.

5-19 The potential energy of a mass m, moving in a plane, is $U(r) = kr^2/2$, that is, a quadratic potential energy of the harmonic oscillator type. Suppose an initial circular orbit is perturbed a bit, a single gentle nudge. Compute in some approximate fashion the shape of the ensuing orbit.

Would you expect a noncircular orbit in the given force field to precess? Why?

Suppose the force were attractive and varied with distance as r^n for some integer n, positive or negative. For which integers would you expect the orbit *not* to precess?

Greater detail about forces that do *not* produce precession may be found in Lowell S. Brown, *Am. J. Phy.*, **46**, 930 (1978).

5-20 Suppose the perturbation in section 5.6 took the position vector or the velocity vector (or both) out of the original plane of motion. Why would the differential equation for δr continue to apply, without change in numerical content, but referred to the new plane of the orbital motion?

5-21 The ingredients are two masses, each of mass m, and a massless spring with spring constant k whose unstressed (no tension) length is ℓ_0. The masses are attached to the ends of the spring, and the system is set into circular motion with a total angular momentum L, computed relative to the center of mass. Without altering the angular momentum, we give the masses a slight kick (symmetrically) so that their separation oscillates about some average value. The situation is sketched in figure P5-21. Assume the oscillation amplitude to be "small," and calculate the frequency of oscillation. (Incidentally, note that we

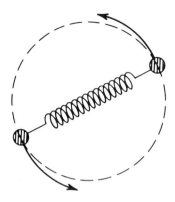

Figure P5-21

have here a classical model for estimating the vibrational spectrum of a rotating diatomic molecule.)

From parameters m, k, and ℓ_0 we can construct a characteristic angular momentum magnitude $m(k/m)^{1/2}\ell_0^2$. What is the limit of your oscillation frequency when L is much smaller than this characteristic size? When much larger?

5-22 The physical context is the same as in the first half of problem 4-6. Suppose you give mass m_2 a sharp but small tug and then let go. What is the immediate effect on mass m_1? What is the ensuing motion like? Be quantitative wherever you can.

5-23 In the northern hemisphere summer is longer than winter; the time from the vernal equinox to the autumnal equinox is about 7 days longer than the time from autumnal to vernal equinoxes. Can you use this information to determine the season in which the earth is closest to the sun? And to estimate the eccentricity of the earth's orbit around the sun? You may assume, as a reasonable approximation, that perihelion passage occurs at one of the solstices. (In actuality, it occurs about 2 weeks later.)

5-24 The sun loses mass at a rate $dM_\odot/dt \simeq -10^{-13}(M_\odot)$ per year because of the radial flux of photons and particles. Calculate the effect on the earth's orbit, taken to be and to remain essentially circular. In particular, what are the fractional changes in the earth's orbital radius and period in 1 year? (Does the radial symmetry imply that angular momentum is conserved?)

5-25 A spacecraft, designed to visit Mars, is launched so that it will drift out to the orbit of Mars while it moves 180° in azimuth, as reckoned from the sun. The velocity during launch is tangential to the earth's orbit (taken to be circular). The spacecraft never passes beyond the orbit of Mars (also taken to be circular).

(a) Sketch the spacecraft's orbit. Determine the values of the eccentricity and the parameter α. (Throughout this problem, you may omit the gravitational pull of the earth and of Mars.)

(b) How much energy (per kilogram of spacecraft) had to be given to the spacecraft during launch? (Remember that the spacecraft initially moved with the earth.)

(c) How much linear momentum (per kilogram) did the rocket engines provide during launch?

(d) Compare the value in part (c) with the momentum that would be required to send the spacecraft *radially* out to the orbit of Mars.

(e) How long does the trip from earth orbit to Mars orbit take (in years)?

Your calculated trajectory is an example of a *transfer orbit*, a relatively inexpensive way to move an object from one circular orbit to another.

CHAPTER SIX

ROTATING FRAMES OF REFERENCE

6.1 Vectors in a rotating frame of reference
6.2 Physics on a rotating table
6.3 The rotating earth
6.4 Foucault pendulum
6.5 The figure of the earth
6.6 A perspective on rotating frames

Were it not for the Coriolis effect, winds on the earth would rush directly from higher-pressure areas to lower-pressure ones, and no strong "highs" or "lows" could develop.... Our weather would be much less changeable than it is.

James McDonald
"The Coriolis Effect," **Scientific American,** May 1952

6.1 VECTORS IN A ROTATING FRAME OF REFERENCE

Rotating objects are everywhere in the universe. To begin with, we live on one. The sun rotates, and so do most other stars. The explosive death of a star often produces a pulsar, believed to be a collapsed, rapidly rotating neutron core, some 10 kilometers in diameter: a gigantic atomic nucleus. For that matter, an ordinary atomic nucleus is a rotating object. Some nuclear isotopes, particularly those of boron, lutetium, and erbium, are highly deformed from spherical shape because of their rotation. The dust grains in interstellar space are in rotation, turning around some 10^4 times per second.

The rotating objects in this brief catalog have fixed shapes and fixed mass distributions, to some good approximation. For purposes of analysis, we may regard each as a collection of mass points with *fixed relative separations*. That "rigidity" condition is most easily implemented if we use a frame of reference that rotates with the body.

That is one reason for studying vectors in a rotating frame of reference. Another reason is simply that our normal, everyday frame of reference—the walls of the room, the sweep of the landscape—is rotating relative to every inertial frame. The phenomena that play themselves out on the surface of the earth are seen by us from a rotating frame. A description that starts from Newton II in an inertial frame will change in appearance when it is couched in the language of the rotating frame. But let's plunge in now and develop that language.

We need to keep in mind two frames of reference: an inertial frame and the rotating frame. Figure 6.1-1 displays the two frames and shows how the location of a single object is described in them. Tersely,

$$\mathbf{r}_I = \mathbf{R} + \mathbf{r}, \tag{6.1-1}$$

where $\mathbf{R}$ denotes the origin of the rotating frame as seen from the inertial frame. The subscript I stands for inertial frame of reference. The rotating frame consists (mathematically) of three orthogonal unit vectors $\mathbf{e}_1$, $\mathbf{e}_2$, and $\mathbf{e}_3$. The position vector as viewed from the rotating frame may be expressed in terms of them:

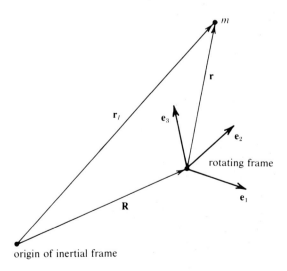

Figure 6.1-1 Descriptions of the location of mass m from both the inertial and the rotating frame.

$$\mathbf{r} = \sum_{i=1}^{3} x_i(t)\mathbf{e}_i(t)$$
$$= x_i \mathbf{e}_i. \qquad (6.1\text{-}2)$$

In the first line we have the components $x_i(t)$ of $\mathbf{r}$ relative to the rotating axes $\mathbf{e}_i(t)$. The second line is just shorthand for the first: whenever a subscript is repeated, we sum the subscript pair over the values 1, 2, and 3. Combining equations (6.1-1) and (6.1-2) gives us

$$\mathbf{r}_I = \mathbf{R} + x_i \mathbf{e}_i. \qquad (6.1\text{-}3)$$

To compute the velocity of the object, we differentiate with respect to time:

$$\frac{d\mathbf{r}_I}{dt} = \frac{d\mathbf{R}}{dt} + \frac{dx_i}{dt}\mathbf{e}_i + x_i \frac{d\mathbf{e}_i}{dt}. \qquad (6.1\text{-}4)$$

The origin of the rotating frame may itself move through space, and $d\mathbf{R}/dt$ accounts for that. The last term needs the most attention; we work on it in two stages.

Uniform Rotation

For a start, consider the special case of uniform rotation. This means rotation at a constant rate about an axis whose orientation is permanently fixed in the inertial frame. To characterize the rotation axis and rate, we may use a vector; the vector's direction specifies the rotation axis, and its magnitude, defined to be $2\pi/$(rotation period), gives the angular rate of rotation. The vector, denoted $\boldsymbol{\omega}$, is called the *angular velocity vector*.

Figure 6.1-2 shows how the rotation carries around a unit vector like $\mathbf{e}_3$. The tip of $\mathbf{e}_3$ moves in a circle about the rotation axis, specified by $\boldsymbol{\omega}$. The tangent to the circle (at the tip of $\mathbf{e}_3$) points in the direction given by $\boldsymbol{\omega} \times \mathbf{e}_3$. As the tip of $\mathbf{e}_3$ is carried around the circle, the change $\Delta\mathbf{e}_3$ in a short time Δt is along the tangent, and so

$$\Delta\mathbf{e}_3 \propto \boldsymbol{\omega} \times \mathbf{e}_3 \, \Delta t. \qquad (6.1\text{-}5)$$

The constant of proportionality is actually just 1; we can deduce that. The part of $\mathbf{e}_3$ that changes is the part perpendicular to $\boldsymbol{\omega}$, the part that lies in the plane of the circle. It has magnitude $\sin(\boldsymbol{\omega}, \mathbf{e}_3)$. In a time Δt, the portion of $\mathbf{e}_3$ in the plane will swing through an angle $2\pi[\Delta t/$(rotation period$)]$. The change $|\Delta\mathbf{e}_3|$ will be the radius, which is $\sin(\boldsymbol{\omega}, \mathbf{e}_3)$, times that infinitesimal angle. That product is precisely the magnitude of $\boldsymbol{\omega} \times \mathbf{e}_3 \, \Delta t$. In equation (6.1-5) we may replace $\propto$ by $=$, divided by Δt, and take the limit:

$$\frac{d\mathbf{e}_3}{dt} = \boldsymbol{\omega} \times \mathbf{e}_3. \qquad (6.1\text{-}6)$$

The expression is a tidy one; we will find its structure preserved even when the rotation is not uniform. To that general situation we turn now.

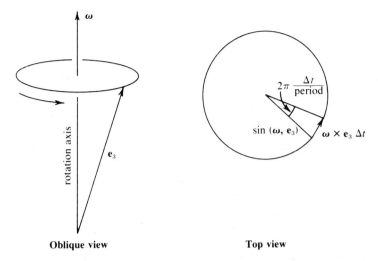

Figure 6.1-2 Uniform rotation.

General Rotation

Even when the rotation is not uniform, we can expect there to be a direction and a rotation rate that provide an *instantaneous* angular velocity $\boldsymbol{\omega}$. Each unit vector will be swung around $\hat{\boldsymbol{\omega}}$ by the rotation. Sketches like those in figure 6.1-2 will apply, at least for an infinitesimal time interval, and such an interval suffices for computing a derivative. Generalizing from equation (6.1-6), we can expect the derivative of $\mathbf{e}_i$, the ith unit vector, to have the form

$$\frac{d\mathbf{e}_i}{dt} = \boldsymbol{\omega} \times \mathbf{e}_i \qquad (6.1\text{-}7)\bigstar$$

always. With the powerful (but abstract) methods of appendix B we can confirm these expectations.

The derivative $d\mathbf{e}_i/dt$ arises as the difference of two vectors, $\mathbf{e}_i(t + \Delta t) - \mathbf{e}_i(t)$, and so it is a vector itself. We may express the derivative in terms of the unit vectors:

$$\frac{d\mathbf{e}_i}{dt} = \lim_{\Delta t \to 0} \frac{\mathbf{e}_i(t + \Delta t) - \mathbf{e}_i(t)}{\Delta t} = \Omega_{ij} \mathbf{e}_j, \qquad (6.1\text{-}8)$$

where the elements Ω_{ij} are functions of time that depend on how the frame is rotating. [Equation (6.1-8) is shorthand for the statement "the vector $d\mathbf{e}_1/dt$ has a component Ω_{11} along $\mathbf{e}_1$, a component Ω_{12} along $\mathbf{e}_2$, and a component Ω_{13} along $\mathbf{e}_3$," together with similar statements for the two other derivatives.] The orthogonality of the unit vectors is preserved by the rotation:

$$\mathbf{e}_i \cdot \mathbf{e}_k = \delta_{ik}. \qquad (6.1\text{-}9)$$

(The Kronecker delta δ_{ik}, defined in appendix B, is 1 if $i = k$ and is 0 if $i \neq k$.) Taking the scalar product of equation (6.1-8) with $\mathbf{e}_k$ and using equation (6.1-9) gives

$$\mathbf{e}_k \cdot \frac{d\mathbf{e}_i}{dt} = \Omega_{ij}\mathbf{e}_k \cdot \mathbf{e}_j = \Omega_{ij}\delta_{kj}$$

$$= \Omega_{ik}. \qquad (6.1\text{-}10)$$

This tells us that the elements Ω_{ik} indeed exist and are determined by the frame's rotation.

More important for us, however, is the *structure* of equation (6.1-8). The preservation of orthogonality and of unit length by the rotation restricts the elements Ω_{ij}. If we differentiate equation (6.1-9) and use equation (6.1-10), we get

$$\frac{d\mathbf{e}_i}{dt} \cdot \mathbf{e}_k + \mathbf{e}_i \cdot \frac{d\mathbf{e}_k}{dt} = 0,$$

$$\Omega_{ik} + \Omega_{ki} = 0.$$

And so we learn that Ω_{ik} is antisymmetric:

$$\Omega_{ki} = -\Omega_{ik}.$$

Consequently there are only three independent elements, for example, Ω_{12}, Ω_{13}, and Ω_{23}. This suggests that we can construct a vectorlike quantity from Ω_{ij} and then use it in place of Ω_{ij}. Let

$$\omega_k \equiv \tfrac{1}{2}\epsilon_{kmn}\Omega_{mn}.$$

(The permutation symbol ϵ_{kmn} is defined in detail in appendix B.) To get back to Ω_{ij}, we multiply by ϵ_{ijk} and use the identity developed in appendix B:

$$\epsilon_{ijk}\omega_k = \tfrac{1}{2}\epsilon_{ijk}\epsilon_{kmn}\Omega_{mn}$$

$$= \tfrac{1}{2}(\delta_{im}\delta_{jn} - \delta_{in}\delta_{jm})\Omega_{mn}$$

$$= \tfrac{1}{2}(\Omega_{ij} - \Omega_{ji}) = \Omega_{ij}.$$

Inserting this form for Ω_{ij} into equation (6.1-8) gives

$$\frac{d\mathbf{e}_i}{dt} = \epsilon_{ijk}\omega_k \mathbf{e}_j$$

$$= (\epsilon_{kij}\omega_k)\,\mathbf{e}_j$$

$$= \boldsymbol{\omega} \times \mathbf{e}_i. \qquad (6.1\text{-}11)$$

The second line suggests a vector product: a vector with components ω_k crossed with a vector whose ith component is 1 and whose other components are 0. Such a vector is $\mathbf{e}_i$, from which the third line follows.

In more detail, the vector $\boldsymbol{\omega}$ is

$$\boldsymbol{\omega} = \omega_k \mathbf{e}_k = \tfrac{1}{2}\epsilon_{kmn}\left(\mathbf{e}_n \cdot \frac{d\mathbf{e}_m}{dt}\right)\mathbf{e}_k \qquad (6.1\text{-}12)$$

and describes the instantaneous axis and rate of rotation. It is the instantaneous angular velocity. Thus, no matter how the axes rotate, the derivative $d\mathbf{e}_i/dt$ always can be written as the vector product of an angular velocity and the unit vector itself. We have a *structure* for the derivative that is thoroughly general.

Illustration

Let us see how the scheme works out in practice. Figure 6.1-3 shows a frame rotating relative to a set of vectors fixed in the inertial frame. The connections are

$$\mathbf{e}_1(t) = \cos\theta\,\hat{\mathbf{x}}_I + \sin\theta\,\hat{\mathbf{y}}_I,$$
$$\mathbf{e}_2(t) = -\sin\theta\,\hat{\mathbf{x}}_I + \cos\theta\,\hat{\mathbf{y}}_I,$$
$$\mathbf{e}_3(t) = \hat{\mathbf{z}}_I.$$

To compute $\boldsymbol{\omega}$, we need

$$\mathbf{e}_2 \cdot \frac{d\mathbf{e}_1}{dt} = \mathbf{e}_2 \cdot (-\sin\theta\,\dot{\theta}\hat{\mathbf{x}}_I + \cos\theta\,\dot{\theta}\hat{\mathbf{y}}_I)$$
$$= (\sin^2\theta + \cos^2\theta)\,\dot{\theta} = \dot{\theta};$$

$$\mathbf{e}_3 \cdot \frac{d\mathbf{e}_1}{dt} = 0;$$

$$\mathbf{e}_2 \cdot \frac{d\mathbf{e}_3}{dt} = 0.$$

Antisymmetry gives the other scalar products in terms of these three elements. Then equation (6.1-12) yields

$$\boldsymbol{\omega} = \tfrac{1}{2}\left[\epsilon_{k12}\dot{\theta} + \epsilon_{k21}(-\dot{\theta})\right]\mathbf{e}_k$$
$$= \dot{\theta}\mathbf{e}_3.$$

The angular velocity emerges just as it should. We may regard equation (6.1-7) as well established and may return to the context of equation (6.1-4).

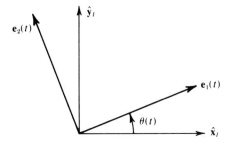

Figure 6.1-3 The axes $\mathbf{e}_1$ and $\mathbf{e}_2$ rotate relative to the inertial axes $\hat{\mathbf{x}}_I$ and $\hat{\mathbf{y}}_I$. The axes $\mathbf{e}_3$ and $\hat{\mathbf{z}}_I$ coincide and are perpendicular to the plane of the paper.

The Primary Derivation Resumed

The last term in equation (6.1-4) may now be written

$$x_i \frac{d\mathbf{e}_i}{dt} = x_i \boldsymbol{\omega} \times \mathbf{e}_i = \boldsymbol{\omega} \times (x_i \mathbf{e}_i)$$

$$= \boldsymbol{\omega} \times \mathbf{r}.$$

Thus equation (6.1-4) becomes

$$\frac{d\mathbf{r}_I}{dt} = \frac{d\mathbf{R}}{dt} + \frac{dx_i}{dt}\mathbf{e}_i + \boldsymbol{\omega} \times \mathbf{r}. \qquad (6.1\text{-}13)$$

For observers in the rotating frame, what is the object's velocity? The observers note that the components x_i change with time, but that is all. For them, the unit vectors $\mathbf{e}_i$ do not change; they are, after all, the observers' very frame of reference. The object's velocity, for them, is just the second term on the right in equation (6.1-13):

$$\mathbf{v} = \dot{x}_i \mathbf{e}_i. \qquad (6.1\text{-}14)$$

The outcome is the form

$$\frac{d\mathbf{r}_I}{dt} = \dot{\mathbf{R}} + \mathbf{v} + \boldsymbol{\omega} \times \mathbf{r}. \qquad (6.1\text{-}15)$$

We should pause to understand intuitively the meaning of the third term. Suppose the components x_i do not change, so that $\mathbf{v} = 0$. Then the object does not move relative to the axes of the rotating frame. Those axes, however, do move relative to the inertial frame, and so the object does move as seen by inertial observers. In short, the angular velocity $\boldsymbol{\omega}$ swings the position vector $\mathbf{r}$ around in space; the term $\boldsymbol{\omega} \times \mathbf{r}$ describes that contribution to the motion.

For Newton II, we need the acceleration as perceived in the inertial frame. Let us differentiate equation (6.1-13) in methodical fashion, invoking equations (6.1-14) and (6.1-7) where appropriate:

$$\frac{d^2\mathbf{r}_I}{dt^2} = \ddot{\mathbf{R}} + \ddot{x}_i \mathbf{e}_i + \dot{x}_i \frac{d\mathbf{e}_i}{dt} + \frac{d\boldsymbol{\omega}}{dt} \times \mathbf{r} + \boldsymbol{\omega} \times \left(\dot{x}_i \mathbf{e}_i + x_i \frac{d\mathbf{e}_i}{dt}\right)$$

$$= \ddot{\mathbf{R}} + \ddot{x}_i \mathbf{e}_i + \boldsymbol{\omega} \times \mathbf{v} + \frac{d\boldsymbol{\omega}}{dt} \times \mathbf{r} + \boldsymbol{\omega} \times (\mathbf{v} + \boldsymbol{\omega} \times \mathbf{r})$$

$$= \ddot{\mathbf{R}} + \mathbf{a} + 2\boldsymbol{\omega} \times \mathbf{v} + \boldsymbol{\omega} \times (\boldsymbol{\omega} \times \mathbf{r}) + \frac{d\boldsymbol{\omega}}{dt} \times \mathbf{r}. \qquad (6.1\text{-}16)$$

For observers in the rotating frame, the object's acceleration is $\ddot{x}_i \mathbf{e}_i$, and that is what $\mathbf{a}$ denotes.

Now we can write Newton II for a small object of mass m. We must start off in the inertial reference frame:

$$m \frac{d^2\mathbf{r}_I}{dt^2} = \mathbf{F}.$$

The equation of motion from the viewpoint of the rotating frame follows when we substitute from equation (6.1-16). If we rearrange to isolate **a**, the substitution yields

$$\mathbf{a} = \frac{\mathbf{F}}{m} - 2\boldsymbol{\omega} \times \mathbf{v} - \boldsymbol{\omega} \times (\boldsymbol{\omega} \times \mathbf{r}) - \ddot{\mathbf{R}} - \frac{d\boldsymbol{\omega}}{dt} \times \mathbf{r}. \qquad (6.1\text{-}17) \bigstar$$

We can best see the implications of this equation by looking at some examples.

6.2 PHYSICS ON A ROTATING TABLE

Let's imagine ourselves on a table that rotates uniformly with respect to an inertial frame. (The kind of merry-go-round that playgrounds often have works nicely. In lieu of that, you can paint a cartesian coordinate grid on a disk about 1 meter in diameter and rotate it; the grid greatly helps in putting yourself mentally into the rotating frame.) Figure 6.1-3 helps us specify the relations between unit vectors. The origins of the two frames coincide permanently, implying $\mathbf{R} = 0$ and hence $\ddot{\mathbf{R}} = 0$. The angular velocity has the constant value

$$\boldsymbol{\omega} = \omega \mathbf{e}_3 = \omega \hat{\mathbf{z}}_I, \qquad (6.2\text{-}1)$$

and so equation (6.1-17) reduces to

$$\mathbf{a} = \frac{\mathbf{F}}{m} - 2\boldsymbol{\omega} \times \mathbf{v} - \boldsymbol{\omega} \times (\boldsymbol{\omega} \times \mathbf{r}). \qquad (6.2\text{-}2)$$

For the object m, take a ball rolling on the table. Then both **r** and **v** lie in the plane of the table, perpendicular to $\boldsymbol{\omega}$. The term in **v**,

$$-2\boldsymbol{\omega} \times \mathbf{v} = -2\omega \mathbf{e}_3 \times \mathbf{v},$$

always generates an acceleration perpendicular to **v**. In this instance, the acceleration is "to the right" if we look along the ball's path, following the motion. We will call $-2\boldsymbol{\omega} \times \mathbf{v}$ the *Coriolis term,* honoring the French engineer Gaspard Coriolis, who recognized in 1835 that such a term must be included when the equation of motion is expressed in a rotating frame.

The last term in equation (6.2-2) may be expanded by the vector identity (B.12):

$$-\boldsymbol{\omega} \times (\boldsymbol{\omega} \times \mathbf{r}) = -[(\boldsymbol{\omega} \cdot \mathbf{r})\boldsymbol{\omega} - \omega^2 \mathbf{r}]$$

$$= +\omega^2 \mathbf{r} \qquad (6.2\text{-}3)$$

when **r** is perpendicular to $\boldsymbol{\omega}$, as it is here. The term generates a radially outward acceleration. To keep the ball at rest in the rotating frame, that is, to ensure $\mathbf{v} = 0$ and $\mathbf{a} = 0$, we need, according to equations (6.2-2) and (6.2-3), a radially inward force **F**. Of course. If we view things from the inertial frame, the ball "at rest" is moving uniformly in a circle, and an honest inward force is needed to maintain such motion.

210 ROTATING FRAMES OF REFERENCE

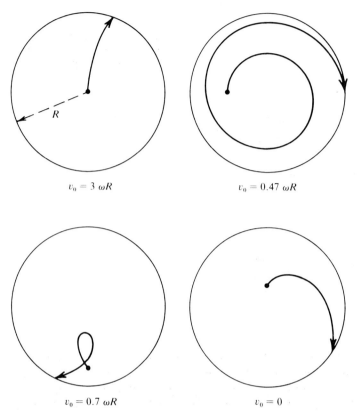

Figure 6.2-1 Force-free motion as seen in the rotating frame. The initial speed is indicated below each drawing. As seen from the inertial frame, every one of these trajectories is a straight line!

Figure 6.2-1 shows some trajectories based on equation (6.2-2) when $\mathbf{F} = 0$. There is no substitute, however, for rolling a ball yourself.

6.3 THE ROTATING EARTH

Suppose now that we want to describe motion at some spot on the earth's surface, for example, in the state of Connecticut. A natural set of directions for the unit vectors would be these:

$\mathbf{e}_1$ points south,

$\mathbf{e}_2$ points east, (6.3-1)

$\mathbf{e}_3$ points radially from the earth's center.

More specifically, $\mathbf{e}_3$ lies along the line from the earth's center to the Old State

House in Hartford, a central location, and south and east are as seen from there.

For almost every purpose, we may regard the earth's rotation rate ω as constant. Often we may ignore the orbital acceleration toward the sun, so that the earth's center may be regarded as a fixed point in an inertial frame. A glance at equation (6.1-17) tells us that the last term, the one in $d\omega/dt$, will vanish. If we want $\ddot{\mathbf{R}}$ to vanish also, we merely take the origin of our {south, east, radial} frame to be at the earth's center, as sketched in figure 6.3-1. The equation of motion then reduces to

$$\mathbf{a} = \frac{\mathbf{F}}{m} - 2\boldsymbol{\omega} \times \mathbf{v} - \boldsymbol{\omega} \times (\boldsymbol{\omega} \times \mathbf{r}). \tag{6.3-2}$$

The unit vectors rotate with the earth's angular velocity ω, whose magnitude is

$$\omega = \frac{2\pi}{60 \times 60 \times 24} = 7.3 \times 10^{-5} \text{ radians/second.}$$

The Plumb Bob

A heavy iron nut on the end of a string makes a good improvised plumb bob, the simple instrument used by masons and carpenters to build a wall that is vertical. Two forces act on the bob: the tension $\mathbf{T}$ from the string and $\mathbf{F}_{\text{gravity}}$, produced by the earth's mass distribution. When the bob is hanging at rest, equation (6.3-2) becomes

$$0 = \frac{\mathbf{T}}{m} + \frac{\mathbf{F}_{\text{gravity}}}{m} - \boldsymbol{\omega} \times (\boldsymbol{\omega} \times \mathbf{r}). \tag{6.3-3}$$

The orientation of the string *defines* the local direction of $\mathbf{g}$. For the magnitude of $\mathbf{g}$, there are two equivalent definitions: the magnitude of the acceleration from rest (when the string is snipped) or the force per unit mass that the

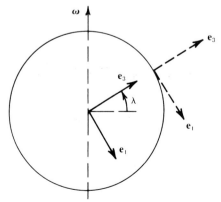

Figure 6.3-1 Natural directions for the unit vectors on the rotating earth. The vector $\mathbf{e}_2$ points east, perpendicularly into the paper. Shown also is the latitude λ of the point where a radial line along $\mathbf{e}_3$ emerges from the earth's surface.

string tension must supply to balance **g**. Either way, equation (6.3-3) tells us that the local, empirical **g** is given by the sum of the second and third terms:

$$\mathbf{g} = \frac{\mathbf{F}_{\text{gravity}}}{m} - \boldsymbol{\omega} \times (\boldsymbol{\omega} \times \mathbf{r}). \tag{6.3-4}$$

The vectors are sketched in figure 6.3-2. If the earth were a perfect sphere, then $\mathbf{F}_{\text{gravity}}/m$ would be purely radial and equal to $-(Gm_\oplus/r^2)\mathbf{e}_3$, but even then **g** would not be purely radial. The contribution to **g** from the earth's rotation has its maximum magnitude at the equator, where it is

$$|\boldsymbol{\omega} \times (\boldsymbol{\omega} \times \mathbf{r})|_{\text{max}} = \omega^2 R_\oplus$$
$$= 3.4 \times 10^{-2} \text{ newtons/kilogram}$$
$$\text{or meters/second}^2.$$

Since the average value of g is 9.8 in these units, the maximum effect is about 0.3 percent. The earth's aspherical shape also alters g as we proceed from pole to equator.

Motion near the Earth's Surface

Now consider objects in motion near the earth's surface. If air resistance is negligible and if there are no forces of propulsion—the rocket engine is off, the ball has left the hand—then the equation of motion follows from equations (6.3-2) and (6.3-4):

$$\mathbf{a} = \mathbf{g} - 2\boldsymbol{\omega} \times \mathbf{v}. \tag{6.3-5}$$

For motion within Connecticut, say, the direction and magnitude of **g** are essentially constant. Then we can integrate equation (6.3-5), term by term, once with respect to time:

$$\mathbf{v} = \mathbf{v}_0 + \mathbf{g}t - 2\boldsymbol{\omega} \times (\mathbf{r} - \mathbf{r}_0), \tag{6.3-6}$$

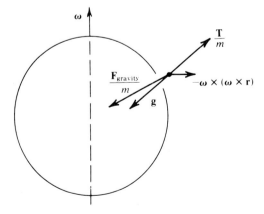

Figure 6.3-2 A plumb bob at rest in the rotating frame. The magnitude of $-\boldsymbol{\omega} \times (\boldsymbol{\omega} \times \mathbf{r})$ is ω^2 times the perpendicular distance from the rotation axis to the mass. The direction is perpendicular to the rotation axis and radially outward. [These properties always hold for the vector $-\boldsymbol{\omega} \times (\boldsymbol{\omega} \times \mathbf{r})$.]

where $\mathbf{v}_0$ and $\mathbf{r}_0$ are the initial velocity and position, respectively. That is, however, the only easy, exact integration. To integrate again, we would need to know $\mathbf{r} - \mathbf{r}_0$ as a function of time, which is precisely what we are trying to compute. The recalcitrant term in equation (6.3-6) is already of order ω, however, and so we may proceed by successive approximations. First we integrate equation (6.3-6) to zeroth order in ω to get

$$\mathbf{r} - \mathbf{r}_0 = \mathbf{v}_0 t + \tfrac{1}{2}\mathbf{g}t^2 + \mathcal{O}(\omega),$$

where the symbol $\mathcal{O}(\omega)$ has been written because the terms that we cannot evaluate are of order ω. Second, we substitute this lowest-order expression for $\mathbf{r} - \mathbf{r}_0$ back into equation (6.3-6):

$$\mathbf{v} = \mathbf{v}_0 + \mathbf{g}t - 2\boldsymbol{\omega} \times [\mathbf{v}_0 t + \tfrac{1}{2}\mathbf{g}t^2 + \mathcal{O}(\omega)].$$

Now we can integrate again:

$$\mathbf{r} - \mathbf{r}_0 = \mathbf{v}_0 t + \tfrac{1}{2}\mathbf{g}t^2 - 2\boldsymbol{\omega} \times (\tfrac{1}{2}\mathbf{v}_0 t^2 + \tfrac{1}{6}\mathbf{g}t^3) + \mathcal{O}(\omega^2). \tag{6.3-7}$$

Here we have $\mathbf{r} - \mathbf{r}_0$ correct to first order in ω.

If we need greater accuracy, we may substitute the current result into equation (6.3-6) and integrate again. It may seem too easy to be true, but the process is valid. Not unconditionally, of course. The expansion cannot literally be in powers of ω because ω is not dimensionless. If we compare the two terms in $\mathbf{v}_0$ in equation (6.3-7), we see that the second term is of order ωt times the first term. The same is true for the two terms in $\mathbf{g}$. Thus the expansion is in powers of the dimensionless product ωt. If the time of flight is 10 minutes or less, then $\omega t \leq 4.4 \times 10^{-2}$. The neglected term in equation (6.3-7), which is of order $(\omega t)^2$, represents a fractional error of order 10^{-3}, at most. Anyway, solving by successive approximations works as long as ωt is less than unity.

Let's be content with motion calculated correctly through order ωt; then we may use equation (6.3-7) as it stands. Two examples will illustrate its implications.

Fall from rest If we drop an object from rest, so that $\mathbf{v}_0 = 0$, the displacement as a function of time is

$$\mathbf{r} - \mathbf{r}_0 = \tfrac{1}{2}\mathbf{g}t^2 - \tfrac{1}{3}\boldsymbol{\omega} \times \mathbf{g}t^3 + \cdots.$$

The earth's rotation induces a deflection in the direction $-\boldsymbol{\omega} \times \mathbf{g}$. If we look at figure 6.3-2, we can work out that the deflection is eastward. What is the size of the deflection for a reasonable falling distance, 30 meters, say? We need the flight time t_F as a function of distance fallen h. Knowing that in zeroth order suffices:

$$h = \tfrac{1}{2}gt_F^2 + \mathcal{O}(\omega t_F \times gt_F^2),$$

and so $t_F \simeq (2h/g)^{1/2}$. Since $\mathbf{g} \simeq g(-\mathbf{e}_3)$, we can use figure 6.3-1 to deduce that

$$|\boldsymbol{\omega} \times \mathbf{g}| \simeq \omega g \, |\hat{\boldsymbol{\omega}} \times \mathbf{e}_3| = \omega g \cos \lambda.$$

214 ROTATING FRAMES OF REFERENCE

Now the deflection follows as

$$\text{Deflection eastward} \simeq \frac{1}{3}\omega g \left(\frac{2h}{g}\right)^{3/2} \cos \lambda. \tag{6.3-8}$$

For $h = 30$ meters and $\lambda = 41°45'58''$ N, the latitude of the Old State House in Hartford, Connecticut, whence $\cos \lambda = 0.746$, the eastward deflection is 2.68×10^{-3} meter, that is, 2.68 millimeters.

If the fall occurs at the equator, the eastward deflection is not difficult to understand. Let us look at the motion from an inertial frame centered on the earth's center. At the instant of release, the object has an angular momentum $L = m(R_\oplus + h)^2 \omega$ because it started at a distance $R_\oplus + h$ and was rotating with the earth's angular velocity ω. Since the gravitational force is purely radial, the angular momentum is conserved in time; that is, $mr^2 \dot\theta = \text{const}$. As the mass falls and r decreases, the angular velocity must increase: $\dot\theta > \omega$. This increase carries the object eastward relative to the line from the earth's center through the starting location (which rotates at the rate ω).

Lateral deflection of a projectile If the object has an initial velocity $\mathbf{v}_0$, then equation (6.3-7) gives us two terms of order ω. The second one, proportional to $-\boldsymbol{\omega} \times \mathbf{g}$, gives an eastward displacement. The first, proportional to $-\boldsymbol{\omega} \times \mathbf{v}_0$, is perpendicular to both $\boldsymbol{\omega}$ and $\mathbf{v}_0$ and so changes direction if $\mathbf{v}_0$ is changed. We should examine two instances.

If a snowball is thrown from the north pole, then $\mathbf{v}_0$ points south and upward, and so $-\boldsymbol{\omega} \times \mathbf{v}_0$ points westward (relative to the initial velocity). The term $\boldsymbol{\omega} \times \mathbf{g}$ is zero at the pole. Thus the snowball's path is deflected westward. From the viewpoint of the inertial frame, that makes sense: the earth rotates eastward under the snowball, and so, of course, the snowball lands somewhere to the west of the initial line of sight.

If a projectile is launched eastward at some middle northern latitude, say, then $-\boldsymbol{\omega} \times \mathbf{v}_0$ produces a deflection southward (and influences the vertical motion). To compute the southward deflection, we can take the scalar product of equation (6.3-7) with $\mathbf{e}_1$, the unit vector in the southerly direction:

$$\mathbf{e}_1 \cdot (\mathbf{r} - \mathbf{r}_0) = 0 + (\simeq 0) - \mathbf{e}_1 \cdot (\boldsymbol{\omega} \times \mathbf{v}_0) \, t^2 + 0. \tag{6.3-9}$$

The first zero arises because $\mathbf{v}_0$ points eastward and upward; the second, because the component of $\mathbf{g}$ in a southerly direction is of order ω^2. The last zero arises because the term in $-\boldsymbol{\omega} \times \mathbf{g}$ points eastward and so may influence the total distance traveled, but does not "deflect" a projectile launched eastward.

The triple product is unchanged under a cyclic permutation (as proved in appendix B), and so we can tidy up equation (6.3-9) by writing

$$\begin{aligned}
\mathbf{e}_1 \cdot (\mathbf{r} - \mathbf{r}_0) &= -\mathbf{v}_0 \cdot (\mathbf{e}_1 \times \boldsymbol{\omega}) t^2 \\
&= -\mathbf{v}_0 \cdot (-\omega \sin \lambda \, \mathbf{e}_2) t^2 \\
&= (\mathbf{e}_2 \cdot \mathbf{v}_0) \, \omega \sin \lambda \, t^2.
\end{aligned} \tag{6.3-10}$$

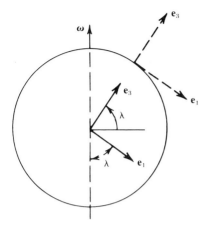

Figure 6.3-3 The angle between e_1 and $-\omega$ is also λ.

Figure 6.3-3 helps us see how the $\sin \lambda$ factor arises.

We need the time of flight to zeroth order only. If the projectile is launched at an angle α with respect to the horizontal, then $t_F = (2v_0 \sin \alpha)/g$, a result we derived in section 1.3. Inserting this into equation (6.3-10) and noting that $e_2 \cdot v_0 = v_0 \cos \alpha$ gives

$$\text{Southerly deflection} = \frac{4\omega v_0^3}{g^2} \sin \lambda \cos \alpha \sin^2 \alpha. \qquad (6.3\text{-}11)$$

A projectile fired eastward at the equator should—by symmetry—experience no lateral deflection. The limit of equation (6.3-11) as the latitude λ goes to zero is indeed zero, a reassuring check on the calculation.

But we had set out to calculate the deflection in middle northern latitudes. During World War I, the Germany army shelled Paris with its Big Bertha cannon from a distance of 76 miles. The shell left the cannon at a speed of 1700 meters/second, about 5 times the speed of sound. Thus the product $\omega v_0^3/g^2$ was 3.7 kilometers. The latitude of Paris is 49°, and the cannon was fired at a large elevation, $\alpha = 55°$. For the product of the trigonometric functions in equation (6.3-11), these values give 0.29, and so the entire right-hand side is 4 kilometers, in round figures. To be sure, the cannon was not fired due east, and air resistance may not be ignored. Still, under the conditions in the field, the deflection was about a mile. The ballistics experts had neatly included the effect in their calculations.

Since the times of Archimedes and Leonardo da Vinci, physics has been brought to bear on war. Sometimes it has been on the side of justice, at other times not. The legacy is a mixture of pride and shame. But probably we should not look at the history as a legacy of the profession. Rather, it is the history of individuals who happen to share an intellectual pursuit. As with all large groups of individuals, some members are admirable and others are not.

The Wind

If the earth were not rotating, the wind would blow directly from high pressure to low. As things really are, the pressure gradient does push the air toward low pressure, but the earth's rotation deflects the air to the right (in the northern hemisphere). The mathematical expression of this is the Coriolis term in equation (6.3-2), the term in $-\omega \times v$. The deflection builds up a circulating pattern, as sketched in figure 6.3-4.

If we begin with just the pressure gradient and the Coriolis effect, the steady-state motion emerges as circular. The Coriolis effect turns the wind perpendicular to the pressure gradient; the two effects cancel, except for a residual inward force to keep the air in circular motion around the low-pressure region.

In reality, there is also friction between the wind and the earth's surface. This dissipates the kinetic energy of the wind; to maintain a steady speed, the wind must head somewhat toward low pressure, picking up energy from the pressure gradient. The net result—a balance among pressure gradient, friction, and Coriolis effect—is the pattern sketched in figure 6.3-4.

In a sense, the Coriolis effect delays the motion of air from high to low pressure. The delay enables the processes that produce the pressure difference in the first place—primarily heating at the earth's surface and condensation—to build up a substantial pressure difference. The result can be a storm of great violence or at least major (and often welcome) changes in the weather.

Very close to the equator, the delay cannot occur. There the rotation vector ω is almost parallel to the earth's surface, and so a vector product with a surface wind produces a Coriolis effect that is predominantly vertical (either upward or downward). No delay means no chance for pressure differences to grow large, and hence little change in the weather can be initiated. The equatorial belt where so little occurs was well known to the sailors of bygone days: the Doldrums.

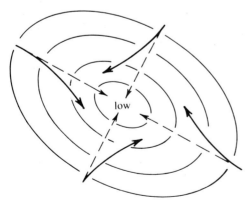

Figure 6.3-4 How the earth's rotation builds a circulating wind pattern. The light lines are contours of constant pressure; the heavy lines denote the actual wind pattern. The dashed lines indicate how the wind would blow if there were no Coriolis effect.

We think of hurricanes and typhoons as tropical storms. Yes, but they are not from the *very* equatorial tropics. These storms always originate outside an equatorial belt of some ±5° in latitude. The Coriolis effect is essential to their formation.

For a moment, let us go back to the beginning of this subsection. The most ideal wind pattern of all would be this: the Coriolis effect swings the wind around perpendicular to the pressure gradient and then balances the gradient with no remainder, so that the motion is in a straight line (on the rotating earth). The meteorologists call such an ideal wind the *geostrophic* wind, a euphonious name and an apt one, for the Greek roots mean "turned by the earth."

At altitudes of 700 meters and more, the friction with the earth's surface ceases to have a major effect. If the wind flows with a radius of curvature that is large—not near the eye of a hurricane—the wind is nearly geostrophic. Rather than blowing from high pressure to low, a steady wind moves at right angles, following the contours of constant pressure.

Analysis in detail relies on Newton II when it is couched in the language of the rotating frame, the contribution made by Gaspard Coriolis in 1835. A qualitative understanding goes back much further, to 1700 or so. George Hadley, a London lawyer, recognized the influence of the earth's rotation. All portions of the surface are carried around at the same angular rate. Those farthest from the rotation axis have the highest speed (relative to the inertial frame provided by the stars). Consider now, with Hadley, a wind blowing southward in the northern hemisphere. It is moving toward the equator, the region where the linear velocity of the earth's surface is largest. The earth will move eastward under the wind. From the viewpoint of a sailing ship on the earth, the wind is deflected westward—and becomes part of the trade winds.

6.4 FOUCAULT PENDULUM

For an introduction, we should return to the rotating table in section 6.2. Imagine a stand attached to the table, with an arm that projects over the table's center. From the arm hangs a pendulum, its point of support directly over the rotation axis. This is sketched in figure 6.4-1. First the table is set into uniform motion, and then the hitherto quiescent pendulum bob is given a sharp horizontal push. What is the bob's motion?

From the viewpoint of the inertial frame, the bob merely swings in a fixed plane. After all, the pendulum's point of support (located on the rotation axis) is not moving, and so the pendulum is "unaware" of any rotation.

For observers on the table, the perceived motion is quite different. A Coriolis effect accelerates the bob to the right (as they look along the path), and so the trajectory is a sequence of loops, as shown in figure 6.4-2. (The inertial observer might say to them, "You see that deflection because the table is moving under the swinging bob.") When the pendulum's oscillation frequency ω_0 is

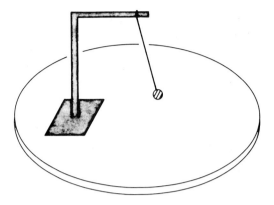

Figure 6.4-1 Pendulum on the rotating table.

high relative to the table's rotation frequency ω, the pendulum makes many swings while the table turns 1°, say. The loops are very narrow, and their orientation changes only slowly. To the observer on the table, the motion appears as a slow precession of the oscillation plane. Otherwise, the pendulum acts perfectly normally.

The rotating table is a fine model for the vicinity of the earth's north pole. From the table, we have a prediction of how a pendulum at the north pole would behave: The plane of oscillation would drift westward (as seen by us) while the earth moved eastward under it (as seen from an inertial frame).

Away from the pole, the analysis is more difficult and presents a challenge to the methods that we have developed. The forces on the pendulum bob are gravity and the tension **T** in the string. Equations (6.3-2) and (6.3-4) give us the

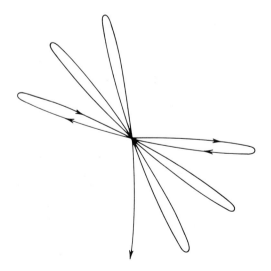

Figure 6.4-2 Looking downward in the rotating frame. The table, when viewed from the inertial frame, rotates counterclockwise.

6.4 FOUCAULT PENDULUM

equation of motion for an observer on the earth:

$$\mathbf{a} = \mathbf{g} + \frac{\mathbf{T}}{m} - 2\boldsymbol{\omega} \times \mathbf{v}. \tag{6.4-1}$$

The tension is directed along the string, away from the bob. Figure 6.4-3 shows why we may write

$$\mathbf{T} = + \frac{\mathbf{r}_s - \mathbf{r}}{\ell} T,$$

where $\mathbf{r}_s$ is the fixed upper terminus of the string and ℓ is the length of the pendulum.

With the unit vectors of equation (6.3-1), we can describe the bob's location by

$$\mathbf{r} = x\mathbf{e}_1 + y\mathbf{e}_2 + z\mathbf{e}_3.$$

The Coriolis term requires

$$\boldsymbol{\omega} \times \mathbf{v} = \begin{vmatrix} \mathbf{e}_1 & \mathbf{e}_2 & \mathbf{e}_3 \\ \omega_1 & 0 & \omega_3 \\ \dot{x} & \dot{y} & \dot{z} \end{vmatrix}$$

$$= \mathbf{e}_1(-\omega_3 \dot{y}) + \mathbf{e}_2(\omega_3 \dot{x} - \omega_1 \dot{z}) + \mathbf{e}_3(\omega_1 \dot{y}).$$

The angular velocity $\boldsymbol{\omega}$ lies in the $(\mathbf{e}_1, \mathbf{e}_3)$ plane, and so $\omega_2 = \mathbf{e}_2 \cdot \boldsymbol{\omega} = 0$.

To determine the precession of the pendulum's oscillation plane, we need to find the (x, y) components of motion. Taking the scalar product of equation (6.4-1) with $\mathbf{e}_1$ gives

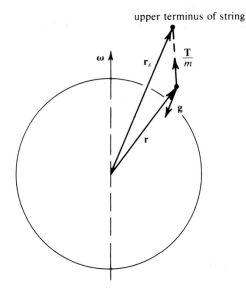

Figure 6.4-3 The tension is along the vector difference $\mathbf{r}_s - \mathbf{r}$, which has length ℓ, the pendulum's length.

$$\ddot{x} = (\simeq 0) + \mathbf{e}_1 \cdot \frac{\mathbf{r}_s - \mathbf{r}}{\ell} \frac{T}{m} + 2\omega_3 \dot{y}.$$

Because **g** is almost purely along $-\mathbf{e}_3$, having only a portion of order ω^2 along $\mathbf{e}_1$, the term $\mathbf{e}_1 \cdot \mathbf{g}$ is of order ω^2 and may be ignored here. The support vector $\mathbf{r}_s$ is along $\mathbf{e}_3$, and so $\mathbf{e}_1 \cdot \mathbf{r}_s = 0$, but $\mathbf{e}_1 \cdot \mathbf{r} = x$. The upshot is

$$\ddot{x} = -\frac{x}{\ell}\frac{T}{m} + 2\omega_3 \dot{y}, \tag{6.4-2}$$

correct through first order in ω.

The scalar product with $\mathbf{e}_2$ gives

$$\ddot{y} = 0 + \mathbf{e}_2 \cdot \frac{\mathbf{r}_s - \mathbf{r}}{\ell}\frac{T}{m} - 2(\omega_3 \dot{x} - \omega_1 \dot{z})$$

$$\simeq -\frac{y}{\ell}\frac{T}{m} - 2\omega_3 \dot{x}, \tag{6.4-3}$$

because the vertical motion $\dot{z}$ is small relative to a horizontal component like $\dot{x}$, provided that the amplitude of swing is small relative to the length ℓ.

Equations (6.4-2) and (6.4-3) are coupled equations in x and y—plus tension T. If the pendulum bob were at rest, we would have $T/m = g$; this was part of our definition of **g** in section 6.3. When the pendulum swings with small amplitude, the tension need vary only a little from the static value. To an adequate approximation, we may use $T/m = g$ in equations (6.4-2) and (6.4-3). They become the pair

$$\ddot{x} = -\omega_0^2 x + 2\omega_3 \dot{y}, \tag{6.4-4a}$$

$$\ddot{y} = -\omega_0^2 y - 2\omega_3 \dot{x}, \tag{6.4-4b}$$

where $\omega_0^2 = g/\ell$.

Now to solve these equations. If the terms in ω_3 were absent, the solution would be oscillation along a straight line. We start the pendulum so that such motion will ensue. We could write such a solution as

$$x = A_x \cos(\omega_0 t + \varphi_0), \tag{6.4-5a}$$

$$y = A_y \cos(\omega_0 t + \varphi_0), \tag{6.4-5b}$$

so that

$$x\mathbf{e}_1 + y\mathbf{e}_2 = (A_x \mathbf{e}_1 + A_y \mathbf{e}_2) \cos(\omega_0 t + \varphi_0). \tag{6.4-6}$$

Figure 6.4-4 shows how the position in the plane would oscillate along a fixed line.

In equations (6.4-4a and b), the ratio of the terms that depend on ω_3 to the linear restoring terms is of order

$$\frac{2\omega_3 \dot{y}}{\omega_0^2 x} \simeq \frac{2\omega_3 \omega_0 y}{\omega_0^2 x} \simeq \frac{2\omega_3}{\omega_0} \tag{6.4-7}$$

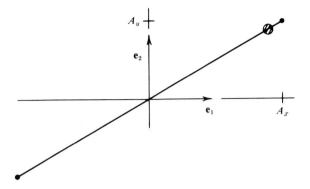

Figure 6.4-4 If the earth did not rotate, equation (6.4-6) would describe the motion. The bob's position along the diagonal line is proportional to $\cos(\omega_0 t + \varphi_0)$.

because the velocities are of order ω_0 times the displacements and the x and y displacements are typically similar in size. A pendulum length of $\ell = 10$ meters is common. For such a pendulum, $\omega_0 = \sqrt{g/\ell} \simeq 1$ radian/second, and the ratio in equation (6.4-7) is of order 10^{-4}. The ratio is small: the Coriolis terms can shift the plane of oscillation only a tiny amount during any one oscillation period. To solve for the precession, perhaps we can use the averaging method.

We allow the amplitudes and the phase constant in equations (6.4-5a and b) to vary with time:

$$x = \mathcal{A}_x(t) \cos[\omega_0 t + \varphi(t)],$$
$$y = \mathcal{A}_y(t) \cos[\omega_0 t + \varphi(t)].$$

Using equation (C.8a) with $\mathcal{A}_x$ in place of $\mathcal{A}$ yields

$$\frac{d\mathcal{A}_x}{dt} = -\frac{1}{\omega_0} \langle 2\omega_3(-\omega_0 \mathcal{A}_y \sin[\cdots])\sin[\cdots]\rangle$$
$$= \omega_3 \mathcal{A}_y \qquad (6.4\text{-}8)$$

because the average of $\sin^2[\cdots]$ is 1/2. Similarly,

$$\frac{d\mathcal{A}_y}{dt} = -\omega_3 \mathcal{A}_x. \qquad (6.4\text{-}9)$$

To generate an equation for φ, we may take $\mathcal{A}_x$ as $\mathcal{A}$ in equation (C.8b):

$$\frac{d\varphi}{dt} = -\frac{1}{\omega_0}\left\langle 2\omega_3(-\omega_0 \mathcal{A}_y \sin[\cdots])\frac{\cos[\cdots]}{\mathcal{A}_x}\right\rangle$$
$$= 0$$

because $2\sin[\cdots]\cos[\cdots] = \sin 2[\cdots]$, and that averages to zero. The phase function does not change, but the amplitudes do, in a coupled fashion.

To solve for $\mathcal{A}_x$ and $\mathcal{A}_y$, we can differentiate equation (6.4-8) and eliminate $d\mathcal{A}_y/dt$ via equation (6.4-9):

$$\frac{d^2 \mathcal{Q}_x}{dt^2} = -\omega_3^2 \mathcal{Q}_x.$$

A general solution to this equation is

$$\mathcal{Q}_x = A \cos(\omega_3 t + \alpha).$$

The corresponding solution for $\mathcal{Q}_y$ comes algebraically from equation (6.4-8):

$$\mathcal{Q}_y = \frac{1}{\omega_3} \frac{d\mathcal{Q}_x}{dt} = -A \sin(\omega_3 t + \alpha).$$

Constants A and α can be determined from the initial conditions, but we do not need to bother.

The averaging method produces the approximate solution

$$x = A \cos(\omega_3 t + \alpha) \cos[\omega_0 t + \varphi(0)],$$
$$y = -A \sin(\omega_3 t + \alpha) \cos[\omega_0 t + \varphi(0)],$$

which is best looked at as

$$x\mathbf{e}_1 + y\mathbf{e}_2 = [\cos(\omega_3 t + \alpha) \mathbf{e}_1 - \sin(\omega_3 t + \alpha) \mathbf{e}_2] A \cos[\omega_0 t + \varphi(0)].$$

The quantity in square brackets is a unit vector that rotates with angular velocity ω_3. Figure 6.4-5 provides a sketch and shows that the vector drifts westward (when $\omega_3 > 0$, as in the northern hemisphere). We deduce a westward precession of the oscillation plane; the angular rate is

$$\omega_3 = \mathbf{e}_3 \cdot \boldsymbol{\omega} = \omega \sin \lambda \tag{6.4-10}$$

for a pendulum at latitude λ. The precession rate is maximum at the north pole, is zero at the equator, and varies smoothly in between. Our methods have met the challenge.

A chance observation led Foucault to the pendulum experiment that bears his name. While building a better driving mechanism for a telescope, he clamped a long flexible rod in a lathe. The rod vibrated readily and maintained

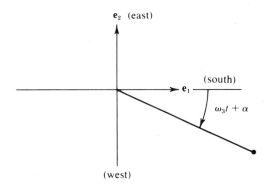

Figure 6.4-5 How the plane of oscillation precesses.

its plane of oscillation, more or less, as the rod was turned very slowly in the lathe. That observation gave Foucault the idea. In his cellar, he hung a massive bob from a wire 2 meters long, set it into motion, and at 2 a.m. on January 8, 1851, watched its plane of oscillation rotate relative to the cellar walls. The pendulum was soon scaled up to 11 meters in the Paris observatory, and the experiment quickly spread across the world.

A little caution should be exercised when interpreting Foucault's experiment. The pendulum's plane of oscillation precesses (relative to a grid on the floor, say), and we infer that the earth rotates. But relative to what? Our theory tells us that the earth must be rotating relative to an inertial frame of reference. The stars in our neighborhood of the Milky Way provide a good approximation to such a frame. It would be unwarranted, however, to infer that the earth rotates—period. That inference would require a satisfactory definition of "absolute" rotation, and no operational definition, at least, is yet available. The classic critique of absolute motion is Ernst Mach's, as expounded in *The Science of Mechanics,* first published in 1883. A century later, it is still provocative reading.

6.5 THE FIGURE OF THE EARTH

The earth's rotation induces an equatorial bulge. How large is it?

When Newton studied the question, he imagined two tunnels bored through the earth, as in figure 6.5-1, and filled with water. Since the polar and equatorial oceans already communicate with one another via the surface water, the tunnels would not disturb the equilibrium or alter any shape. The pressure at the top of each water column is atmospheric pressure; since the columns join at the bottom, their pressures are equal there. The values of g along the equatorial column (at $\theta = \pi/2$) differ from those along the polar column (at $\theta = 0$)

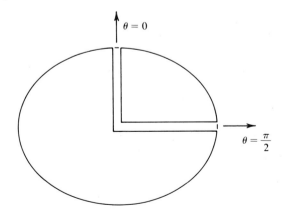

Figure 6.5-1 Newton's strategem.

because of the earth's rotation and bulge. To get the same change in pressure between center and surface, the two columns must be of different height. We can turn that observation into a calculation of the bulge.

First we need the connection between pressure and g. Imagine a short cylinder of water in one column. The force pushing upward is $p(r)A$, where $p(r)$ is the pressure at radial distance r and A is the cylinder's cross-sectional area. The downward force is the pressure at the cylinder's top, which is $p(r + \Delta r)A$, and the effect of g, namely $g\rho A \, \Delta r$, where ρ is the fluid density. The column is in equilibrium, and so we must have

$$p(r + \Delta r)A + g\rho A \, \Delta r = p(r)A.$$

We may cancel A throughout, transfer $p(r)$ to the left and $g\rho \, \Delta r$ to the right, divide by Δr, and pass to the limit:

$$\frac{dp}{dr} = -g\rho. \tag{6.5-1}$$

The basic requirement is that the pressure difference between bottom and top be the same for both columns. That means the integral of equation (6.5-1) from bottom to top must be the same for both columns, because the integral of the left-hand side is $p_{\text{top}} - p_{\text{bottom}}$. Since we only imagine the tunnels, we may imagine them filled with fluid of uniform density. Then ρ is a constant factor, and the basic requirement is translated into the equality

$$\int_0^{R_p} g(r, 0) \, dr = \int_0^{R_{\text{eq}}} g\left(r, \frac{\pi}{2}\right) dr, \tag{6.5-2}$$

where R_p and R_{eq} are the polar and equatorial radii, respectively. Since g varies with angle from the north pole, as well as with radius, we have to distinguish $g(r, 0)$ along the polar axis from $g(r, \pi/2)$ along the equatorial tunnel.

Equation (6.5-2) is exact, but we cannot infer the difference in radii, $\Delta R = R_{\text{eq}} - R_p$, from it as it stands. The simplest approximation that will give us an answer is this: We include the rotation term $\omega^2 r$ in g, but ignore the influence of the bulge itself on g. In short, we write

$$g\left(r, \frac{\pi}{2}\right) \simeq g(r, 0) - \omega^2 r. \tag{6.5-3}$$

Now we can express the right-hand side of equation (6.5-2) as

$$\int_0^{R_p} g(r, 0) \, dr + \int_{R_p}^{R_{\text{eq}}} g(r, 0) \, dr - \int_0^{R_{\text{eq}}} \omega^2 r \, dr.$$

The first term here cancels with the term on the left in equation (6.5-2). We can write the outcome as

$$g(R, 0) \, \Delta R \simeq \frac{\omega^2 R^2}{2},$$

where R stands for either R_p or R_{eq} or their average when the distinction is unimportant. The fractional difference in radii is

$$\frac{\Delta R}{R} \simeq \frac{\omega^2 R}{2g(R, 0)} \simeq 1.5 \times 10^{-3}, \tag{6.5-4}$$

given $R = 6.38 \times 10^3$ kilometers. The absolute size of ΔR is then about 10 kilometers. This estimate is correct in order of magnitude—that is the measure of our success—but it is too small by about a factor of 2.

Newton's estimate was better because he included the influence of the bulge itself. Let us represent the earth gravitationally by a sphere plus a mass ring at the equator. The sphere makes the same contribution to $g(r, 0)$ as to $g(r, \pi/2)$. That is not so for the ring. When we studied a mass ring in section 1.6, we reasoned that the ring would exert a pull inward along the symmetry axis. That translates into a positive contribution to $g(r, 0)$. In the plane of the ring, however, we found a force radially outward; that becomes a negative contribution to $g(r, \pi/2)$. Thus the ring enhances $g(r, 0)$ and diminishes $g(r, \pi/2)$. Since the $\omega^2 r$ term also diminishes $g(r, \pi/2)$, the bulge and rotation reinforce each other: we should expect a $\Delta R/R$ ratio larger than our early estimate, and that is correct.

Geopotential

With another line of approach, we can improve our estimate. To begin, note that the surface of the ocean is perpendicular to the local $\mathbf{g}$ vector. (We have to average the waves, of course, and the tides.) If there were a component of $\mathbf{g}$ parallel to the surface, it would cause water to flow, which is contrary to equilibrium. Next, recall that the gradient of a function is perpendicular to contours along which the function is constant; we worked that out in section 1.4. If we can find a function whose gradient is $\mathbf{g}$, then a contour of "function = constant" will describe the ocean's surface. This will provide the difference in radii and a lot more, too.

Equation (6.3-4) gives us the structure of $\mathbf{g}$:

$$\mathbf{g} = \frac{\mathbf{F}_{\text{gravity}}}{m} - \boldsymbol{\omega} \times (\boldsymbol{\omega} \times \mathbf{r}).$$

Figure 6.5-2 shows how we can express the double cross product in terms of the problem's natural unit vectors, $\hat{\mathbf{r}}$ and $\hat{\boldsymbol{\theta}}$. Thus

$$\mathbf{g} = \frac{\mathbf{F}_{\text{gravity}}}{m} + \omega^2 r \sin\theta \, (\sin\theta \, \hat{\mathbf{r}} + \cos\theta \, \hat{\boldsymbol{\theta}}). \tag{6.5-5}$$

A gradient form for the first term certainly exists:

$$\frac{\mathbf{F}_{\text{gravity}}}{m} = -\text{grad}\, \frac{U}{m} = -\text{grad}\, \tilde{U},$$

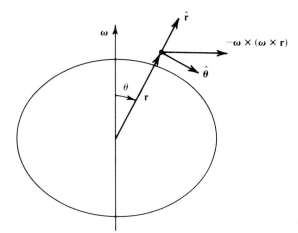

Figure 6.5-2 Expressing $-\boldsymbol{\omega} \times (\boldsymbol{\omega} \times \mathbf{r})$ in terms of $\hat{\mathbf{r}}$ and $\hat{\boldsymbol{\theta}}$.

where $\tilde{U}$ denotes the gravitational potential energy per unit mass. For the portion in ω^2, we need a function $f(r, \theta)$ such that

$$\frac{\partial f}{\partial r} = \omega^2 r \sin^2 \theta,$$

$$\frac{1}{r} \frac{\partial f}{\partial \theta} = \omega^2 r \sin \theta \cos \theta.$$

Integrating the first equation with respect to r implies

$$f = \tfrac{1}{2}\omega^2 r^2 \sin^2 \theta + \text{(function of } \theta \text{ only)}.$$

Substitution in the second equation tells us that the supplementary function may be only a constant. Since the constant disappears when the gradient is formed, we may set it to zero to start with. The conclusion is that equation (6.5-5) may be written as

$$\mathbf{g} = -\text{grad}\,(\tilde{U} - \tfrac{1}{2}\omega^2 r^2 \sin^2 \theta).$$

The quantity in parentheses is what we are after. It is an effective potential energy per unit mass, and so we denote it by

$$\tilde{U}_{\text{eff}}(\mathbf{r}; \omega) = \tilde{U}(\mathbf{r}) - \tfrac{1}{2}\omega^2 r^2 \sin^2 \theta. \tag{6.5-6}$$

Its more formal name is the *geopotential*.

All we need now is $\tilde{U}(\mathbf{r})$ or a suitable form for it. The oblate earth acts gravitationally much as a sphere plus an equatorial ring, and so we may adopt the corresponding form for $\tilde{U}$. Looking at equation (5.5-2) and replacing solar constants by terrestrial ones gives us

$$\tilde{U}_{\text{eff}}(\mathbf{r}; \omega) = -\frac{GM_\oplus}{r}\left[1 + \frac{M_{\text{ring}}}{M_\oplus} \frac{R^2}{4r^2}(1 - 3\cos^2\theta)\right] - \tfrac{1}{2}\omega^2 r^2 \sin^2\theta.$$

$$\tag{6.5-7}$$

Although the ring portion was derived under the specification $r \gg R$, it retains the correct structure—the correct dependence on r and θ—to describe an oblate earth for any $r \geq R$.

The reasoning that introduced this subsection implies that $\tilde{U}_{\text{eff}}$ will remain constant as **r** is moved mentally from point to point on the ocean's surface. If we evaluate $\tilde{U}_{\text{eff}}$ at the north pole, say, then the right-hand side of equation (6.5-7) equals that value at all points on the surface. The equality gives us r_{surface} as a function of θ and thus describes the shape of the earth's oceans.

To get at ΔR, we need only equate the values of $\tilde{U}_{\text{eff}}$ at $\theta = 0$ and $\theta = \pi/2$:

$$-\frac{GM_\oplus}{R_p}\left[1 + \frac{M_{\text{ring}}}{M_\oplus}\frac{1}{4}(-2)\right] = -\frac{GM_\oplus}{R_p + \Delta R}\left(1 + \frac{M_{\text{ring}}}{M_\oplus}\frac{1}{4}\right) - \frac{\omega^2 R^2}{2}.$$

In the ring and ω^2 terms, it suffices to use just R. Now multiply through by $-R_p/(GM_\oplus)$; expand on the right as

$$\frac{R_p}{R_p + \Delta R} = \left(1 + \frac{\Delta R}{R_p}\right)^{-1} \simeq 1 - \frac{\Delta R}{R};$$

cancel the leading terms; and solve for

$$\frac{\Delta R}{R} \simeq \frac{\omega^2 R/2}{GM_\oplus/R^2} + \frac{3}{4}\frac{M_{\text{ring}}}{M_\oplus}. \tag{6.5-8}$$

Since $GM_\oplus/R^2 = g$ at the earth's surface (to adequate approximation here), the first term reproduces our earlier result. The ring term enters positively, enhancing ΔR. That is what we needed and expected. Good.

It is tempting to estimate M_{ring} in terms of ΔR, as we did for the solar oblateness, and then solve equation (6.5-8) for ΔR. Let's try. The ring mass comes from a band of circumference $2\pi R$, thickness ΔR, and width of order R. Thus

$$\frac{M_{\text{ring}}}{M_\oplus} \simeq \frac{2\pi R \cdot R \cdot \Delta R \, \rho_s}{(4\pi/3) R^3 \langle\rho\rangle} = \frac{3}{2}\frac{\rho_s}{\langle\rho\rangle}\frac{\Delta R}{R} = \#\frac{\Delta R}{R},$$

where ρ_s denotes the density near the earth's surface and $\langle\rho\rangle$ is the average density. All we can say honestly is that the mass ratio is some number of order unity, denoted by $\#$, times $\Delta R/R$. Inserting this into equation (6.5-8) and solving gives

$$\frac{\Delta R}{R} \simeq \frac{\omega^2 R/(2g)}{1 - \frac{3}{4}\#}.$$

The denominator is the difference of two similar numbers and hence is very sensitive to the value of $\#$. By this route we cannot get a reliable estimate of ΔR.

Fortunately for us, and for geophysics in general, the earth's bulge has a marked influence on satellite orbits. The bulge produces a noncentral force, the satellite's angular momentum need not remain constant, and the orbital plane drifts through space. The function $\tilde{U}$ can be expanded through higher inverse powers of r, with associated angular functions and numerical coefficients. A comparison of satellite tracking data and theoretical orbits generates values for

228 ROTATING FRAMES OF REFERENCE

the numerical coefficients. In particular, the satellite data imply

$$\frac{M_{\text{ring}}}{M_\oplus} = 2.17 \times 10^{-3}; \tag{6.5-9}$$

that is, we need to adopt this value in our expression for $\tilde{U}$ if satellite orbits computed from $\tilde{U}$ are to be consistent with observation.

Now we can return to equation (6.5-8) and evaluate it as

$$\frac{\Delta R}{R} \simeq 1.73 \times 10^{-3} + 1.62 \times 10^{-3} = 3.35 \times 10^{-3}.$$

This ratio is within 1 percent of the internationally accepted value. It implies a radial difference $R_{\text{eq}} - R_p$ of 21.4 kilometers. For a comparison, the summit of Mt. Everest is 8.8 kilometers above sea level.

6.6 A PERSPECTIVE ON ROTATING FRAMES

In section 6.1 we wrote

$$\mathbf{r}_I = \mathbf{R} + x_i \mathbf{e}_i$$

to describe an object's position as observed from the inertial frame. That position will change if $\mathbf{R}$ changes, if the components x_i change, or if the unit vectors $\mathbf{e}_i$ change. For the observer in the rotating frame, however, the only variables are the components x_i. If the object moves relative to the observer's triad of reference axes, the x_i change, and then $\mathbf{v} \equiv \dot{x}_i \mathbf{e}_i$ is the object's velocity, as far as she or he is concerned.

For the inertial observer, however, that is only a portion of the story. Changes in $\mathbf{R}$ and in the $\mathbf{e}_i$ also change $\mathbf{r}_I$ and contribute to the velocity $d\mathbf{r}_I/dt$. It is the need to include these changes that makes the calculation complicated—and makes the acceleration $d^2\mathbf{r}_I/dt^2$ even worse.

Our aim was to write the equations of motion in variables that the rotating observer finds natural. Newton II demands, however, that we start off in an inertial reference frame, using the acceleration perceived there. Our strategy was this: Into Newton II we inserted $d^2\mathbf{r}_I/dt^2$ as written in equation (6.1-16), and then we moved all terms except $\mathbf{a}$ to the other side of the equation, which produced equation (6.1-17):

$$\mathbf{a} = \frac{\mathbf{F}}{m} - 2\boldsymbol{\omega} \times \mathbf{v} - \boldsymbol{\omega} \times (\boldsymbol{\omega} \times \mathbf{r}) - \ddot{\mathbf{R}} - \frac{d\boldsymbol{\omega}}{dt} \times \mathbf{r}.$$

So much for the equations of motion. We must specify, of course, precisely how the frame rotates relative to the inertial frame. The angular velocity vector $\boldsymbol{\omega}$ conveys this information. And whenever we need to know how the unit vectors change with time, we can refer to equation (6.1-7):

$$\frac{d\mathbf{e}_i}{dt} = \boldsymbol{\omega} \times \mathbf{e}_i.$$

The rotation swings the vector $\mathbf{e}_i$ around in an arc tangent to the direction defined by $\boldsymbol{\omega} \times \mathbf{e}_i$.

Problem solving in the rotating frame is little different from the problem solving that we did earlier. To be sure, in equation (6.1-17) there are new terms on the "force side." We can reason out their influence on the motion, or we can treat them as just part of the mathematics.

WORKED PROBLEM

WP6-1 Some large biological molecules are spun in a centrifuge. Each molecule has mass $m = 5 \times 10^{-21}$ kilogram and (roughly spherical) volume $V = 3.85 \times 10^{-24}$ meter³. The centrifuge tube is rotated in a horizontal circle at 15,000 revolutions/minute, as indicated in figure WP6-1. The molecules start at an initial radius $r_i = 0.1$ meter and sediment outward to $r_f = 0.18$ meter, taking about 10 hours to make the trip. Their motion through the liquid is resisted by a drag force: $-\gamma$ times velocity relative to the liquid, where $\gamma = 2 \times 10^{-10}$ newton · second/meter. The liquid has a density like that of water, $\rho = 10^3$ kilograms/meter.

Problem. Work out the motion of a molecule as seen in a frame rotating with the centrifuge tube. To evaluate the pressure gradient dp/dr, you may want to consider the equilibrium for the liquid itself in that frame. If you can justify dropping the Coriolis term (to simplify an equation), do so.

Two forces act on each molecule. One is a drag force $-\gamma \mathbf{v}$, where $\mathbf{v}$, the molecule's velocity as seen in the rotating frame, is also its velocity relative to the liquid. The other force arises from the pressure gradient in the liquid and is $-V$ grad $p = -V(dp/dr)\hat{\mathbf{r}}$, where $\hat{\mathbf{r}}$ is a radially directed unit vector. (Thinking of the molecule as cylinder-shaped and recalling the analysis at the start of section 6.5 will confirm this expression.) Appeal to Newton II, as expressed in equa-

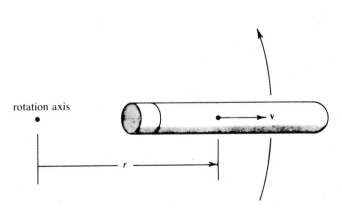

Figure WP6-1 Looking down on the rotating centrifuge tube.

tion (6.1-17), enables us to write

$$\mathbf{a} = -\frac{\gamma}{m}\mathbf{v} - \frac{V}{m}\frac{dp}{dr}\hat{\mathbf{r}} - 2\boldsymbol{\omega}\times\mathbf{v} + \omega^2\mathbf{r}, \tag{1}$$

because we may take $\mathbf{R} = 0$, because the centrifuge rotates with constant angular velocity $\omega = 2\pi(15{,}000)/60 = 1.57 \times 10^3$ radians/second, and because $\mathbf{r}$ is perpendicular to the rotation axis $\boldsymbol{\omega}$, whence the double vector product reduces as in equation (6.2-3).

To evaluate dp/dr, we should note that a volume of fluid at the same location as the molecule would be completely at rest in the rotating frame. Thus

$$0 = -\frac{V}{\rho V}\frac{dp}{dr}\hat{\mathbf{r}} + \omega^2\mathbf{r},$$

and so

$$\frac{dp}{dr} = +\rho\omega^2 r.$$

(The pressure is higher on the large-r side of the tiny fluid volume to provide the net inward force that keeps the fluid in uniform circular motion, as seen from the inertial frame.)

Now we may express equation (1) as

$$\mathbf{a} = -\frac{\gamma}{m}\mathbf{v} - 2\boldsymbol{\omega}\times\mathbf{v} + \left(1 - \frac{\rho V}{m}\right)\omega^2\mathbf{r}. \tag{2}$$

Because $1 - \rho V/m = 0.23$, the ratio of the Coriolis term to the last term is of order $v/(\omega r)$. We estimate the speed v from the data as $v \simeq 0.08$ meter/(10 hours $\times$ 3600 seconds/hour) $= 2 \times 10^{-6}$ meter/second in order of magnitude. This is to be compared with $\omega r \simeq 2 \times 10^2$ meters/second, giving a ratio of $v/(\omega r) \simeq 10^{-8}$. Only if we were interested in minute deviations from purely radial motion would there be any point in retaining the Coriolis term.

After we have dropped the Coriolis term, we may presume purely radial motion and may write equation (2) as

$$\ddot{r} = -\frac{\gamma}{m}\dot{r} + \left(1 - \frac{\rho V}{m}\right)\omega^2 r. \tag{3}$$

The structure is similar to what we met in section 2.1 on the harmonic oscillator: terms linear in the dependent variable and with constant coefficients. There will be exponential solutions:

$$r = A_+ e^{p_+ t} + A_- e^{p_- t},$$

where

$$p_\pm \equiv \left\{-1 \pm \left[1 + 4\left(1 - \frac{\rho V}{m}\right)\omega^2\left(\frac{m}{\gamma}\right)^2\right]^{1/2}\right\}\frac{\gamma}{2m}. \tag{4}$$

The term in ω^2 has the numerical value 1.4×10^{-15}, and so we may use the binomial expansion, retaining only the first term dependent on ω. That approximation yields

$$p_+ \simeq \left(1 - \frac{\rho V}{m}\right)\omega^2\frac{m}{\gamma} = \frac{+1}{7.1 \times 10^4 \text{ seconds}},$$

$$p_- \simeq -\frac{\gamma}{m} = \frac{-1}{2.5 \times 10^{-11} \text{ second}}.$$

The decay time for the A_- term is so short that the term vanishes almost instantly. To excellent approximation we may retain only the A_+ term and write

$$r(t) \simeq r_i \exp\left[\left(1 - \frac{\rho V}{m}\right)\omega^2\frac{m}{\gamma}t\right].$$

An aside. If we imagine inserting the final approximate solution into the right-hand side of equation (3), we see that it corresponds to saying $\ddot{r} \simeq 0$. The damping coefficient γ/m is so large, relatively speaking, that the molecule quickly approaches a "terminal speed"

$$\dot{r}_{\text{terminal}} \equiv \left(1 - \frac{\rho V}{m}\right)\omega^2\frac{m}{\gamma}r.$$

Thereafter, as r increases slowly, a gradual increase in $\dot{r}$ will suffice to keep the molecule near the terminal speed, and so the acceleration is almost zero.

Sometimes one makes the outer fluid regions more dense than the inner, to stabilize the fluid against convective motion. Then both ρ and the viscous damping coefficient γ are functions of r. Setting the right-hand side of equation (3) equal to zero produces an equation of the form $dr/dt = f(r)$ that can still be integrated without great difficulty.

PROBLEMS

6-1 A satellite passes over the north pole, headed south toward Greenwich, England, along the zero of longitude. What is the satellite's longitude when it passes the equator? (You may assume that the satellite just skims the earth's surface, being in free fall at radius $r_\oplus = 6.4 \times 10^6$ meters. Try for a simple method.)

6-2 A mass m moves on a frictionless, uniformly rotating table in a circle concentric with the rotation axis ω. As seen from the rotating frame, the mass travels in uniform circular motion with speed v_0 and radius r_0.

(*a*) What is the acceleration **a**?

(*b*) Use the machinery of sections 6.1 and 6.2 to compute the force that must be acting on the mass.

(*c*) The mass may be traveling around the circle in the same sense as the table rotates (relative to an inertial frame) or in the opposite sense. Try to explain why the answer in part (*b*) depends on the sense of travel.

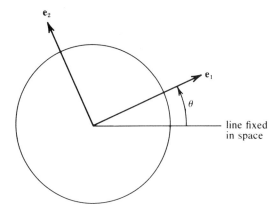

Figure P6-3

6-3 A table rotates about an axis fixed in an inertial frame of reference. A set of orthogonal axes $\mathbf{e}_i$ rotates with the table. The $\mathbf{e}_3$ axis coincides with the fixed rotation axis; the $\mathbf{e}_1$ and $\mathbf{e}_2$ axes rotate relative to the inertial frame as indicated in figure P6-3. The angle θ changes with time as $\theta = \dot{\Omega} t^2 / 2$, where $\dot{\Omega}$ = constant > 0, with units of radians per second^2. Thus the motion is one of constant angular *acceleration*. A small object of mass m lies on the frictionless surface of the table.

  (*a*) First situation: The object is permanently at rest at $\mathbf{r} = A\mathbf{e}_1$. (Any forces needed to achieve this are provided by an external agent.) Compute the object's acceleration as perceived in the inertial frame. Do this for general $t \geq 0$ and then evaluate explicitly for $t = 0$ and $t = 3$ seconds. You may refer vectorial results here and in part (*b*) to the axes that rotate with the table.

  (*b*) Second situation: The object moves such that $\mathbf{r}(t) = v_0 t \mathbf{e}_1$. That is, the motion is of uniform velocity along $\mathbf{e}_1$ as seen in the rotating frame. Determine the force $\mathbf{F}$ required to produce this motion.

T**6-4** The context is section 6.3. Calculate the angle between $\mathbf{g}$ and $-\mathbf{e}_3$ as a function of latitude λ. Where is the deviation angle a maximum? You may take the earth as spherical and work to lowest nonvanishing order in ω. A good way to start is with the vector product $|(-\mathbf{e}_3) \times \mathbf{g}|$. Can you see why?

6-5 The context is the plumb bob in section 6.3. Adopt the viewpoint of the inertial frame in which the bob moves in uniform circular motion. Calculate the tension $\mathbf{T}$ in the string. (You may take the earth as perfectly spherical.) If you define $\mathbf{g}$ by $-\mathbf{T}/m$, can you cast your expression for $\mathbf{g}$ into the form in equation (6.3-4)?

6-6 In section 6.3 we found that an object dropped from rest experiences an eastward deflection. Suppose an object is tossed vertically upward (along $-\hat{\mathbf{g}}$). Where does it land relative to the starting location (to first order in ω)?

  Suppose the toss occurs at the equator. Can you explain the deflection qualitatively with an angular momentum argument?

6-7 For an object dropped from rest, extend our calculation in section 6.3 to second order in ω. Estimate the numerical size of the southerly deflection that would appear (under ideal conditions) if you were to try the experiment. (I presume that you would seek to maximize a small effect.) Compare that expected deflection with the wavelength of red light, 6.5×10^{-7} meter.

An historical account of attempts to measure such deflections—including his own attempt—is given by E. H. Hall, *Phys. Rev.* **16**, 246 (1903).

6-8 In section 6.3 we calculated the lateral deflection of a projectile launched due eastward. For fixed initial velocity v_0, what change in the range—the horizontal distance to the landing site—is produced by the earth's rotation? A calculation of the range correct to first order in ω will suffice. Why may you then ignore the "lateral deflection" that we calculated? And why must you compute the correction of order ω to the time of flight?

6-9 Calculate the change in g between the pole and the equator, working things out consistently to first order in $\Delta R/R$ and quantities of like magnitude. What fraction of the change is *directly* due to the earth's rotation? (Before the days of artificial satellites, measurements of how g varies with latitude were a major source of information about the earth's gross mass distribution.)

6-10 If a bucket of water is spun about a vertical axis (which coincides with the bucket's symmetry axis), the surface becomes concave. Can you adapt the geopotential method to this context and determine the surface shape?

6-11 Focus attention on just the earth and moon, taken to be in circular orbit about their common center of mass (which is about $0.7 R_\oplus$ from the earth's center). Bearing in mind that the moon's pull on a kilogram of material—rock or water—will vary with location on (or inside) the earth, develop a theory of the tides. Why should there be *two* high tides (and two low tides) each day?

6-12 A youngster swings a bucket of water in a vertical circle. How large must the angular velocity be so that the water does not fall out of the pail? Work out a solution from a reference frame rotating with the child's arm and the bucket.

6-13 *An effective g.* A circular platform coasts through interstellar space, sweeping up diffuse matter as its goes. The platform's initial mass and velocity are M_0 and v_0, respectively. The diffuse material provides an average mass density in space of value ρ; the platform sweeps out a cylinder whose cross-sectional area is A.

(*a*) Determine the platform's velocity and acceleration as a function of time.

(*b*) If an astronaut were to stand on the platform and toss a ball, what would be the ball's acceleration (as seen from the astronaut's reference frame)? In short, what would be the magnitude and direction of a "$g_{\text{effective}}$"?

6-14 A fragile spacecraft will be pulled out of shape (beyond hope of recovery) if the gravitational force per unit mass differs by 3 newtons/kilogram between any two ends of the spacecraft, separated by 7 meters.

(*a*) Why would a *difference* in the gravitational force per unit mass be crucial (in determining whether the spacecraft remains intact) rather than the magnitude itself of the gravitational force per unit mass?

(b) How close to a neutron star could the spacecraft safely travel? [Typical theoretical values for the mass and radius, respectively, of a neutron star are $m = 1.4 m_\odot \simeq 4.7 \times 10^5 m_\oplus$ and $R = 2 \times 10^{-3} R_\oplus$.]

6-15 A satellite is initially in synchronous orbit around the earth, that is, its circular orbit has a period of 24 hours and lies in the earth's equatorial plane, so that the satellite hovers over some single location on the equator (which we will call "site A"). Suppose a meteor flashes by the satellite, striking an antenna. Describe the subsequent motion of the satellite's center of mass as observed from site A. (Some of your description will have to be purely qualitative, but not all need be.)

6-16 A simple pendulum of length ℓ and mass m hangs initially at equilibrium from a support in the laboratory.

(a) The support is moved upward with constant acceleration $\mathbf{a}_s$ at a small angle θ_s relative to the vertical. Describe the pendulum's response.

(b) Now suppose the experiment is repeated, but the support is accelerated (at rate $\mathbf{a}_s$ and angle θ_s) for a time t_s only. Thereafter, the support moves with constant velocity. Again, describe the pendulum's response.

CHAPTER SEVEN

EXTENDED BODIES IN ROTATION

7.1 Equations for location and orientation
7.2 Simple precession
7.3 How L is related to ω
7.4 A novel pendulum
7.5 L is not necessarily parallel to ω
7.6 Diagonal form for the inertia tensor
7.7 Euler's equations for a rigid body
7.8 Axisymmetric and torque-free
7.9 Chandler wobble
7.10 An interlude on kinetic energy
7.11 The symmetric, supported top
7.12 Precession of the equinoxes
7.13 Survey of the critical notions

Disturbed by Newcomb's suspicions of the earth's irregularities as a timekeeper, I could think of nothing but precession and nutation, and tides and monsoons, and settlements of the equatorial regions, and meltings of the polar ice.

Lord Kelvin
Address to the British Association (1876)

7.1 EQUATIONS FOR LOCATION AND ORIENTATION

Among the multitude of rotating objects—planets, stars, atomic nuclei, and dust grains, to name a few—many have fixed shapes and fixed mass distributions, at least to a good approximation. For purposes of analysis, we may regard such a rotating object as a collection of mass points with *fixed relative*

236 EXTENDED BODIES IN ROTATION

separations. Needed, then, are equations for the location and orientation of the body as a whole.

In deriving those equations from Newton II, we can most easily incorporate the rigidity condition—that the mass points have fixed relative separations—if we use two frames of reference: a frame that rotates with the body and an inertial frame. Figure 7.1-1 shows the two frames and the pertinent vectors. The origin of the "body frame" moves with the body, and unit vectors $\mathbf{e}_i$ rotate with the body. The mass points that constitute the body are labeled by a subscript. The location of mass m_α, as viewed from the inertial frame, is

$$\mathbf{r}_{I\alpha} = \mathbf{R} + \mathbf{r}_\alpha, \qquad (7.1\text{-}1)$$

where

$$\mathbf{r}_\alpha = x_{i\alpha}\mathbf{e}_i(t). \qquad (7.1\text{-}2)$$

The elements $x_{i\alpha}$ describe the location of m_α relative to the origin of the body frame and relative to axes $\mathbf{e}_i$ that are fixed in the body (because they rotate with the body). Because the relative separations are fixed, the elements $x_{i\alpha}$ are constants:

$$x_{i\alpha} = \text{constant}. \qquad (7.1\text{-}3)$$

That is a tremendous boon.

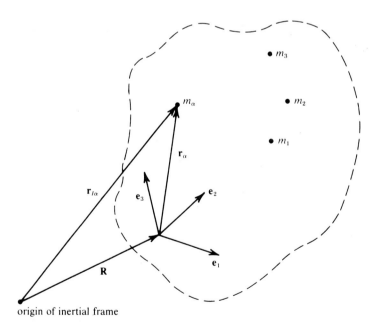

Figure 7.1-1 The body frame and the inertial frame. The dashed line indicates the bounding surface of the extended body.

7.1 EQUATIONS FOR LOCATION AND ORIENTATION

To get the velocity of m_α, we insert equation (7.1-2) into equation (7.1-1) and differentiate:

$$\begin{aligned}
\mathbf{v}_{I\alpha} &= \dot{\mathbf{R}} + \frac{dx_{i\alpha}}{dt}\mathbf{e}_i + x_{i\alpha}\frac{d\mathbf{e}_i}{dt} \\
&= \dot{\mathbf{R}} + 0 + x_{i\alpha}\boldsymbol{\omega} \times \mathbf{e}_i \\
&= \dot{\mathbf{R}} + \boldsymbol{\omega} \times (x_{i\alpha}\mathbf{e}_i) \\
&= \dot{\mathbf{R}} + \boldsymbol{\omega} \times \mathbf{r}_\alpha.
\end{aligned} \quad (7.1\text{-}4)\bigstar$$

For the third term, we use equation (6.1-7) and the definition of the angular velocity $\boldsymbol{\omega}$. The motion of the body frame's origin contributes a common amount $\dot{\mathbf{R}}$, while the rotation of the body itself contributes $\boldsymbol{\omega} \times \mathbf{r}_\alpha$ (because the position vector $\mathbf{r}_\alpha$ is carried around by the angular velocity $\boldsymbol{\omega}$).

Newton II for m_α is

$$\frac{d}{dt}(m_\alpha \mathbf{v}_{I\alpha}) = \mathbf{F}_\alpha,$$

where the force $\mathbf{F}_\alpha$ includes the mutual forces that keep the relative separations fixed.

Equation for Location

To get an equation for the location of the body, we sum Newton II for all the mass points:

$$\sum_\alpha \frac{d}{dt}(m_\alpha \mathbf{v}_{I\alpha}) = \sum_\alpha \mathbf{F}_\alpha.$$

Equation (7.1-4) enables us to write this equation as

$$\frac{d}{dt}\left(\sum_\alpha m_\alpha \dot{\mathbf{R}} + \boldsymbol{\omega} \times \sum_\alpha m_\alpha \mathbf{r}_\alpha\right) = \sum_\alpha \mathbf{F}_\alpha. \quad (7.1\text{-}5)$$

The expression on the left-hand side will simplify if we place the origin of the body frame at the center of mass. The generalization, from two particles to many particles, of the center-of-mass vector $\mathbf{R}_{CM}$ is

$$\begin{aligned}
\mathbf{R}_{CM} &= \frac{1}{M}\sum_\alpha m_\alpha \mathbf{r}_{I\alpha} \\
&= \frac{1}{M}\left[\left(\sum_\alpha m_\alpha\right)\mathbf{R} + \sum_\alpha m_\alpha \mathbf{r}_\alpha\right], \quad (7.1\text{-}6)
\end{aligned}$$

where M denotes the total mass: $M = \Sigma m_\alpha$. If we do place the body frame's origin at the center of mass, then $\mathbf{R} = \mathbf{R}_{CM}$, and equation (7.1-6) implies

$$\sum_\alpha m_\alpha \mathbf{r}_\alpha = 0. \quad (7.1\text{-}7)$$

Most succinctly, the distance from the center of mass to the center of mass is zero. When applied to equation (7.1-5), the zero value in equation (7.1-7) implies that the second sum on the left-hand side is zero. Thus, if we make the choice $\mathbf{R} = \mathbf{R}_{CM}$, equation (7.1-5) becomes

$$\frac{d}{dt}(M\dot{\mathbf{R}}_{CM}) = \sum_\alpha \mathbf{F}_\alpha. \qquad (7.1\text{-}8) \bigstar$$

With regard to the forces that maintain the relative separations, it is fair to assume that they satisfy Newton III and hence cancel (by pairs) in the sum. Only the net external force remains. Equation (7.1-8) tells us that the net external force determines the motion of the center of mass, with no explicit influence from the rotation.

There is nothing novel here (although there is good information), and so we drop the location issue for the remainder of the chapter.

Equation for Orientation

An equation for the orientation of the body is next. A torque often changes an orientation; so let us take torques relative to the origin of the body frame and then sum:

$$\sum_\alpha \mathbf{r}_\alpha \times \frac{d}{dt}(m_\alpha \mathbf{v}_{I\alpha}) = \sum_\alpha \mathbf{r}_\alpha \times \mathbf{F}_\alpha. \qquad (7.1\text{-}9)$$

The next step is to pull the time derivative out in front and to reveal the angular momentum. Thus we write the left-hand side as

$$\frac{d}{dt}\left(\sum_\alpha \mathbf{r}_\alpha \times m_\alpha \mathbf{v}_{I\alpha}\right) - \sum_\alpha \left(\frac{d\mathbf{r}_\alpha}{dt} \times m_\alpha \mathbf{v}_{I\alpha}\right).$$

The first sum is the angular momentum relative to the origin of the body frame. We will denote it by $\mathbf{L}$:

$$\mathbf{L} = \sum_\alpha \mathbf{r}_\alpha \times m_\alpha \mathbf{v}_{I\alpha}. \qquad (7.1\text{-}10) \bigstar$$

With the aid of equation (7.1-4) we can write the second sum as

$$\sum_\alpha m_\alpha(\boldsymbol{\omega} \times \mathbf{r}_\alpha) \times (\dot{\mathbf{R}} + \boldsymbol{\omega} \times \mathbf{r}_\alpha).$$

The second term on the right vanishes because $\boldsymbol{\omega} \times \mathbf{r}_\alpha$ crossed with itself is identically zero. The first term will vanish

If we choose the center of mass as the (7.1-11a)
origin of the body frame, so that
$\Sigma m_\alpha \mathbf{r}_\alpha = 0$, or

If the body has a point spatially fixed (7.1-11b)
in the inertial frame and we place the origin
of the body frame there, so that $\dot{\mathbf{R}} = 0$.

Making either choice will suffice for our purposes. Thus the entire expression vanishes. The equation for orientation becomes

$$\frac{d}{dt}(\mathbf{L}) = \sum_\alpha \mathbf{r}_\alpha \times \mathbf{F}_\alpha. \qquad (7.1\text{-}12) \bigstar$$

On the right-hand side is the torque relative to the origin of the body frame. This equation and the definition of **L** in equation (7.1-10) are the fundamental equations for the chapter.

More about Torque

Before going on, we should note a major simplification in evaluating the torque expression. The force $\mathbf{F}_\alpha$ on mass m_α arises, in general, from two sources: the other masses in the body and agents outside the body (such as gravity or a magnetic field). As before, we may suppose that the forces among the masses satisfy Newton III. Moreover, we may justifiably take them to act along the line joining the masses. Therefore, as reasoned in figure 7.1-2 and its caption, those "internal" forces produce no net torque. In evaluating the right-hand side of equation (7.1-12) we need include only the forces arising outside the body, the "external" forces.

Rotational motion in three dimensions is subtle, and so we proceed by small steps. In section 7.2 we study the motion that arises from a common type of torque. Then we investigate the connection between the angular momentum **L** and the angular velocity **ω** of the body's rotation. Once we have that connection well in hand, we return to dynamics—to equation (7.1-12)—and expand it in a particularly fruitful component form. Examples and applications then occupy the remainder of the chapter, except for a theoretical interlude to develop an expression for the kinetic energy in rotational motion.

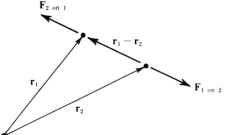

Figure 7.1-2 The forces that particles 1 and 2 exert on each other contribute, to the total torque, an amount

$$\mathbf{r}_1 \times \mathbf{F}_{2\text{ on }1} + \mathbf{r}_2 \times \mathbf{F}_{1\text{ on }2} = (\mathbf{r}_1 - \mathbf{r}_2) \times \mathbf{F}_{2\text{ on }1} = 0;$$

the first step follows from Newton III; the second, because $\mathbf{F}_{2\text{ on }1}$ is along the line joining the two masses. The analysis applies equally to all pairs of masses, and so the forces between the masses produce no net torque.

7.2 SIMPLE PRECESSION

Time and time again an interaction generates a torque with the form

$$\text{(Some vector whose direction is constant)} \times \mathbf{L}. \quad (7.2\text{-}1)$$

Such a torque is perpendicular to both vectors that appear in it. This is the feature to note most especially. When we insert the torque (7.2-1) into equation (7.1-12), our equation for orientation becomes

$$\frac{d\mathbf{L}}{dt} = \mathfrak{F}\hat{c} \times \mathbf{L}, \quad (7.2\text{-}2) \bigstar$$

where $\hat{c}$ denotes a constant unit vector and where $\mathfrak{F}$ denotes some function. When multiplied together, as $\mathfrak{F}\hat{c}$, these two elements describe "some vector whose direction is constant." Separating that vector into direction and magnitude will prove fruitful; it shows clearly the influence of each. We will explore the implications of equation (7.2-2) methodically.

Taking the scalar product of equation (7.2-2) with $\mathbf{L}$ gives

$$\mathbf{L} \cdot \frac{d\mathbf{L}}{dt} = \mathbf{L} \cdot (\mathfrak{F}\hat{c} \times \mathbf{L}),$$

and so

$$\frac{1}{2}\frac{d}{dt}(\mathbf{L} \cdot \mathbf{L}) = 0$$

because the torque is perpendicular to $\mathbf{L}$. We deduce that the magnitude of $\mathbf{L}$ remains constant.

If we take the scalar product with the constant unit vector $\hat{c}$, we find

$$\frac{d}{dt}(\hat{c} \cdot \mathbf{L}) = 0$$

because the torque is perpendicular to $\hat{c}$ also. The component of $\mathbf{L}$ along the constant unit vector does not change.

We can profitably decompose $\mathbf{L}$ into vectors parallel and perpendicular to $\hat{c}$,

$$\mathbf{L} = \mathbf{L}_\parallel + \mathbf{L}_\perp,$$

as sketched on the left in figure 7.2-1. We know that $\mathbf{L}_\parallel$ does not change. Since the magnitude L is constant, the pythagorean theorem tells us that $|\mathbf{L}_\perp|$ must be constant. All that can happen is that $\mathbf{L}_\perp$ precesses about $\hat{c}$.

For a time interval Δt, equation (7.2-2) implies a change $\Delta \mathbf{L}$ equal to

$$\frac{d\mathbf{L}}{dt}\Delta t = \mathfrak{F}\,\hat{c} \times (\mathbf{L}_\parallel + \mathbf{L}_\perp)\,\Delta t$$

$$= \mathfrak{F}\,\hat{c} \times \mathbf{L}_\perp\,\Delta t.$$

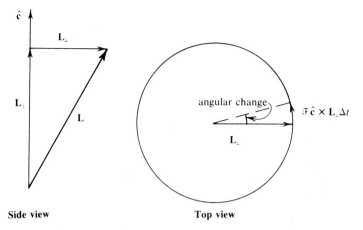

Figure 7.2-1 The vector **L** and its change with time. Because the torque is perpendicular to both **L** and $\hat{c}$, it can only cause **L** to precess about $\hat{c}$.

This change is shown on the right in figure 7.2-1. From the sketch we can extract the angular change in time Δt. We find that $\mathbf{L}_\perp$ swings around at the angular rate

$$\frac{|\mathcal{F}\hat{c} \times \mathbf{L}_\perp \Delta t|}{|\mathbf{L}_\perp| \Delta t} = \mathcal{F}.$$

Note that this rate does not depend on either the magnitude L or the angle between **L** and $\hat{c}$.

Here is a summary. The vector **L**

1. Has a constant magnitude and
2. Precesses about $\hat{c}$ at a constant angle of inclination and with an angular frequency given by $\mathcal{F}$.

Because the torque is perpendicular to both **L** and $\hat{c}$, the torque can only cause **L** to trace a cone centered on $\hat{c}$. That motion is the precession. Figure 7.2-2 shows an oblique view of it.

The equation that we just analyzed has the structure

$$\frac{d}{dt}(\text{vector}) = \begin{pmatrix} \text{some vector} \\ \text{whose direction} \\ \text{is constant} \end{pmatrix} \times \begin{pmatrix} \text{same vector} \\ \text{as on LHS} \end{pmatrix}. \quad (7.2\text{-}3)$$

Whenever this structure arises, the properties just summarized hold, with "vector" in place of **L**.

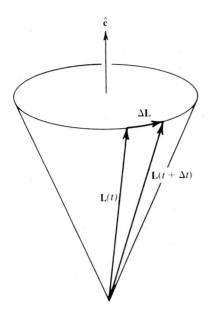

Figure 7.2-2 An oblique view of the precession: L traces a cone centered on ĉ.

The Dust Grain: An Example

An interstellar dust grain is a tiny object, perhaps 5×10^{-7} meter across, roughly the wavelength of the visible light that the grains scatter so well. The grains are in rapid rotation; an angular velocity of 10^4 radians/second is typical. Moreover, usually they are electrically charged. The net charge depends on the free electron density in the grain's interstellar environment and also changes from time to time, as the grain captures or loses electrons. An average value of 20 electrons would be representative. A magnetic field pervades the galaxy, and so that is a part of the grain's environment.

The picture of a charged rotating grain in a magnetic field implies magnetic forces and, typically, a torque. Figure 7.2-3 depicts a charge q being carried around by the spinning grain. In the magnetic field **B**, the magnetic force is

$$\mathbf{F}_{\text{mag}} \propto \mathbf{v}_q \times \mathbf{B},$$

where $\mathbf{v}_q$ is the charge's velocity. The sketch shows the product $\mathbf{v}_q \times \mathbf{B}$ when the charge is at two specific points on the circular path. At each of these points the torque is perpendicular to the $\omega \mathbf{B}$ plane. What about the average torque, computed by averaging over the entire path?

Here is another opportunity to reason effectively by using vectors. When the charge is at location **r** relative to the grain's center, the torque is

$$\text{Torque} \propto \mathbf{r} \times (\mathbf{v}_q \times \mathbf{B})$$
$$\propto \mathbf{r} \times [(\boldsymbol{\omega} \times \mathbf{r}) \times \mathbf{B}],$$

when we recall that the velocity can be written as $\mathbf{v}_q = \boldsymbol{\omega} \times \mathbf{r}$ because the charge is carried around by the rotation ω. After we average this torque expression over all values of **r** along the path, the result can depend on only the two fixed vectors, ω and **B**. The result must be proportional in magnitude to ω and to B, and it must be a vector. Moreover, it must vanish if ω and **B** are parallel; analysis with a sketch will confirm that claim. The only combination of ω and **B** that meets the requirements is

$$\langle \text{Torque} \rangle \propto \omega \times \mathbf{B}, \tag{7.2-4}$$

and so this vector product must emerge as the average torque for the specific charge q. In short, the average torque is perpendicular to the $\omega\mathbf{B}$ plane. (The vector product form may be familiar from the torque on a current loop, to which the charge's motion is equivalent.)

For a typical grain, we need to sum the torques associated with some 20 charges. Let us write the outcome as

$$\langle \text{Total torque} \rangle = \gamma \mathbf{L} \times \mathbf{B}, \tag{7.2-5}$$

where **L** is the grain's angular momentum about its center of mass and γ is a constant. (Implicitly we have taken $\omega \propto \mathbf{L}$; the proportionality is a good approximation, which we investigate in section 7.8.)

Equation (7.1-12) tells us that the torque generates the change in angular momentum, and so we have

$$\frac{d\mathbf{L}}{dt} = \gamma \mathbf{L} \times \mathbf{B}.$$

If we reverse the locations of **L** and **B** on the right, at the cost of a minus sign, so

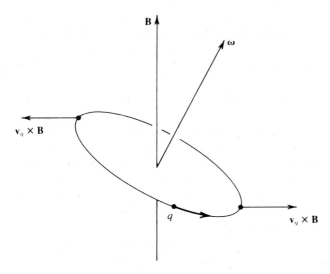

Figure 7.2-3 How a torque arises when a spinning grain carries around a charge q.

that

$$\frac{d\mathbf{L}}{dt} = (-\gamma \mathbf{B}) \times \mathbf{L},$$

we find the form displayed in equation (7.2-2). The magnetic torque, we find, will cause the grain's angular momentum to precess. Thus the spinning grain will precess bodily about the magnetic field.

The galactic magnetic field has a magnitude of order 3×10^{-6} gauss, equivalent to 3×10^{-10} tesla. The product $|-\gamma B|$ works out to roughly 10^{-11} radian/second. That implies one precessional revolution in 20,000 years. In interstellar space, where there may be only 20 hydrogen atoms per cubic centimeter and one dust grain in a volume the size of a civic auditorium, that time is not so long as to be irrelevant. It is, in fact, short enough that precession determines the average orientation of the dust grains, an average that lies along the local magnetic field.

7.3 HOW L IS RELATED TO ω

We turn now to the connection between the angular momentum $\mathbf{L}$ and the angular velocity ω of the body's rotation. The angular momentum, computed relative to the body frame's origin, was displayed earlier in equation (7.1-10):

$$\mathbf{L} = \sum_\alpha \mathbf{r}_\alpha \times m_\alpha \mathbf{v}_{I\alpha}. \qquad (7.3\text{-}1)$$

Here $\mathbf{v}_{I\alpha}$ denotes the velocity of mass m_α as seen from the inertial frame of reference. Equation (7.1-4) expressed that velocity as

$$\mathbf{v}_{I\alpha} = \dot{\mathbf{R}} + \omega \times \mathbf{r}_\alpha. \qquad (7.3\text{-}2)$$

The motion of the body frame's origin contributes a common amount $\dot{\mathbf{R}}$, while the rotation of the body itself contributes $\omega \times \mathbf{r}_\alpha$ (because the position vector $\mathbf{r}_\alpha$ is carried around by the angular velocity ω). Inserting the expression for $\mathbf{v}_{I\alpha}$ into equation (7.3-1) gives us

$$\mathbf{L} = \sum_\alpha m_\alpha \mathbf{r}_\alpha \times (\dot{\mathbf{R}} + \omega \times \mathbf{r}_\alpha).$$

The first thing to note is that the vector product containing $\dot{\mathbf{R}}$ makes no net contribution. We are working under stipulations (7.1-11a and b). If we have chosen the center of mass as the origin of the body frame, then $\Sigma m_\alpha \mathbf{r}_\alpha = 0$, and $\dot{\mathbf{R}}$ drops out. If the body has a point spatially fixed and we have chosen that point as the origin of the body frame, then $\dot{\mathbf{R}}$ equals zero. Either way, under our stipulations, the angular momentum reduces to

$$\mathbf{L} = \sum_\alpha m_\alpha \mathbf{r}_\alpha \times (\omega \times \mathbf{r}_\alpha). \qquad (7.3\text{-}3)$$

7.3 HOW L IS RELATED TO ω

To see the connection with ω most clearly, we must expand the double vector product and factor out the angular velocity as best we can. Let us drop the index α for now and use the notation developed in appendix B. Identity (B.12) will get us started:

$$\mathbf{r} \times (\boldsymbol{\omega} \times \mathbf{r}) = (\mathbf{r} \cdot \mathbf{r})\boldsymbol{\omega} - (\mathbf{r} \cdot \boldsymbol{\omega})\mathbf{r}$$
$$= r^2 \omega_i \mathbf{e}_i - x_j \omega_j x_i \mathbf{e}_i$$
$$= (r^2 \delta_{ij} - x_i x_j) \omega_j \mathbf{e}_i.$$

The succeeding steps factor out the unit vectors $\mathbf{e}_i$ and the components ω_j of the angular velocity vector $\boldsymbol{\omega}$. These steps isolate the portion that depends on only the location of a specific mass (relative to the body axes). For equation (7.3-3) we need the sum

$$I_{ij} \equiv \sum_{\alpha} m_{\alpha} (\delta_{ij} r_{\alpha}^2 - x_{i\alpha} x_{j\alpha}). \qquad (7.3\text{-}4) \bigstar$$

The elements I_{ij} are a set of *constants* that describe how mass is distributed throughout the rigid body. The array is called the *moment of inertia tensor*.

The array has the pictorial form

$$I_{ij} = \begin{pmatrix} \sum m_{\alpha}(y_{\alpha}^2 + z_{\alpha}^2) & -\sum m_{\alpha} x_{\alpha} y_{\alpha} & -\sum m_{\alpha} x_{\alpha} z_{\alpha} \\ -\sum m_{\alpha} y_{\alpha} x_{\alpha} & \sum m_{\alpha}(x_{\alpha}^2 + z_{\alpha}^2) & -\sum m_{\alpha} y_{\alpha} z_{\alpha} \\ -\sum m_{\alpha} z_{\alpha} x_{\alpha} & -\sum m_{\alpha} z_{\alpha} y_{\alpha} & \sum m_{\alpha}(x_{\alpha}^2 + y_{\alpha}^2) \end{pmatrix}, \qquad (7.3\text{-}5)$$

where the index i labels rows and j labels columns.

Now we can go back to equation (7.3-3) and write succinctly

$$\mathbf{L} = I_{ij} \omega_j \mathbf{e}_i. \qquad (7.3\text{-}6) \bigstar$$

Familiar Ground

Equation (7.3-6) is a generalization of something familiar. To make the connection, let us examine a ring of radius R and choose $\mathbf{e}_3$ to lie along the symmetry axis, the central perpendicular to the plane of the ring. The ring is sketched in figure 7.3-1. The origin of the body frame is at the ring's center, its center of mass. Therefore element I_{33} becomes

$$I_{33} = \sum_{\alpha} m_{\alpha} (r_{\alpha}^2 - z_{\alpha}^2)$$
$$= \sum_{\alpha} m_{\alpha}(x_{\alpha}^2 + y_{\alpha}^2)$$
$$= M_{\text{ring}} R^2$$

because each mass m_{α} lies at the radial distance R from $\mathbf{e}_3$. The expression $M_{\text{ring}} R^2$ is familiar as the moment of inertia about the symmetry axis.

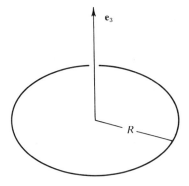

Figure 7.3-1 The unit vector e_3 lies along the symmetry axis of the metal ring.

An element such as

$$I_{12} = -\sum_\alpha m_\alpha x_\alpha y_\alpha$$

equals zero by symmetry in the plane of the ring (because for every positive y_α at fixed x_α there is a canceling negative y_α). Symmetry sets I_{13} and I_{23} equal to zero (or we can reason that the ring has vanishing thickness, and so $z_\alpha = 0$ for all the mass elements). Finally, symmetry ensures that I_{11} equals I_{22}. The sum of I_{11} and I_{22} is easy to evaluate:

$$(I_{11} + I_{22})_{\text{ring}} = \sum_\alpha m_\alpha (x_\alpha^2 + y_\alpha^2 + 2z_\alpha^2)$$

$$= M_{\text{ring}} R^2.$$

Then we take half of this result for each element individually.

With all the elements in hand, we can display the array:

$$(I_{ij})_{\text{ring}} = M_{\text{ring}} R^2 \begin{pmatrix} \frac{1}{2} & 0 & 0 \\ 0 & \frac{1}{2} & 0 \\ 0 & 0 & 1 \end{pmatrix}. \tag{7.3-7}$$

The general expression for the ring's angular momentum follows from the display and from equation (7.3-6). Because the array I_{ij} is diagonal, the double sum (over the indices i and j) yields

$$\mathbf{L} = \tfrac{1}{2} M_{\text{ring}} R^2 (\omega_1 \mathbf{e}_1 + \omega_2 \mathbf{e}_2) + M_{\text{ring}} R^2 \omega_3 \mathbf{e}_3. \tag{7.3-8}$$

If we specify that the rotation is purely around $\mathbf{e}_3$, then $\boldsymbol{\omega} = \omega \mathbf{e}_3$, implying $\omega_3 = \omega$ and $\omega_1 = \omega_2 = 0$. Equation (7.3-8) reduces to

$$\mathbf{L} = M_{\text{ring}} R^2 \omega \mathbf{e}_3,$$

probably familiar as the angular momentum when a ring rotates about its symmetry axis.

Now suppose the rotation is about a spatially fixed axis perpendicular to e_3, so that a diameter of the ring lies along ω. Without loss of generality, we may take the body axis e_1 to lie along that direction; then $\omega_1 = \omega$ and $\omega_2 = \omega_3 = 0$. The angular momentum is now

$$\mathbf{L} = \tfrac{1}{2} M_{\text{ring}} R^2 \omega \mathbf{e}_1.$$

The magnitude of **L** is smaller than before, and the reason is not hard to find. When the ring rotates about its symmetry axis, all mass points are a distance R from the rotation axis. When the ring rotates about a diameter, most mass points lie at distances less than R from the rotation axis. There is, in fact, a continuum of distances, from the value R to zero. Because most mass points lie at a smaller distance, they contribute less to the angular momentum.

In general, equation (7.3-6) tells us that I_{ij} answers the question, How much angular momentum along the direction $\mathbf{e}_i$ is produced by a rotation rate ω_j along the direction $\mathbf{e}_j$?

7.4 A NOVEL PENDULUM

To see how the machinery of sections 7.1 and 7.3 functions, let us work out an example. Here is a novel pendulum: A mass m_2 is fixed at the end of a rod of length ℓ, and a mass m_1 is located at the midpoint, as sketched on the left in figure 7.4-1. If the pendulum swings in the plane of the sketch, how does the oscillation frequency depend on the mass ratio m_1/m_2?

The pendulum has a point spatially fixed in an inertial frame, its pivot point. We can profitably put the origin of the body axes there, orient $\mathbf{e}_1$ along the rod, and orient $\mathbf{e}_3$ perpendicular to the plane of oscillation, as shown on the right in figure 7.4-1. The angular velocity is then

$$\boldsymbol{\omega} = \omega_3 \mathbf{e}_3 = \dot{\theta} \mathbf{e}_3. \qquad (7.4\text{-}1)$$

The coordinates of the masses, referred to the body axes, are

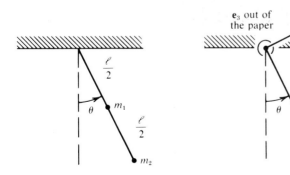

Figure 7.4-1 The pendulum.

$$x_{i1} = \frac{\ell}{2}\delta_{i1},$$

$$x_{i2} = \ell\delta_{i1}.$$

If the rod has negligible mass, as we will suppose, the inertia tensor is

$$I_{ij} = m_1\left[\delta_{ij}\left(\frac{\ell}{2}\right)^2 - \left(\frac{\ell}{2}\right)^2\delta_{i1}\delta_{j1}\right] + m_2[\delta_{ij}\ell^2 - \ell^2\delta_{i1}\delta_{j1}]$$

$$= \left(m_1\frac{\ell^2}{4} + m_2\ell^2\right)(\delta_{ij} - \delta_{i1}\delta_{j1}).$$

The array has the pictorial form

$$I_{ij} = \left(\frac{m_1}{4} + m_2\right)\ell^2\begin{pmatrix} 0 & 0 & 0 \\ 0 & 1 & 0 \\ 0 & 0 & 1 \end{pmatrix},$$

where, as before, i labels rows and j labels columns. The angular momentum is

$$\mathbf{L} = I_{ij}\omega_j\mathbf{e}_i = I_{i3}\dot{\theta}\mathbf{e}_i$$

$$= \left(\frac{m_1}{4} + m_2\right)\ell^2\dot{\theta}\mathbf{e}_3.$$

For the torque, we have

$$\sum_\alpha \mathbf{r}_\alpha \times \mathbf{F}_\alpha = \frac{\ell}{2}\mathbf{e}_1 \times (m_1\mathbf{g}) + \ell\,\mathbf{e}_1 \times (m_2\mathbf{g})$$

$$= \left(\frac{m_1}{2} + m_2\right)\ell g\,(-\sin\theta\,\mathbf{e}_3).$$

The force $\mathbf{F}_1$ on m_1 arises from gravity and from a stress in the rod. We need evaluate, however, only the contribution from gravity, the external force, because the contribution from the stress, as an internal force, will cancel in the sum. The same is true for the force on m_2.

Appeal to equation (7.1-12) produces

$$\frac{d}{dt}\left[\left(\frac{m_1}{4} + m_2\right)\ell^2\dot{\theta}\mathbf{e}_3\right] = -\left(\frac{m_1}{2} + m_2\right)\ell g \sin\theta\,\mathbf{e}_3.$$

In the present situation, $\mathbf{e}_3$ does not move relative to the inertial frame, and so differentiation gives

$$\ddot{\theta}\mathbf{e}_3 = -\frac{\frac{1}{2}m_1 + m_2}{\frac{1}{4}m_1 + m_2}\frac{g}{\ell}\sin\theta\,\mathbf{e}_3.$$

The coefficients of $\mathbf{e}_3$ must be equal. If the amplitude of swing is small, so that $\sin\theta \simeq \theta$, we have an equation of harmonic oscillator form. We can read off the square of the angular frequency:

$$\omega_0^2 = \frac{1 + m_1/(2m_2)}{1 + m_1/(4m_2)} \frac{g}{\ell}. \tag{7.4-2}$$

If m_1/m_2 goes to zero, we recover the frequency for a simple pendulum of length ℓ. If m_1/m_2 goes to infinity, then ω_0^2 approaches $g/(\ell/2)$; that is correct. For a general value of the mass ratio, equation (7.4-2) gives a nonlinear interpolation between these two limits.

Fixed Axis of Rotation

Our pendulum swings about a line fixed in space, the perpendicular to the plane of the sketch. The angular velocity $\boldsymbol{\omega}$ always points along that line. To be sure, $\boldsymbol{\omega}$ reverses direction after each half swing, but $\boldsymbol{\omega}$ never wanders in orientation from the fixed line. The pendulum has a spatially fixed axis of rotation. An aside to examine such a situation in general is worth our while.

When a body rotates about a spatially fixed axis, we may always take one body axis to coincide with the rotation axis. To see why, note that mass points that lie (instantaneously) along $\boldsymbol{\omega}$ are not moved through space by the rotation. Thus those mass points continue to lie along $\boldsymbol{\omega}$. They can serve as guides when we choose the orientation of the body axes. To be specific, we may choose body axis $\mathbf{e}_3$ to pass through those mass points and thus to lie permanently along the rotation axis. In terms of the body axes, the angular velocity takes the simple form $\boldsymbol{\omega} = \omega_3 \mathbf{e}_3$.

The angular momentum **L** follows now from equation (7.3-6) as

$$\mathbf{L} = I_{ij}\omega_j \mathbf{e}_i$$
$$= I_{i3}\omega_3 \mathbf{e}_i$$

because $\omega_1 = \omega_2 = 0$. If the moment of inertia tensor is diagonal, then **L** simplifies further:

$$\mathbf{L} = I_{33}\omega_3 \mathbf{e}_3.$$

We should remember that because $\mathbf{e}_3$ points along the spatially fixed rotation axis, its time derivative is zero. A torque will change the value of ω_3, but nothing else.

Our pendulum typifies this simplified situation. Another example is provided by a disk hung from a stiff wire that lies along the disk's symmetry axis. A twist in the wire produces a restoring torque, so that the disk oscillates: a torsional oscillator.

7.5 L IS NOT NECESSARILY PARALLEL TO $\boldsymbol{\omega}$

Imagine four equal masses glued to the surface of a thin plastic spherical shell, as sketched in figure 7.5-1. The masses are located at opposite ends of two mutually perpendicular diameters. The shell rotates with constant angular velocity

250 EXTENDED BODIES IN ROTATION

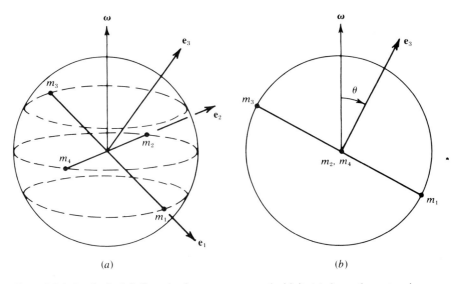

Figure 7.5-1 A spherical shell carries four masses around with it. (*a*) shows the system in perspective. The unit vector $\mathbf{e}_3$ points perpendicular to the plane defined by the four masses. (*b*) shows the masses at a special instant and in projection. At the instant when masses m_1 and m_3 lie in the plane of the paper, mass m_2 is a distance R directly beneath that plane, and mass m_4 is directly above.

$\boldsymbol{\omega}$, carrying with it the four masses. (Some bearings provide whatever torque is needed to ensure uniform rotation.)

We will reckon the angular momentum **L** from the system's center of mass, located at the center of the shell. Here is the central question: Does the angular momentum point along $\boldsymbol{\omega}$?

First we calculate **L** by summing $\mathbf{r}_\alpha \times (m_\alpha \mathbf{v}_{I\alpha})$ for the four masses. Take m_2 first. Velocity $\mathbf{v}_{I2}$ lies in the plane perpendicular to $\boldsymbol{\omega}$, and $\mathbf{r}_2$ lies in that plane. Thus the contribution to **L**, the quantity $\mathbf{r}_2 \times (m_2 \mathbf{v}_{I2})$, points along $\boldsymbol{\omega}$. Mass m_4 moves with the same speed as m_2 but in precisely the opposite direction. Because $\mathbf{r}_4$ is the negative of $\mathbf{r}_2$, the contribution to **L** is the same as that for m_2. Both m_2 and m_4 contribute along $\boldsymbol{\omega}$.

Mass m_1 moves in a plane perpendicular to $\boldsymbol{\omega}$, but $\mathbf{r}_1$ is *not* perpendicular to $\boldsymbol{\omega}$. The vector product $\mathbf{r}_1 \times (m_1 \mathbf{v}_{I1})$ points along the direction of the unit vector $\mathbf{e}_3$ in the sketch. Because mass m_3 has a velocity and a location that are the negative of those for mass m_1, it contributes to **L** precisely as m_1 does—along $\mathbf{e}_3$.

Thus the sum giving **L** is composed of two vectors along $\boldsymbol{\omega}$ and two along $\mathbf{e}_3$. Since $\mathbf{e}_3$ is not parallel to $\boldsymbol{\omega}$, the sum is not parallel to $\boldsymbol{\omega}$. Therefore **L** is not parallel to $\boldsymbol{\omega}$. That may be a shock, but it is true.

Now let us calculate **L** via the inertia tensor. We use the center of mass as the origin of the body axes: $\mathbf{e}_1$ points toward mass m_1, $\mathbf{e}_2$ points toward mass m_2, and $\mathbf{e}_3$ is perpendicular to the plane containing all four masses.

Mass m_1, which lies along the direction $\mathbf{e}_1$, will contribute to I_{ij} just as mass m_1 did in section 7.4. We need only replace $\ell/2$ by the distance R. Mass m_3 lies along $-\mathbf{e}_1$. The moment-of-inertia tensor depends on position components only as binary products, however, and so m_3 makes precisely the same contribution as m_1. Masses m_2 and m_4 lie along $\mathbf{e}_2$ and $-\mathbf{e}_2$, respectively. Their contribution must be like that of m_1 and m_3 but with the index 1 replaced by 2.

Putting together the pieces, we get

$$I_{ij} = 2mR^2(\delta_{ij} - \delta_{i1}\delta_{j1}) + 2mR^2(\delta_{ij} - \delta_{i2}\delta_{j2})$$

$$= 2mR^2 \begin{pmatrix} 1 & 0 & 0 \\ 0 & 1 & 0 \\ 0 & 0 & 2 \end{pmatrix}. \tag{7.5-1}$$

Again we find a diagonal array. The double sum in equation (7.3-6) reduces to

$$\mathbf{L} = 2mR^2(\omega_1 \mathbf{e}_1 + \omega_2 \mathbf{e}_2) + 4mR^2\omega_3 \mathbf{e}_3.$$

To make the direction of $\mathbf{L}$ more apparent, we can add $2mR^2\omega_3\mathbf{e}_3$ to the first terms, so that they become proportional to $\boldsymbol{\omega}$, and then compensate by subtracting the same amount from the last term:

$$\mathbf{L} = 2mR^2\boldsymbol{\omega} + 2mR^2\omega_3\mathbf{e}_3. \tag{7.5-2}$$

We can extract the value of ω_3 from figure 7.5-1. It is $\omega \cos\theta$, a positive value. Thus $\mathbf{L}$ is composed of positive contributions along $\boldsymbol{\omega}$ and $\mathbf{e}_3$, just as we found earlier.

More importantly, both calculations show us that $\mathbf{L}$ is not parallel to $\boldsymbol{\omega}$.

In computing an angular momentum, we must always choose some fiducial point for reckoning the position $\mathbf{r}_\alpha$ in the vector product $\mathbf{r}_\alpha \times (m_\alpha \mathbf{v}_{I\alpha})$. Whenever a position vector $\mathbf{r}_\alpha$ is not perpendicular to $\boldsymbol{\omega}$, the contribution to $\mathbf{L}$ will not be along $\boldsymbol{\omega}$. Sometimes the net result of such contributions will still—by cancellations—be along $\boldsymbol{\omega}$, but we cannot count on that. The connection between $\mathbf{L}$ and $\boldsymbol{\omega}$ is necessarily more complicated than a proportionality. Only an expression such as

$$\mathbf{L} = I_{ij}\omega_j \mathbf{e}_i$$

can describe the connection in full generality.

7.6 DIAGONAL FORM FOR THE INERTIA TENSOR

Later we will need the moment of inertia tensor for a sphere. We work it out here, in part as another example of how to compute the array I_{ij}.

Symmetry requires that the inertia tensor for a sphere be diagonal and have equal diagonal elements. We can use that property to advantage:

$$(I_{11} + I_{22} + I_{33})_{\text{sphere}} = \sum_{\text{sphere}} m_\alpha [3r_\alpha^2 - (x_\alpha^2 + y_\alpha^2 + z_\alpha^2)]$$

$$= \sum_{\text{sphere}} m_\alpha 2 r_\alpha^2.$$

To evaluate the final sum, we first convert it to an integral. We imagine decomposing the sphere into nested shells: radius r, thickness dr, surface area $4\pi r^2$, and hence volume $4\pi r^2\, dr$. The mass associated with such a spherical shell is the mass density times the volume. Our sum becomes the integral

$$\int_0^R 2r^2\, \rho(r)\, 4\pi r^2\, dr,$$

where $\rho(r)$ is the mass density and R is the sphere's radius. If the density is everywhere the same, the integral gives

$$\frac{8\pi}{5} \rho R^5 = 3 \cdot \tfrac{2}{5} M_{\text{sphere}} R^2.$$

Each diagonal element is equal to one-third of this value, and so we find

$$(I_{ij})_{\substack{\text{homogeneous}\\\text{sphere}}} = \tfrac{2}{5} M_{\text{sphere}} R^2 \begin{pmatrix} 1 & 0 & 0 \\ 0 & 1 & 0 \\ 0 & 0 & 1 \end{pmatrix}. \tag{7.6-1}$$

Whenever the distribution of mass is smooth, we may replace a sum over masses by an integral. What we have denoted by m_α is replaced by the mass density times an infinitesimal volume element. The summation is then replaced by integration between endpoints set by the physical extent of the body.

Knowing something about the inertia tensor for a disk will be useful. A disk has the same symmetry as a ring. If we put the origin of the body axes at the disk's center of mass and if we orient $\mathbf{e}_3$ along the symmetry axis, the inertia tensor must have, qualitatively, the form we found for a ring in equation (7.3-7):

$$(I_{ij})_{\text{disk}} = \begin{pmatrix} I_* & 0 & 0 \\ 0 & I_* & 0 \\ 0 & 0 & I_{33} \end{pmatrix}, \tag{7.6-2}$$

where I_* denotes the equal values of I_{11} and I_{22}. How does I_{33} compare with I_*? We should write the sums that define I_{33} and I_{11}:

$$I_{33} = \sum_\alpha m_\alpha (r_\alpha^2 - z_\alpha^2),$$

$$I_{11} = \sum_\alpha m_\alpha (r_\alpha^2 - x_\alpha^2).$$

For a disk that is thin relative to its radius, the average of z_α^2 is less than the

7.6 DIAGONAL FORM FOR THE INERTIA TENSOR

average of x_α^2, and so I_{33} is greater than I_{11}. This knowledge will suffice for our purposes. (A still more explicit evaluation is provided in WP7-3.)

The few times that we have calculated the moment-of-inertia tensor we found a diagonal array. That favorable outcome is typical of the shapes we meet or choose to employ. Those shapes have symmetries, like the axial symmetry of a disk, and as long as we align the body axes along the natural directions specified by the symmetries, a diagonal array emerges automatically.

A tennis racket provides a new example. It has no axis of rotational symmetry, but its does possess two reflection symmetries. Provided we align e_1 along the handle's central axis, set e_2 perpendicular to the strings, and lay e_3 in the strings' plane (or use some permutation of these choices), we find a diagonal array. The moral here is to recognize symmetries when you see them and to use them as a guide in choosing directions.

To be sure, even if the object has no symmetries, a judicious choice of the body axes will produce a diagonal array. That property is not obvious, but a proof can be built on the symmetry of the general expression for I_{ij}, equation (7.3-4). We will not construct a proof here; rather, we will develop a plausibility argument.

Suppose we have chosen a set of axes in the body (the set e_i), have computed I_{ij}, and have found that it is not diagonal. Can we choose a new set of axes (the set e_i') so that the new moment of inertia tensor I'_{ij}, constructed from the position components $x'_{i\alpha}$, is diagonal?

We need to ensure that three sums are zero:

$$\sum_\alpha m_\alpha x'_\alpha y'_\alpha, \quad \sum_\alpha m_\alpha x'_\alpha z'_\alpha, \quad \sum_\alpha m_\alpha y'_\alpha z'_\alpha.$$

That will suffice precisely because the moment-of-inertia tensor is symmetric. In getting all the terms on one side of the diagonal to be zero, we simultaneously annihilate all the terms on the other side.

Now we ask, How much freedom (or flexibility) do we have in choosing a new set of axes? We may produce the new axes by rotating the old axes around some specific direction. To specify the direction of the rotation axis, we can use latitude and longitude on a unit sphere. (The two angles in spherical polar coordinates would work equally well.) Thus two parameters are needed for the direction and give us 2 degrees of freedom. We also may rotate about the chosen axis by a small angle or a large one. The amount of rotation may be specified freely and gives us another degree of freedom.

In short, we have 3 degrees of freedom in choosing the new axes, and we need to achieve only three conditions, the three zero sums. This equality in the counting makes it plausible that we can achieve our goal—a diagonal form for I'_{ij}—and indeed we can.

Any set of body axes for which the moment-of-inertia tensor is diagonal is called a set of *principal axes*. Our plausibility argument says that we can always choose principal axes, and we do so whenever it provides a simplification.

7.7 EULER'S EQUATIONS FOR A RIGID BODY

We can return now to dynamics, to the way in which a torque changes an angular momentum and an orientation. If we write **L** in equation (7.1-12) in terms of $\boldsymbol{\omega}$, the equation becomes

$$\frac{d}{dt}(I_{ij}\omega_j \mathbf{e}_i) = \sum_\alpha \mathbf{r}_\alpha \times \mathbf{F}_\alpha. \tag{7.7-1}$$

At times an expanded form of this equation is useful. If we perform the differentiation, the left-hand side becomes

$$I_{ij}\dot\omega_j \mathbf{e}_i + I_{ij}\omega_j \frac{d\mathbf{e}_i}{dt}.$$

Because

$$\frac{d\mathbf{e}_i}{dt} = \boldsymbol{\omega} \times \mathbf{e}_i, \tag{7.7-2}$$

equation (7.7-1) can be written as

$$I_{ij}\dot\omega_j \mathbf{e}_i + \boldsymbol{\omega} \times \mathbf{e}_i \, I_{ij}\omega_j = \sum_\alpha \mathbf{r}_\alpha \times \mathbf{F}_\alpha. \tag{7.7-3}$$

Now we specify that principal axes have been chosen, so that I_{ij} is diagonal. The first double sum equals

$$I_{11}\dot\omega_1 \mathbf{e}_1 + I_{22}\dot\omega_2 \mathbf{e}_2 + I_{33}\dot\omega_3 \mathbf{e}_3. \tag{7.7-4a}$$

The second double sum—the sum with the vector product—generates six terms:

$$(\omega_3 \mathbf{e}_2 - \omega_2 \mathbf{e}_3) I_{11}\omega_1 + (-\omega_3 \mathbf{e}_1 + \omega_1 \mathbf{e}_3) I_{22}\omega_2 + (\omega_2 \mathbf{e}_1 - \omega_1 \mathbf{e}_2) I_{33}\omega_3. \tag{7.7-4b}$$

To reduce equation (7.7-3) to component form, first we insert the expansions that we have written out as expressions (7.7-4a and b) and then we take scalar products with unit vectors $\mathbf{e}_1$, $\mathbf{e}_2$, and $\mathbf{e}_3$. Each scalar product gives an equation; together, they generate the following three equations:

$$\begin{aligned} I_{11}\dot\omega_1 - (I_{22} - I_{33})\omega_2\omega_3 &= \mathbf{e}_1 \cdot \sum \mathbf{r}_\alpha \times \mathbf{F}_\alpha, \\ I_{22}\dot\omega_2 - (I_{33} - I_{11})\omega_3\omega_1 &= \mathbf{e}_2 \cdot \sum \mathbf{r}_\alpha \times \mathbf{F}_\alpha, \\ I_{33}\dot\omega_3 - (I_{11} - I_{22})\omega_1\omega_2 &= \mathbf{e}_3 \cdot \sum \mathbf{r}_\alpha \times \mathbf{F}_\alpha. \end{aligned} \tag{7.7-5} \bigstar$$

These equations are called *Euler's equations for a rigid body*.

Recall that $\boldsymbol{\omega}$ gives the rotation rate of the body axes relative to the inertial frame. Euler's equations describe how the components of that rotation rate change with time. If we can solve for the components $\omega_j(t)$, then we can go to equation (7.7-2) and determine the vectors $\mathbf{e}_i(t)$. Because the vectors $\mathbf{e}_i$ rotate with the body, once we know them, we have the orientation of the body as a function of time. That is fine in principle, but the program is difficult to carry out in practice. We usually look for some stratagem that will give us an answer without all the labor of the direct approach.

This is not to dismiss Euler's equations. In fact, they prove useful in section 7.8. But Euler's equations play an indispensable role primarily in very complex problems, such as the motion of the moon about its center of mass, where no simple method will suffice.

7.8 AXISYMMETRIC AND TORQUE-FREE

The earth is a rotating body with an axis of symmetry, at least until you look very closely. The same is true for other planets, especially fluid ones such as Jupiter. Because of the equatorial bulge, none of these objects is spherical, but each does have an axis of rotational symmetry. To some extent—or for some purposes—all planets are free from torques (relative to their center of mass). These examples are already sufficient reason to study the symmetric disk shown in figure 7.8-1.

We begin with a sphere rotating uniformly about a fixed axis. The sphere is made of three sections. At $t = 0$, the upper and lower sections are lifted away, leaving behind a disk; its orientation at that instant is shown in the figure. The disk is the object that we will study. The sphere as a whole serves merely to establish unambiguously the disk's angular velocity at $t=0$; it is the vector $\boldsymbol{\omega}(0)$ in the figure. Note especially that $\boldsymbol{\omega}(0)$ does not lie along the disk's symmetry

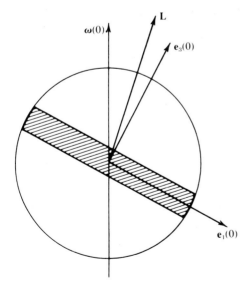

Figure 7.8-1 Until $t = 0$, the disk and the two caps rotated as a single sphere about the axis $\boldsymbol{\omega}(0)$. The vector $\mathbf{e}_1(0)$ lies in the plane defined by $\boldsymbol{\omega}(0)$ and the symmetry axis $\mathbf{e}_3(0)$.

axis, nor does it lie in the plane of the disk. Rather, $\boldsymbol{\omega}(0)$ points in a "general" direction relative to those special orientations.

We place the origin of the body frame at the center of mass and specify no torques from external forces. The internal forces that hold the mass points in their fixed relative separations generate no net torque. Thus the right-hand side of equation (7.1-12) is zero, which implies

$$\mathbf{L} = I_{ij}\omega_j \mathbf{e}_i = \text{constant vector.} \tag{7.8-1}$$

Let us evaluate the double sum at $t = 0$.

Vector $\mathbf{e}_3$ is perpendicular to the plane of the disk; vectors $\mathbf{e}_1$ and $\mathbf{e}_2$ lie in the disk. Symmetry gives to the inertia tensor the form that we worked out in equation (7.6-2):

$$I_{ij} = \begin{pmatrix} I_* & 0 & 0 \\ 0 & I_* & 0 \\ 0 & 0 & I_{33} \end{pmatrix}.$$

We have also the inequalities

$$I_{33} > I_* > 0.$$

With this much information, we can go back to equation (7.8-1) and assert that

$$\mathbf{L} = I_*(\omega_1 \mathbf{e}_1 + \omega_2 \mathbf{e}_2) + I_{33}\omega_3 \mathbf{e}_3 \tag{7.8-2a}$$

at all times and that

$$\mathbf{L} = I_*\omega_1(0)\mathbf{e}_1(0) + I_{33}\omega_3(0)\mathbf{e}_3(0) \tag{7.8-2b}$$

when the right-hand side is evaluated at $t = 0$. Since $I_{33} \neq I_*$, we find that $\mathbf{L}$ is not parallel to $\boldsymbol{\omega}(0)$. The situation is geometrically similar to what we studied in section 7.5, where we found $\mathbf{L}$ not parallel to $\boldsymbol{\omega}$. A difference in directions need not surprise us.

To see where $\mathbf{L}$ points, we first write equation (7.8-2b) as

$$\mathbf{L} = I_*[\omega_1(0)\mathbf{e}_1(0) + \omega_3(0)\mathbf{e}_3(0)] + (I_{33} - I_*)\omega_3(0)\mathbf{e}_3(0).$$

Because $I_{33} > I_*$, we have $\mathbf{L}$ as a linear combination of $\boldsymbol{\omega}(0)$ and $\mathbf{e}_3(0)$ with positive coefficients. Thus the sum that gives $\mathbf{L}$ will lie between $\boldsymbol{\omega}(0)$ and $\mathbf{e}_3(0)$, as shown in figure 7.8-1.

This completes the preliminaries; we can go on to investigate the disk's motion.

Evolution of ω

What can we learn about ω from Euler's equations for a rigid body? Taking the first equation in the set (7.7-5) and bearing in mind our special form for I_{ij}, we get

$$I_*\dot{\omega}_1 - (I_* - I_{33})\omega_2\omega_3 = 0,$$

which we can rearrange as

$$\dot{\omega}_1 = -\frac{I_{33} - I_*}{I_*}\omega_3\omega_2. \tag{7.8-3}$$

The second of Euler's equations gives a similar result:

$$\dot{\omega}_2 = +\frac{I_{33} - I_*}{I_*}\omega_3\omega_1. \tag{7.8-4}$$

Last, the third equation gives

$$I_{33}\dot{\omega}_3 - (I_* - I_*)\omega_1\omega_2 = 0,$$

and so

$$\dot{\omega}_3 = 0.$$

We find that ω_3 retains its original value:

$$\omega_3 = \omega_3(0).$$

The constancy of ω_3 enables us to write equations (7.8-3) and (7.8-4) as

$$\dot{\omega}_1 = -\Omega\omega_2, \tag{7.8-5a}$$

$$\dot{\omega}_2 = +\Omega\omega_1, \tag{7.8-5b}$$

where

$$\Omega \equiv \frac{I_{33} - I_*}{I_*}\omega_3. \tag{7.8-6}$$

As defined here, Ω is a constant with the dimensions of an angular velocity. To solve the pair of coupled equations, we differentiate the first and then use the second to get

$$\ddot{\omega}_1 = -\Omega\dot{\omega}_2 = -\Omega^2\omega_1.$$

In this equation we can recognize a familiar form: the harmonic oscillator, once again. The solution that meets our initial conditions is

$$\omega_1 = \omega_1(0)\cos\Omega t.$$

Then ω_2 follows algebraically from equation (7.8-5a):

$$\omega_2 = \frac{1}{\Omega}\dot{\omega}_1 = \omega_1(0)\sin\Omega t.$$

The summary form is

$$\boldsymbol{\omega} = \omega_1(0)[\cos\Omega t\, \mathbf{e}_1 + \sin\Omega t\, \mathbf{e}_2] + \omega_3(0)\mathbf{e}_3. \tag{7.8-7}\bigstar$$

The vector in square brackets is a unit vector; it rotates with angular frequency Ω relative to axes $\mathbf{e}_1$ and $\mathbf{e}_2$. The implication is this: $\boldsymbol{\omega}$ precesses about the symmetry axis $\mathbf{e}_3$ with angular frequency Ω *as viewed from the body frame*. In section 7.9 we will see how this precession applies to the earth's rotation.

Orientation of the Disk in Space

A question of more intuitive interest is this: How does the orientation of the disk change (relative to the inertial frame)? The question amounts to asking, How does the orientation of the symmetry axis e_3 change with time? After recalling two facts, we can make rapid progress to the answer.

First, we do have one spatially fixed vector in the problem, the angular momentum **L**. Any motion of the symmetry axis ultimately must be motion relative to **L**.

Second, the general equation (6.1-7) applies to e_3:

$$\frac{d\mathbf{e}_3}{dt} = \boldsymbol{\omega} \times \mathbf{e}_3. \tag{7.8-8}$$

Our strategy is to combine these two observations.

We need the connection between **L** and $\boldsymbol{\omega}$. Let us go back to equation (7.8-2a), add $I_*\omega_3 \mathbf{e}_3$ to get $I_*\boldsymbol{\omega}$, and compensate in the last term:

$$\mathbf{L} = I_*\boldsymbol{\omega} + (I_{33} - I_*)\omega_3 \mathbf{e}_3. \tag{7.8-9}\bigstar$$

We can solve this equation for $\boldsymbol{\omega}$:

$$\boldsymbol{\omega} = \frac{\mathbf{L}}{I_*} - \frac{(I_{33} - I_*)}{I_*} \omega_3 \mathbf{e}_3. \tag{7.8-10}$$

When we insert this expression for $\boldsymbol{\omega}$ into equation (7.8-8) and remember that $\mathbf{e}_3 \times \mathbf{e}_3 = 0$, we get

$$\frac{d\mathbf{e}_3}{dt} = \frac{\mathbf{L}}{I_*} \times \mathbf{e}_3. \tag{7.8-11}\bigstar$$

This equation has precisely the structure we studied in section 7.2 and described verbally in equation (7.2-3). The general results of that section give us our answer: The symmetry axis e_3 precesses about **L** with angular frequency L/I_*.

Surprisingly, there is a commonplace example of this precession: a poorly thrown U.S. football. The ball spins, but the ends wobble around in the air. The symmetry axis is precessing around the fixed direction of the angular momentum.

$\boldsymbol{\omega}$ as Seen from the Inertial Frame

Our analysis with Euler's equations gave us the motion of $\boldsymbol{\omega}$ as viewed from the body frame. How does the evolution of $\boldsymbol{\omega}$ appear when it is viewed from the inertial frame?

The answer comes readily. In equation (7.8-10) we have $\boldsymbol{\omega}$ expressed as a linear combination of **L** and $\mathbf{e}_3$. Thus $\boldsymbol{\omega}$ lies in the plane that **L** and $\mathbf{e}_3$ instantaneously define. As $\mathbf{e}_3$ precesses around the spatially fixed vector **L**, so must $\boldsymbol{\omega}$, and it must precess with precisely the same angular frequency, L/I_*. Because the linear combination of **L** and $\mathbf{e}_3$ in equation (7.8-10) is formed with constant coefficients, **ω** has a constant magnitude and maintains a constant angle of inclination relative to **L**.

Summary

This paragraph and figure 7.8-2 summarize what we derived in this section. The $\boldsymbol{\omega}\mathbf{L}\mathbf{e}_3$ plane precesses around **L** with angular frequency L/I_*. Thus **ω** traces the cone centered on **L** (called the *space cone*). Simultaneously, **ω** precesses about $\mathbf{e}_3$ with respect to the body axes; the angular frequency is $\Omega = (I_{33} - I_*)\omega_3/I_*$. Thus **ω** traces the cone centered on $\mathbf{e}_3$ and fixed in the body. (It is called the *body cone*.) The vector **ω** traces the two cones simultaneously. The points of the body that lie (instantaneously) along the direction of **ω** are at rest (relative to the center of mass) because a rotation does not move points on the very axis of rotation. Thus the body cone "rolls without slipping" on the space cone.

The Dust Grain Again

When we studied a spinning dust grain in section 7.2, we took $\boldsymbol{\omega} \propto \mathbf{L}$ but noted that we should later examine the proposition. The disk of this section tells us that we should not expect the proportionality to be literally true. Nonetheless,

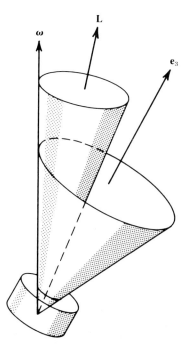

Figure 7.8-2 The body cone (centered on $\mathbf{e}_3$) rolls without slipping on the space cone (centered on **L**). The line of contact gives **ω** as a function of time. Because the body cone is fixed in the disk, the movement of that cone through space shows how the disk's orientation changes.

the proportionality holds in an average sense that is amply sufficient to justify our work. Here is why.

The magnetic torque on the grain is tiny. The characteristic time for it to change **L** is thousands of years. Over an hour, say, **L** is essentially constant. The angular velocity ω is of order 10^4 radians/second, with a period of $2\pi/10^4 \simeq 10^{-3}$ second. Any precession of ω relative to **L** will have a similar characteristic time scale. (We may approximate the shape of a dust grain by a thick disk. Then our work earlier in this section tells us that ω precesses with angular frequency L/I_*. For a cylinder whose height is about equal to its diameter, L is approximately $I_*\omega$, in order of magnitude, and so the precession frequency is approximately ω.) In an hour's time, the precession of ω relative to **L** will give an average ω that lies along **L** and is proportional to L in magnitude. Using the average in a torque expression that has a time scale of 10,000 years is sufficient, and so the step from equation (7.2-4) to equation (7.2-5) follows as an amply justified approximation.

7.9 CHANDLER WOBBLE

Let us ignore for now the small torques that the sun and moon exert on the earth. Then the results of section 7.8 apply. To see them in some detail, we need to estimate the earth's inertia tensor. We may approximate the oblate earth by a sphere plus an equatorial ring of mass. The summation form in equation (7.3-4) implies that the earth's inertia tensor will be composed additively of $(I_{ij})_{\text{sphere}}$ and $(I_{ij})_{\text{ring}}$. We have worked them out in equations (7.6-1) and (7.3-7), respectively. Their sum gives us an approximation to the earth's inertia tensor:

$$(I_{ij})_{\text{earth}} \simeq \tfrac{2}{5} M_{\text{sphere}} R^2 \begin{pmatrix} 1+\tilde{H} & 0 & 0 \\ 0 & 1+\tilde{H} & 0 \\ 0 & 0 & 1+2\tilde{H} \end{pmatrix},$$

where
$$\tilde{H} = \frac{5}{4} \frac{M_{\text{ring}}}{M_{\text{sphere}}}.$$

According to equations (7.8-6) and (7.8-7), the earth's angular velocity ω will precess about the symmetry axis $\mathbf{e}_3$ at the rate

$$\Omega = \frac{I_{33} - I_*}{I_*} \omega_3 = \frac{\tilde{H}}{1+\tilde{H}} \omega_3$$

$$\simeq \frac{5}{4} \frac{M_{\text{ring}}}{M_{\text{sphere}}} \omega. \tag{7.9-1}$$

The approximation in the second line follows because $\tilde{H}$ is much less than 1 and because ω is always very close to the symmetry axis, so that $\omega_3 \mathbf{e}_3$ is almost the whole of ω. This precession is sketched in figure 7.9-1. In section 6.5 we

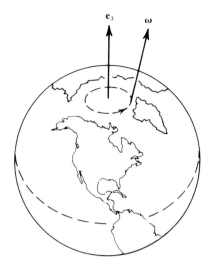

Figure 7.9-1 How ω moves as seen from the body frame, that is, from the viewpoint of us on earth. The size of the precession circle is much exaggerated.

adopted a value $M_{\text{ring}}/M_\oplus = 2.17 \times 10^{-3}$, based on satellite tracking data. If we use this value in equation (7.9-1), we infer

$$\Omega \simeq \tfrac{1}{368}\, \omega,$$

which implies a precessional period of 368 days. We may not take this as more than an estimate because the "spherical part" of the earth is not homogeneous (as in our approximation) nor is the bulge concentrated in an equatorial ring of negligible cross section. Nonetheless, we ought to have a good order-of-magnitude estimate, and indeed we do. Observations that are independent of the motion immediately before us give

$$\frac{I_{33} - I_*}{I_{33}} = \frac{1}{306}.$$

This ratio and the ratio of moments in equation (7.9-1) can differ only by tenths of a percent, and so a period of essentially 306 days is predicted.

The observed period, however, is markedly different. The ω axis moves relative to the surface of the earth with a period of some 427 days. This much longer period was discovered in 1891 by S. C. Chandler, an associate of the Harvard College Observatory. Chandler presented his discovery, meticulously substantiated, in a series of papers in the *Astronomical Journal*. Even as the papers were appearing—literally between installments 3 and 4—Simon Newcomb proposed an explanation for the longer period. In the *Astronomical Journal* of December 23, 1891, Newcomb wrote:

> Mr. Chandler's remarkable discovery, that the apparent variations in terrestrial latitudes may be accounted for by supposing a revolution of the axis of rotation of the earth around that of

figure, in a period of 427 days, is in such disaccord with the received theory of the earth's rotation that, at first, I was disposed to doubt its possibility. But I am now able to point out a *vera causa* which affords a complete explanation of this period.

Newcomb recapitulates the kind of calculation we have made and cites the prediction: a period of 306 days. Then he continues, as follows:

> The question now arises whether Mr. Chandler's result can be reconciled with dynamic theory. I answer that it can, because the theory which assigns 306 days as the time of revolution is based on the hypothesis that the earth is an absolutely rigid body. But, as a matter of fact, the fluidity of the ocean plays an important part in the phenomenon, as does also the elasticity of the earth.

Newcomb goes on to say that when the rotation axis departs from the symmetry axis e_3 of the earth, it carries some of the bulge with it. The solid earth stretches, and the oceans flow a little. The shift in shape is equivalent to reducing the difference $I_{33} - I_*$ and hence lengthens the period.

The motion of ω relative to the earth's surface carries a curiously informal name, *the Chandler wobble*. More precisely, the motion with the 427-day period carries that name; there are variations with other periods, too. For example, there is an annual variation, believed to be caused by a shift in air mass, from over the oceans in summer to over Siberia in winter.

From his stellar observations, Chandler inferred "a revolution of the earth's [rotation axis ω] in a period of 427 days, from west to east, with a radius of thirty feet, measured at the earth's surface." The modern value for the period is a trifle longer, perhaps 2 percent longer, with an uncertainty of similar size. The radius varies, by a factor of 2, in an irregular manner. The motion is damped on a time scale of perhaps 25 years, although that estimate is thoroughly uncertain. Equally uncertain is the energy source that rejuvenates the motion. Earthquakes, as the source, have been proposed by some geophysicists and dismissed by others. Although Newcomb pointed out the essential element—an earth that flexes and flows—a great share of the puzzle remains unsolved.

Although the geophysical problem may remain a puzzle, we can understand what Newcomb meant when he spoke of "the apparent variations in terrestrial latitudes." Understanding that will also help us understand the precessional motions as a whole.

Chandler's observations implied that ω and e_3 are separated by some 10 meters where they poke through the arctic ice. The corresponding separation between ω and L is much less, as we calculate here. An efficient way to calculate the angle between two nearly parallel vectors is to form their vector product. That product is proportional to the sine of the angle between the two vectors, and for a small angle, the sine is approximately equal to the angle itself. Using equation (7.8-9), we get

$$\omega \times L = (I_{33} - I_*)\omega_3\, \omega \times e_3$$

because $\boldsymbol{\omega} \times \boldsymbol{\omega} = 0$. If we divide by the magnitudes ω and L and then form absolute values, we get

$$|\hat{\boldsymbol{\omega}} \times \hat{\mathbf{L}}| = \frac{I_{33} - I_*}{L/\omega_3} |\hat{\boldsymbol{\omega}} \times \mathbf{e}_3|.$$

Because $\mathbf{L}$ is almost along $\mathbf{e}_3$, the ratio L/ω_3 is approximately $(I_{33}\omega_3)/\omega_3$. Thus on the right-hand side we have the ratio $(I_{33} - I_*)/I_{33}$, which we know to be about $\frac{1}{300}$. Therefore the angle between $\boldsymbol{\omega}$ and $\mathbf{L}$ is smaller, by a factor of about $\frac{1}{300}$, than the angle between $\boldsymbol{\omega}$ and $\mathbf{e}_3$. On the earth's surface $\boldsymbol{\omega}$ and $\mathbf{L}$ are separated by only 3 centimeters.

We should summarize the orientational features of the three vectors as noted by an observer in the inertial frame:

1. Vector $\mathbf{L}$ is fixed in space.
2. Vector $\boldsymbol{\omega}$ precesses about $\mathbf{L}$ in a circle of radius 3 centimeters and so is *almost* fixed in space.
3. Vector $\mathbf{e}_3$ precesses about $\mathbf{L}$ in a circle of radius 10 meters.

Consider now the great circle that passes through Cambridge, Massachusetts (where Chandler worked), and through the point where $\mathbf{e}_3$ emerges from the earth's surface. As $\boldsymbol{\omega}$ moves around $\mathbf{e}_3$ (as seen on earth), it crosses the great circle twice. For those two instants, figure 7.9-2 shows the relative

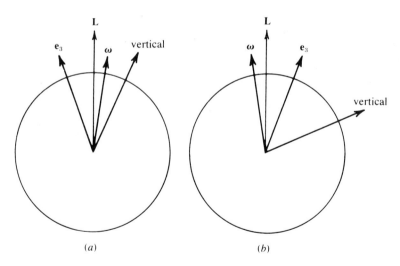

Figure 7.9-2 (*a*) The relative orientations of $\mathbf{L}$, $\boldsymbol{\omega}$, $\mathbf{e}_3$, and the local vertical in Cambridge, Massachusetts, when $\boldsymbol{\omega}$ lies between $\mathbf{e}_3$ and Cambridge on the great circle. (*b*) As in (*a*), but after $\boldsymbol{\omega}$ has precessed around $\mathbf{e}_3$ by 180°, in some $\frac{1}{2}$ (427) days. (To make the diagrams clear, the angles between $\boldsymbol{\omega}$ and $\mathbf{L}$ and between $\mathbf{e}_3$ and $\mathbf{L}$ are greatly exaggerated. The angle between $\mathbf{e}_3$ and the vertical in Cambridge remains constant and is faithfully drawn.)

264 EXTENDED BODIES IN ROTATION

orientations of **L**, **ω**, **e**$_3$, and the vertical at Cambridge. Between the time of one crossing and the other, the earth tips bodily. We may take the direction to the north star to be along the fixed direction **L** (a good approximation here). The tipping changes the angle between the local vertical and the north star. If that angle is taken to define the colatitude, then the colatitude changes. Because the latitude of a city is 90° minus the colatitude, we have here an explanation of why Newcomb spoke of "the apparent variations in terrestrial latitudes."

7.10 AN INTERLUDE ON KINETIC ENERGY

The kinetic energy of a rotating body figures prominently in the theory of molecular energy levels. It is part of the energy conservation law that we will use to analyze a spinning top. Thus an interlude to work out a general expression for such kinetic energy is time well spent.

We need to start with a typical mass m_α in the rigid body and with its velocity as seen from an inertial reference frame:

$$\mathbf{v}_{I\alpha} = \dot{\mathbf{R}} + \boldsymbol{\omega} \times \mathbf{r}_\alpha.$$

The goal is to express the kinetic energy in terms of the two natural velocities—the velocity $\dot{\mathbf{R}}$ of the body frame's origin and the angular velocity of rotation **ω**—together with constants that describe the mass and its distribution throughout the body.

Summing the contributions from all the mass points, we may write the kinetic energy as

$$KE = \sum_\alpha \tfrac{1}{2} m_\alpha \mathbf{v}_{I\alpha} \cdot \mathbf{v}_{I\alpha}$$

$$= \sum_\alpha \tfrac{1}{2} m_\alpha (\dot{\mathbf{R}} + \boldsymbol{\omega} \times \mathbf{r}_\alpha) \cdot (\dot{\mathbf{R}} + \boldsymbol{\omega} \times \mathbf{r}_\alpha). \qquad (7.10\text{-}1)$$

When we expand the scalar product, we get cross terms of the form

$$\sum_\alpha m_\alpha \dot{\mathbf{R}} \cdot (\boldsymbol{\omega} \times \mathbf{r}_\alpha) = \dot{\mathbf{R}} \cdot \boldsymbol{\omega} \times \left(\sum_\alpha m_\alpha \mathbf{r}_\alpha \right).$$

Such terms are numerically zero (1) if we choose the center of mass as the origin of the body frame, so that $\Sigma m_\alpha \mathbf{r}_\alpha = 0$, or (2) if the body has a point spatially fixed in the inertial frame and we place the origin of the body frame there, so that $\dot{\mathbf{R}} = 0$. These conditions appeared in section 7.1; we agreed to impose one or the other thenceforth because they suffice for our purposes (and for most applications). Thus the cross terms are zero for us, and we may drop them.

A few steps simplify the term with the vector products:

$$\sum_\alpha \tfrac{1}{2} m_\alpha (\boldsymbol{\omega} \times \mathbf{r}_\alpha) \cdot (\boldsymbol{\omega} \times \mathbf{r}_\alpha) = \tfrac{1}{2} \sum_\alpha m_\alpha \boldsymbol{\omega} \cdot [\mathbf{r}_\alpha \times (\boldsymbol{\omega} \times \mathbf{r}_\alpha)]$$

$$\begin{aligned}&= \tfrac{1}{2} \boldsymbol{\omega} \cdot \mathbf{L} \\ &= \tfrac{1}{2} \boldsymbol{\omega} \cdot (I_{ij}\omega_j \mathbf{e}_i) \\ &= \tfrac{1}{2} \omega_i I_{ij} \omega_j.\end{aligned} \qquad (7.10\text{-}2)$$

A triple product is unchanged by cyclic permutation of the vectors (as proved in appendix B). If, on the left-hand side, we regard $\boldsymbol{\omega} \times \mathbf{r}_\alpha$, $\boldsymbol{\omega}$, and $\mathbf{r}_\alpha$ as three vectors in a triple product, then we may permute cyclically to get an $\boldsymbol{\omega}$ outside. We can recognize the sum as the angular momentum $\mathbf{L}$, expressed in the form displayed in equation (7.3-3), and so the succinct result $\tfrac{1}{2}\boldsymbol{\omega} \cdot \mathbf{L}$ follows. But we can also expand that result, by using (7.3-6), and thereby arrive at the last line.

Returning now to equation (7.10-1), we may write

$$\text{KE} = \tfrac{1}{2} \left(\sum_\alpha m_\alpha \right) |\dot{\mathbf{R}}|^2 + \tfrac{1}{2} \omega_i I_{ij} \omega_j. \qquad (7.10\text{-}3)\bigstar$$

If we choose the center of mass as the origin of the body frame, we find that the kinetic energy decomposes neatly into the energy of the center-of-mass motion plus the energy of rotation about the center of mass. If we exploit a spatially fixed point and operate under condition 2, then $\dot{\mathbf{R}} = 0$, and the kinetic energy is, of course, purely rotational energy. Section 7.11 provides such an instance, and an application under condition 1 is made in WP7-2.

7.11 THE SYMMETRIC, SUPPORTED TOP

Figure 7.11-1 shows a spinning top. One point is spatially fixed; the tip rests in a slight depression at the apex of the support. We should take the point of support as the origin of the body frame, so that the contact force will drop out of the torque expression. The net torque comes solely from the gravitational forces:

$$\begin{aligned}\sum_\alpha \mathbf{r}_\alpha \times \mathbf{F}_\alpha &= \sum_\alpha \mathbf{r}_\alpha \times (m_\alpha \mathbf{g}) \\ &= \left(\sum_\alpha m_\alpha \mathbf{r}_\alpha \right) \times \mathbf{g}.\end{aligned}$$

We may express the center of mass as

$$\sum_\alpha m_\alpha \mathbf{r}_\alpha = M R_\text{CM} \mathbf{e}_3, \qquad (7.11\text{-}1)$$

where R_CM denotes the distance from the support point to the center of the mass,

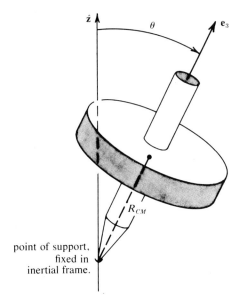

Figure 7.11-1 A spinning top.

which must lie along the symmetry axis $\mathbf{e}_3$. Bearing in mind that $\mathbf{g} = -g\hat{\mathbf{z}}$, we can write the angular momentum equation (7.1-12) in the present context as

$$\frac{d\mathbf{L}}{dt} = -MR_{\text{CM}}g\,\mathbf{e}_3 \times \hat{\mathbf{z}}. \tag{7.11-2}$$

We can solve for the top's motion from this dynamical equation. We will find it easier, however, to use the conserved quantities.

Conserved Quantities

In section 1.7 we developed the consequence of rotational invariance. If an object's potential energy does not change when the object is rotated about some axis, then the component of angular momentum along that axis is conserved in time. We can rotate the top about the z axis (at fixed angle θ, according to the sketch) without changing the gravitational potential energy. That rotational invariance implies

$$\hat{\mathbf{z}} \cdot \mathbf{L} = L_z = \text{constant}. \tag{7.11-3}$$

That is not all, however. The top is symmetric about the axis $\mathbf{e}_3$, and so we may certainly rotate the top about $\mathbf{e}_3$ without changing the potential energy. The implication is

$$\mathbf{e}_3 \cdot \mathbf{L} = L_3 = \text{constant}. \tag{7.11-4}$$

Although the orientation of $\mathbf{e}_3$ may change with time (relative to the inertial frame), the angular momentum maintains a component of constant size along $\mathbf{e}_3$.

Both implications from rotational invariance can be confirmed with equation (7.11-2). They could, of course, be derived from it, but the invariance argument is quicker, at least for L_3.

There is one more conserved quantity: energy. Let us note the kinetic part first. Because the origin of the body frame is fixed in the inertial frame, equation (7.10-3) reduces to $\frac{1}{2}\omega_i I_{ij}\omega_j$, the kinetic energy of rotation.

The gravitational potential energy is Mg times the height of the center of mass, which is $(R_{\text{CM}}\mathbf{e}_3) \cdot \hat{\mathbf{z}}$. We may ignore the potential energy of the forces that hold the top together because that energy does not change. The contact force acts on the point of the body that is spatially fixed, and so it cannot change the energy. In summary, we need include only the gravitational energy.

Energy conservation takes the form

$$\tfrac{1}{2}I_{ij}\omega_i\omega_j + MgR_{\text{CM}}\,\mathbf{e}_3 \cdot \hat{\mathbf{z}} = E. \tag{7.11-5}$$

We now have the three dynamical quantities that are conserved during the top's motion.

How the Top's Orientation Changes

How does the symmetry axis $\mathbf{e}_3$ move? That amounts to asking how the angle θ (between $\mathbf{e}_3$ and $\hat{\mathbf{z}}$) changes and how $\mathbf{e}_3$ precesses about $\hat{\mathbf{z}}$ (at some instantaneous θ). Let us recall that $\boldsymbol{\omega}$ describes how the orientation changes with time and thus must contain the information we need. We can usefully decompose $\boldsymbol{\omega}$ into three orthogonal pieces, as sketched in figure 7.11-2. There is a portion along the symmetry axis, just $\omega_3\mathbf{e}_3$. There is a portion perpendicular to the $\hat{\mathbf{z}}\mathbf{e}_3$ plane, written $\omega_\perp\hat{\mathbf{u}}_\perp$. And there is a portion lying in that plane, denoted $\omega_\|\hat{\mathbf{u}}_\|$. Thus

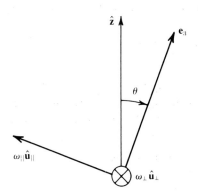

Figure 7.11-2 The decomposition of $\boldsymbol{\omega}$ into three natural orthogonal pieces. The portion $\omega_\perp\hat{\mathbf{u}}_\perp$ points into the paper, perpendicular to the $\hat{\mathbf{z}}\mathbf{e}_3$ plane. The symbols $\hat{\mathbf{u}}_\perp$ and $\hat{\mathbf{u}}_\|$ denote unit vectors.

268 EXTENDED BODIES IN ROTATION

$$\boldsymbol{\omega} = \omega_1\mathbf{e}_1 + \omega_2\mathbf{e}_2 + \omega_3\mathbf{e}_3$$
$$= \omega_\perp\hat{\mathbf{u}}_\perp + \omega_\|\hat{\mathbf{u}}_\| + \omega_3\mathbf{e}_3.$$

The first line gives $\boldsymbol{\omega}$ in terms of cartesian unit vectors; the second, in terms of the unit vectors $\hat{\mathbf{u}}_\perp$, $\hat{\mathbf{u}}_\|$, and $\mathbf{e}_3$.

The derivative $d\theta/dt$ is given by

$$\dot{\theta} = \omega_\perp \tag{7.11-6}$$

because that part of the angular velocity will swing $\mathbf{e}_3$ toward $\hat{\mathbf{z}}$ or away from it. To begin with, can we extract $\dot{\theta}$ from the conservation laws?

The top has the same symmetry as the disk we studied in section 7.8, and so its inertia tensor must have the structure displayed in equation (7.6-2). Energy conservation appears as

$$\tfrac{1}{2}I_*(\omega_1^2 + \omega_2^2) + \tfrac{1}{2}I_{33}\omega_3^2 + MgR_{\text{CM}}\cos\theta = E,$$

while the angular momentum takes the form

$$\mathbf{L} = I_*(\omega_1\mathbf{e}_1 + \omega_2\mathbf{e}_2) + I_{33}\omega_3\mathbf{e}_3.$$

The conservation of $\mathbf{e}_3 \cdot \mathbf{L}$ becomes

$$\mathbf{e}_3 \cdot \mathbf{L} = I_{33}\omega_3 = L_3,$$

and so

$$\omega_3 = \frac{L_3}{I_{33}}. \tag{7.11-7}$$

Thus ω_3 is constant in time, and we may replace it in the energy expression with the ratio L_3/I_{33}. The energy contains also the sum $\omega_1^2 + \omega_2^2$, which we may rewrite as

$$\omega_1^2 + \omega_2^2 = \omega_\perp^2 + \omega_\|^2. \tag{7.11-8}$$

The first term on the right is $\dot{\theta}^2$, the quantity that we are looking for, and the second term we can get from L_z, as follows.

Figure 7.11-2 shows the essential vectors. The conservation of $\hat{\mathbf{z}} \cdot \mathbf{L}$ takes the form

$$\hat{\mathbf{z}} \cdot \mathbf{L} = I_* \hat{\mathbf{z}} \cdot (\omega_\perp\hat{\mathbf{u}}_\perp + \omega_\|\hat{\mathbf{u}}_\|) + I_{33}\omega_3 \hat{\mathbf{z}} \cdot \mathbf{e}_3 = L_z.$$

The sketch tells us that

$$\hat{\mathbf{z}} \cdot \hat{\mathbf{u}}_\perp = 0$$

and

$$\hat{\mathbf{z}} \cdot \hat{\mathbf{u}}_\| = \sin\theta.$$

Solving for $\omega_\|$, we get

$$\omega_\| = \frac{L_z - L_3 \cos\theta}{I_* \sin\theta}. \tag{7.11-9}$$

Now we may use equations (7.11-6) through (7.11-9) in the energy expression to arrive at

$$\frac{1}{2}I_*\left[\dot{\theta}^2 + \left(\frac{L_z - L_3\cos\theta}{I_*\sin\theta}\right)^2\right] + \frac{L_3^2}{2I_{33}} + MgR_{CM}\cos\theta = E. \quad (7.11\text{-}10)$$

We can solve this equation for $d\theta/dt$ as a function of θ and then integrate to find θ as a function of time. We come back to this equation shortly; let us go on for now.

The next question is this: How does $\mathbf{e}_3$ precess about $\hat{\mathbf{z}}$ (at some instantaneous θ)? Figure 7.11-2 is again a help: the portion $\omega_\parallel\hat{\mathbf{u}}_\parallel$ will swing $\mathbf{e}_3$ around in the precessional sense. (And that portion of $\boldsymbol{\omega}$ is the entire story: The part of $\boldsymbol{\omega}$ along $\mathbf{e}_3$ does not move $\mathbf{e}_3$ itself, and the third orthogonal part of $\boldsymbol{\omega}$ is what gives $\dot{\theta}$.) Thus some of what we need is ready for us in equation (7.11-9). In a time Δt, the tip of $\mathbf{e}_3$ will swing around $\hat{\mathbf{z}}$ by a distance $\omega_\parallel \Delta t$. Really the part of $\mathbf{e}_3$ perpendicular to $\hat{\mathbf{z}}$ does the precessing; its magnitude is $\sin\theta$. The angle through which it swings is $\omega_\parallel \Delta t/\sin\theta$. (A top view would be similar to figure 7.2-1.) Dividing the angle by Δt, we get

$$\text{Precessional angular frequency} = \frac{L_z - L_3\cos\theta}{I_*\sin^2\theta}. \quad (7.11\text{-}11)\bigstar$$

Extracting the Implications

Now we have the necessary equations; we can commence extracting their implications.

We may construe equation (7.11-10) as giving kinetic and effective potential energies for a "particle" of "mass" I_* and "coordinate location" θ:

$$\tfrac{1}{2}I_*\dot{\theta}^2 + U_{\text{eff}}(\theta; L_z, L_3) = E, \qquad \bigstar$$

where

$$U_{\text{eff}}(\dot{\theta}; L_z, L_3) = \frac{(L_z - L_3\cos\theta)^2}{2I_*\sin^2\theta} + \frac{L_3^2}{2I_{33}} + MgR_{CM}\cos\theta. \quad (7.11\text{-}12)$$

[margin note: there shouldn't be a dot]

This tactic will help us to think about a complicated equation.

Even so, U_{eff} is difficult to graph for general L_z and L_3. For a start, consider the situation when $L_z = 0$. We can achieve that experimentally by starting the top with $\mathbf{e}_3$ perpendicular to $\hat{\mathbf{z}}$ and with the top spinning purely about $\mathbf{e}_3$, there being no other motion initially. (Some forms of support will permit such a starting angle.) The function U_{eff} reduces to

$$U_{\text{eff}}(\theta; 0, L_3) = \frac{L_3^2}{2I_*}\frac{\cos^2\theta}{\sin^2\theta} + \frac{L_3^2}{2I_{33}} + MgR_{CM}\cos\theta. \quad (7.11\text{-}13)$$

The three contributions and their sum are plotted in figure 7.11-3. Because $\sin\theta$ goes to zero as θ approaches 0 or π, the effective potential energy rises steeply toward infinity at the ends of the angular range.

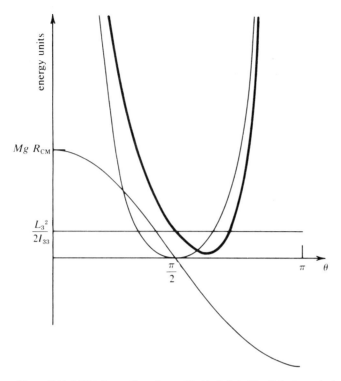

Figure 7.11-3 The heavy line shows $U_{\text{eff}}(\theta; 0, L_3)$. The light lines depict the three contributions in equation (7.11-13).

For the initial conditions described a few lines above, we have $\theta(0) = \pi/2$ and $\dot\theta(0) = 0$. The total energy E is then merely $L_3^2/(2I_{33})$. The "coordinate" θ for the "particle" must oscillate between two turning points, $\theta = \pi/2$ and the point $\theta_{\max}$ where the U_{eff} curve again intersects the total energy line, which is near 0.6π in the figure. The symmetry axis drops at first—that makes sense—but then recovers and oscillates between $\theta = \pi/2$ and $\theta_{\max}$. The oscillation is called *nutation*, from the Latin word for "nodding."

That is not the whole story. Equation (7.11-11) tells us how the symmetry axis precesses about $\hat{z}$. Since $L_z = 0$ in the present situation, the equation reduces to

$$\text{Precessional angular frequency} = \frac{-L_3 \cos \theta}{I_* \sin^2 \theta}. \qquad (7.11\text{-}14)$$

If we start the top spinning positively around $\mathbf{e}_3$, so that $L_3 > 0$, then the factor multiplying $\cos \theta$ is negative. As θ drops from $\pi/2$ toward $\theta_{\max}$, $\cos \theta$ goes negative and precession commences. The precession will be in the positive sense around $\hat{z}$. The precession rate has a fixed sign (when not equal to zero) but a variable magnitude (as θ oscillates). The combined effect of precession and nutation is sketched on the left in figure 7.11-4.

Our expression for the precession rate came from the $\hat{z} \cdot \mathbf{L}$ conservation law. That law can give us a qualitative picture of why precession must occur. As the top drops away from $\theta = \pi/2$, the spin about its symmetry axis gives some angular momentum along $\hat{z}$, really, along $-\hat{z}$. But $\hat{z} \cdot \mathbf{L}$ is zero to start with. Hence the top must move bodily around $\hat{z}$ to keep $\hat{z} \cdot \mathbf{L} = 0$ at all times. That bodily motion is the precession.

Why does the top not fall all the way to $\theta = \pi$? The interplay of angular momentum and energy conservation prevents that from happening. Gravity pulls down on the top; when released, the top commences to fall over. Gravita-

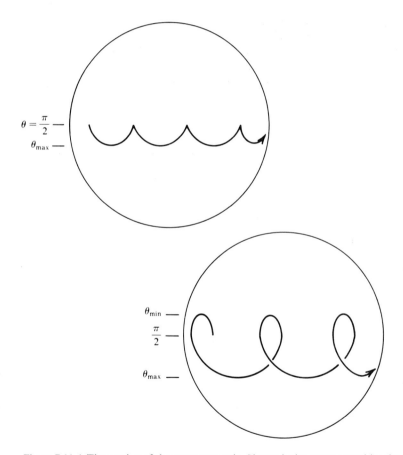

Figure 7.11-4 The motion of the symmetry axis. Shown is the curve traced by the tip of $\mathbf{e}_3$ on the unit sphere. (*a*) The initial conditions are $\theta(0) = \pi/2$, $\dot{\theta}(0) = 0$, and no motion other than spin about $\mathbf{e}_3$. (*b*) Now $\dot{\theta}(0) < 0$, but still $\theta(0) = \pi/2$, and there is no azimuthal motion initially.

tional potential energy is converted to kinetic energy, but the constancy of $\hat{z} \cdot \mathbf{L}$ shunts much of that energy into precessional motion. The situation is then similar (only) to a rubber stopper swung more or less horizontally at the end of a string (as in figure P4-6); the support force plays the role of the string tension.

Now imagine starting the top a bit differently. We will have $\theta(0) = \pi/2$, as before, but we give the top a push upward, so that $\dot{\theta}(0) < 0$. As before, there is no azimuthal or precessional motion at $t = 0$, and so $L_z = 0$ still holds. Because $\dot{\theta}(0) \neq 0$, the total energy is larger than before. Figure 7.11-3 tells us that θ will now oscillate between $\theta_{\min} < \pi/2$ and a new $\theta_{\max}$, somewhat larger than before. Because θ now crosses $\pi/2$ in its oscillations, the cosine in equation (7.11-14) will change sign regularly. The precession will be in the positive sense when $\theta > \pi/2$, as before, and in the negative sense when $\theta < \pi/2$. The ensuing motion of the symmetry axis is sketched on the right in figure 7.11-4.

What we have described so far is the ideal situation. When we try the experiment, we find the nutation and the precession, but the nutation often damps out after a dozen oscillations or so. That is due to friction at the support point. Nutation appears as a transient while the top settles toward uniform precession. But friction at the support also nibbles away L_3, the angular momentum about the symmetry axis. The term in $U_{\text{eff}}(\theta; 0, L_3)$ with the $1/\sin^2\theta$ factor is proportional to L_3^2 and will diminish. That term is crucial in setting the size of $\theta_{\max}$. As L_3 decreases, $\theta_{\max}$ increases. Sooner or later the real support will fail in the role so easily filled by the ideal support: keeping the tip of the top spatially fixed. And so the top slides off the support—but only then "falls down."

If L_z is not zero, the graph of U_{eff} versus θ is qualitatively the same as in figure 7.11-3. The first term in equation (7.11-12) sends U_{eff} to infinity at each end of the angular range, and the curve has only a single minimum. (That minimum may now lie at $\theta < \pi/2$.) To avoid difficulties with the support, we may want to release the top with θ small, 30°, say. If we do that with $\dot{\theta}(0) = 0$ and no azimuthal motion initially, the top will nutate and precess as sketched in the left portion of figure 7.11-4, except that the turning points in θ will be at smaller values.

The situation when $L_z = L_3$ is special: U_{eff} no longer goes to infinity as θ goes to zero, the vertical orientation. The implications are explored in problem 7-20.

7.12 PRECESSION OF THE EQUINOXES

The earth's bulge is tilted relative to the earth's orbital plane, as sketched in figure 7.12-1. The sun pulls more strongly on the nearer portion of the bulge than on the more distant part and thus exerts a net torque on the earth. Certainly the earth is not an object "with one point spatially fixed in an inertial frame," and so we should take torques relative to the earth's center of mass. Doing that in the sketch indicates a net torque perpendicular to the plane of the page and

7.12 PRECESSION OF THE EQUINOXES

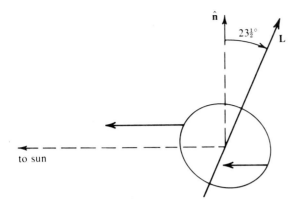

Figure 7.12-1 The sun exerts a torque on the earth's bulge. (The unit vector $\hat{n}$ denotes the perpendicular to the orbital plane.)

toward us. Such a torque will cause **L** to precess around the perpendicular to the orbital plane. Sketches for other locations of the earth in its orbit will show torques that induce precession in the same sense. (There are two exceptions: on the first day of spring and of fall, **L** is perpendicular to the sun-earth line, and symmetry sets the torque to zero.)

We can readily estimate the torque and the precession rate, at least in order of magnitude. The solar gravitational force per unit mass, as it acts on the bulge, is

$$\simeq \frac{GM_\odot}{(r \pm R)^2} \simeq \frac{GM_\odot}{r^2}\left(1 \mp 2\frac{R}{r}\right).$$

where r and R are the earth's orbital and surface radii, respectively. The far and near parts of the bulge differ in distance by roughly $\pm R$. Previously, we regarded the bulge as a ring with mass M_{ring}. Continuing with that description, we can estimate the net solar torque as "R times the solar force on the near part of the ring" minus "R times the solar force on the far part of the ring." This difference leads to the estimate

$$\text{Net solar torque} \simeq R\frac{GM_\odot}{r^2}\frac{R}{r}M_{\text{ring}};$$

factors of 2 and the like are ignored.

In figure 7.12-1, the torque is perpendicular to $\hat{\mathbf{L}}$ and to $\hat{\mathbf{n}}$, the perpendicular to the orbital plane; in short, the torque is proportional to $-\hat{\mathbf{n}} \times \hat{\mathbf{L}}$. If we average the torque over the orbit (that is, over 1 year), the average direction must be similarly expressible in terms of $\hat{\mathbf{L}}$ and $\hat{\mathbf{n}}$ (because r has been averaged). Thus we must be able to write

$$\text{Average solar torque} \simeq -\frac{GM_\odot M_{\text{ring}}R^2}{r^3}\,\hat{\mathbf{n}} \times \hat{\mathbf{L}}. \qquad (7.12\text{-}1)$$

This form is particularly useful when we ask about the precession rate. We have here the torque structure that we studied in section 7.2. If we compare equation (7.12-1) with equation (7.2-2) and recall the conclusions of that section, we can write immediately that

$$\text{Precessional angular frequency} \simeq \frac{GM_\odot M_{ring}R^2}{r^3 L}.$$

For the earth's angular momentum (relative to its center of mass), we can make the estimate $L \simeq M_\oplus R^2 \omega$, based on a spherical approximation to the shape and with all the subtleties of the Chandler wobble ignored. Inserting this, we get

$$\text{Precession frequency} \simeq \frac{GM_\odot M_{ring}R^2}{r^3 M_\oplus R^2 \omega}. \tag{7.12-2}$$

We can simplify this. Since the earth's orbit is essentially circular, Newton II implies [via equation (5.8-7)] that

$$M_\oplus r \omega_{orbit}^2 \simeq \frac{GM_\odot M_\oplus}{r^2}.$$

Thus we can eliminate several factors on the right in equation (7.12-2) in terms of the earth's orbital angular frequency ω_{orbit}:

$$\text{Precession frequency} \simeq \frac{M_{ring}}{M_\oplus} \frac{\omega_{orbit}}{\omega} \omega_{orbit}$$

$$\simeq (2 \times 10^{-3}) \tfrac{1}{365} \omega_{orbit}$$

$$\simeq \frac{1}{2 \times 10^5} \omega_{orbit}.$$

The estimate of $M_{ring}/M_\oplus$ comes from equation (6.5-9).

The precession period, as estimated so far, is of order 2×10^5 years. The moon also exerts a torque on the earth's bulge. The average of that torque is about twice the solar average. The full average is then 3 times as large as the torque we used, which reduces the period by a factor of 3.

The numerical factors that we implicitly set to 1 in the order-of-magnitude calculation work out to a similar reduction. The earth's angular momentum is indeed observed to precess with a period of 26,000 years. (The discovery itself goes back to Hipparchus, in the second century B.C.)

Currently the north star coincides in direction with **L**, but the precession is slowly carrying **L** away from Polaris. Because the tilt (shown in figure 7.12-1) is as large as $23\tfrac{1}{2}°$, the earth's axis will move—in several thousand years—far from the north star. Indeed, in 13,000 years, the direction on the sky will have shifted by more than the size of the constellation that forms the Big Dipper.

As **L** creeps around, the two locations in the earth's orbit where **L** is perpendicular to **r** creep also. They are the locations of equal daytime and nighttime, the equinoxes. The location of the equinoxes precesses as **L** precesses and provides the historical name for the entire phenomenon: the precession of the equinoxes.

7.13 SURVEY OF THE CRITICAL NOTIONS

To describe a rotating extended body, we need equations for location and orientation. This chapter focuses on the latter. Computing the orientation as it evolves in time is complicated because several vectors play essential roles:

1. The orthogonal unit vectors $\mathbf{e}_i$ tell us the instantaneous orientation.
2. The angular velocity $\boldsymbol{\omega}$ describes how the orientation changes with time.
3. The angular momentum $\mathbf{L}$ is the dynamical quantity that changes in response to torques.

In chapter 6 we learned how the rotation alters the body's orientation:

$$\frac{d\mathbf{e}_i}{dt} = \boldsymbol{\omega} \times \mathbf{e}_i.$$

The vectors $\mathbf{L}$ and $\boldsymbol{\omega}$ are not necessarily parallel. The shape and mass distribution of the body determine how much angular momentum along the direction $\mathbf{e}_i$ is produced by a rotation rate ω_j along the direction $\mathbf{e}_j$. The essential information is summarized in the moment-of-inertia tensor:

$$I_{ij} = \sum_\alpha m_\alpha (\delta_{ij} r_\alpha^2 - x_{i\alpha} x_{j\alpha}).$$

The relation between $\mathbf{L}$ and $\boldsymbol{\omega}$ is expressed by equation (7.3-6):

$$\mathbf{L} = I_{ij} \omega_j \mathbf{e}_i.$$

Equation (7.1-12) says that the time rate of change of the angular momentum equals the net torque:

$$\frac{d\mathbf{L}}{dt} = \sum_\alpha \mathbf{r}_\alpha \times \mathbf{F}_\alpha.$$

All orientation equations in the chapter follow from equations (7.1-12) and (7.3-6).

Equations for the time derivatives of components ω_j are worked out in section 7.7. Euler's equations are quite generally applicable, but we can often do better with $\mathbf{L}$ and conservation laws.

For applications that you are likely to meet after a course in mechanics, the equation for simple precession and its solution are probably the most significant. Equation (7.2-2) displayed the structure of the equation:

$$\frac{d\mathbf{L}}{dt} = \mathcal{F} \hat{\mathbf{c}} \times \mathbf{L},$$

276 EXTENDED BODIES IN ROTATION

where $\hat{c}$ is a constant unit vector and $\mathcal{F}$ is some function. The consequences are these:

1. The vector **L** has constant magnitude.
2. It precesses about $\hat{c}$ at a constant angle of inclination and with an angular frequency given by $\mathcal{F}$.

Sometimes a torque will take the form $\mathcal{F}\hat{c} \times \mathbf{L}$ only after an approximation has been made. Computing an average often suffices.

Second in importance is likely to be an expression for the kinetic energy of rotation and translation. We worked out such an expression in section 7.10. When no point in the body is spatially fixed, the natural origin for the body frame is the center of mass. The kinetic energy splits neatly into two contributions:

$$\tfrac{1}{2}\left(\sum_\alpha m_\alpha\right)|\dot{\mathbf{R}}|^2, \quad \text{energy of center-of-mass motion;}$$

$$\tfrac{1}{2}\omega_i I_{ij}\omega_j, \quad \text{energy of rotational motion about center of mass.}$$

There are two kinematic quantities, the velocities $\dot{\mathbf{R}}$ and $\boldsymbol{\omega}$. With each is associated some characterization of the body's mass: the total mass Σm_α with $\dot{\mathbf{R}}$ and the moment of inertia tensor I_{ij} with $\boldsymbol{\omega}$. The way in which the total mass is distributed throughout the body is crucial for the rotational energy.

WORKED PROBLEMS

WP7-1 The stars in an oblate elliptical galaxy provide a mass distribution that we can approximate by a uniform sphere of mass plus an equatorial ring. For a dust grain in orbit within the galaxy, the potential energy is the sum of two contributions. According to figure 1.5-2, the sphere contributes

$$U_{\text{sphere}} = \tfrac{1}{2}kr^2 - \text{const}$$

for some appropriate positive constant k (meant to look like a spring constant because the associated gravitational force is a linear restoring force). For the equatorial ring, we can turn to equation (1.6-6) and abbreviate its contribution as

$$U_{\text{ring}} = -\mathcal{C}(r^2 - 3z^2) - \text{const}',$$

where $\mathcal{C}$ is a positive constant proportional to the ring's mass.

The dust grain is, let us suppose, in almost circular orbit about the galactic center.

Problem. How does the orientation of the grain's orbit change with time (over the long run)? An order-of-magnitude response will suffice.

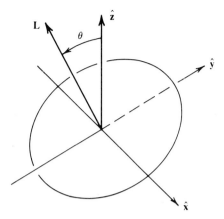

Figure WP7-1 The orbit relative to the galactic equatorial plane.

An orientation question calls for the angular momentum **L**. If a torque tips **L**, the orbital plane will change, too. The dust grain experiences a force

$$\mathbf{F} = -\text{grad } U = -k\mathbf{r} + \mathcal{C}(2\mathbf{r} - 6z\hat{\mathbf{z}}).$$

The radial portions produce no torque (relative to the galactic center). To see what the portion $-6\mathcal{C}z\hat{\mathbf{z}}$ does, let us look at figure WP7-1. That force pushes down where the orbit rises above the galactic equatorial plane; it pushes up when the orbit is below. We can infer an average torque directed along the unit vector $(\mathbf{L} \times \hat{\mathbf{z}})/(L \sin \theta)$.

What is the size of the average torque, in order of magnitude? When z has its maximum value, namely $r \sin \theta$, the magnitude of $\mathbf{r} \times \mathbf{F}$ is $6 \mathcal{C} r \sin \theta \, (r \cos \theta)$ because $|\mathbf{r} \times \hat{\mathbf{z}}| = r \cos \theta$. The average torque is a significant fraction of this product.

For the angular momentum equation, we may write

$$\frac{d\mathbf{L}}{dt} \simeq 6\mathcal{C}r^2 \sin \theta \cos \theta \, \frac{\mathbf{L} \times \hat{\mathbf{z}}}{L \sin \theta},$$

which is valid as an average over one orbital period. The structure is akin to the "simple precession" form we studied in section 7.2. [A transposition of **L** and $\hat{\mathbf{z}}$ in the vector product, which costs only a minus sign, puts the present equation in the form of equation (7.2-2).] The angular momentum will precess about $\hat{\mathbf{z}}$ with a precession frequency given, in order of magnitude, by $(6\mathcal{C}r^2 \cos \theta)/L$.

Because the orbital plane is perpendicular to **L**, the orbital plane will precess about $\hat{\mathbf{z}}$ also.

WP7-2 A molecule consisting of a carbon atom and a sulfur atom rotates about an axis perpendicular to the line connecting the two atoms, as sketched in figure WP7-2. The center of mass is at rest, and the square of the angular momentum is $L^2 = 2.22 \times 10^{-68}$ joule $\cdot$ meter^2 $\cdot$ kilogram. The carbon-sulfur

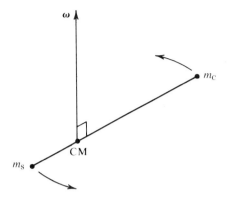

Figure WP7-2 The rotating carbon-sulfur molecule.

separation is $\ell = 1.54 \times 10^{-10}$ meter. To adequate accuracy, the masses are $m_C = 12 m_p$ and $m_S = 32 m_p$, where m_p is the proton's mass.

Problem. What is the kinetic energy of rotation?

To answer the question, we can use equation (7.10-3), provided we have the connection between **L** (whose square is given) and **ω**. A calculation of the moment-of-inertia tensor can follow the pattern in section 7.4. If $\mathbf{e}_1$ points from the center of mass toward the carbon atom along the symmetry axis, then

$$x_{iC} = \frac{m_S}{m_S + m_C} \ell \delta_{i1},$$

and

$$x_{iS} = \frac{-m_C}{m_S + m_C} \ell \delta_{i1}.$$

Using these expressions in equation (7.3-4) leads to

$$I_{ij} = I \begin{pmatrix} 0 & 0 & 0 \\ 0 & 1 & 0 \\ 0 & 0 & 1 \end{pmatrix},$$

where

$$I = \frac{m_C m_S}{m_C + m_S} \ell^2.$$

(We can recognize the combination of masses as the reduced mass.)

The problem specifies that **ω** is perpendicular to the symmetry axis $\mathbf{e}_1$, and so $\omega_1 = 0$. Thus, by equation (7.3-6), we have

$$\mathbf{L} = I_{ij} \omega_j \mathbf{e}_i$$
$$= I(\omega_2 \mathbf{e}_2 + \omega_3 \mathbf{e}_3) = I\boldsymbol{\omega}.$$

Equation (7.10-3) now implies

$$KE = \tfrac{1}{2}I(\omega_2{}^2 + \omega_3{}^2) = \tfrac{1}{2}I\omega^2$$
$$= \frac{1}{2}\frac{L^2}{I}.$$

Numerical evaluation gives $I = 3.46 \times 10^{-46}$ kilogram · meter^2 and $KE = 3.21 \times 10^{-23}$ joule.

The problem specified L^2 because that quantity would determine the allowed rotational energies in a quantum mechanical treatment of the carbon-sulfur molecule.

WP7-3 A disk of constant density ρ has a radius R and a height h.

Problem. Compute the moment-of-inertia tensor. The origin of the body axes is to be placed at the center of mass, and the unit vector e_3 is to lie along the disk's symmetry axis. Qualitatively, a disk can be thin, like a pancake, or elongated, like a cigar. What value of the ratio h/R separates these two regimes?

The moment-of-inertia tensor for the disk has the form displayed in equation (7.6-2). We need compute only I_{33} and I_*.

We may decompose the disk into annuli and add their contributions. Because I_{ij} is defined by a sum [in equation (7.3-4)], such a superposition principle holds. The "summation" is best done by integration in cylindrical coordinates. We need to distinguish the radius variable in cylindrical coordinates from the position magnitude r. Let us write

$$r_{\text{cyl}} \equiv \sqrt{x^2 + y^2}.$$

The element I_{33} follows as

$$I_{33} = \sum_\alpha m_\alpha (x_\alpha{}^2 + y_\alpha{}^2) = \int_{z=-h/2}^{+h/2} \int_{r_{\text{cyl}}=0}^{R} r_{\text{cyl}}{}^2 \rho \, 2\pi r_{\text{cyl}} \, dr_{\text{cyl}} \, dz$$

$$= 2\pi\rho \frac{R^4}{4} h = \tfrac{1}{2} M_{\text{disk}} R^2.$$

Because the elements I_{11} and I_{22} must be equal, we can use the same trick that we used for a ring in section 7.3. We work out the sum:

$$I_{11} + I_{22} = \int_{z=-h/2}^{+h/2} \int_{r_{\text{cyl}}=0}^{R} (r_{\text{cyl}}{}^2 + 2z^2) \rho \, 2\pi r_{\text{cyl}} \, dr_{\text{cyl}} \, dz$$

$$= 2\pi\rho \left[\frac{R^4}{4} h + 2\frac{R^2}{2} \frac{2}{3}\left(\frac{h}{2}\right)^3 \right]$$

$$= \frac{1}{2} M_{\text{disk}} R^2 \left(1 + \frac{1}{3}\frac{h^2}{R^2}\right).$$

Therefore

$$I_* = \frac{1}{4} M_{\text{disk}} R^2 \left(1 + \frac{1}{3}\frac{h^2}{R^2}\right).$$

Now that we have expressions for I_{33} and I_* in terms of h and R, we can compare their magnitudes. The ratio I_*/I_{33} is

$$\frac{I_*}{I_{33}} = \frac{1}{2}\left(1 + \frac{1}{3}\frac{h^2}{R^2}\right).$$

This ratio is less than 1 if $h/R < \sqrt{3} = 1.73$.

A kinematic distinction between a thin disk and an elongated disk can be extracted from equation (7.8-9). If I_{33} is greater than I_*, then **L** lies between **ω** and **e**$_3$. If I_{33} is less than I_*, then **L** lies outside. Thus the ratio $I_*/I_{33} = 1$ forms the boundary between the two regimes. Our calculation says that the boundary is equivalent to the ratio $h/R = \sqrt{3}$.

PROBLEMS

7-1 The nucleus of the helium isotope $^3_2$He acts as a tiny spinning bar magnet. In an externally imposed magnetic field **B**, the nucleus experiences a torque γ**L** × **B**, where **L** is the nuclear angular momentum and the constant γ is -2.04×10^8 per tesla · second. In a typical laboratory field of 0.6 tesla, how long does it take for the nucleus to precess 90°?

7-2 If the polar ice caps were to melt, the level of the seas would rise, the increase in height being $\Delta h \simeq 60$ meters. Estimate the change that would ensue in the length of the day. (Describe your simplifying assumptions as you make them.)

7-3 A satellite circles the earth, moving close to its surface. The orbit is inclined at 45° to the earth's equatorial plane. The earth's equatorial bulge will exert a torque on the satellite. Use a sketch to help you to determine the direction of the average torque (i.e., the torque averaged over one round trip). Next, use dimensional analysis to estimate the magnitude of the average torque. At what rate, expressed symbolically, does the satellite's angular momentum precess? If **L** precesses through a full cycle in 7 weeks, how massive is the earth's equatorial bulge (relative to $m_\oplus$), in order of magnitude?

7-4 How would the results in section 7.4 change if the rod had a nonnegligible mass $m_{\rm rod}$? That mass is distributed uniformly along length ℓ.

7-5 A piece of sheet metal, in the shape of an equilateral triangle, is pivoted at one vertex and swings in the plane of the triangle. Why must the frequency of small oscillations be some numerical multiple of $(g/\ell)^{1/2}$, where ℓ is the length of a side? Determine the numerical factor.

7-6 A disk is suspended by a stiff wire, aligned along the disk's symmetry axis. The wire, when twisted, produces a restoring torque proportional to the angular displacement θ from the equilibrium orientation: torque $= -k'\theta$, where k' is a torsional constant. The disk has a radius of 0.1 meter and a mass of 0.2 kilogram. The measured oscillation period is 6 seconds.

(a) Use the data to calculate the torsional constant k'.

(*b*) Now a U.S. football is suspended vertically below the disk, so that the two objects rotate as a single rigid body. The oscillation period increases to 9.82 seconds. What is the U.S. football's moment of inertia for rotation about its symmetry axis?

(*c*) When the U.S. football is suspended horizontally, the period increases again, to 11.70 seconds. What is I_{33}/I_* for the football?

7-7 To teach circular motion in an introductory laboratory, one often swings a rubber stopper around at the end of a string. The string traces a conical surface. Suppose one replaces the string with a rigid rod (but a massless one), pivoted at one end, as sketched in figure P7-7. The rotating system consists of the massless rod (of length ℓ) and the concentrated mass m. At the instant shown, the velocity of the mass is into the paper.

(*a*) Describe or calculate the angular momentum **L** of the system and the gravitational torque on it.

(*b*) Calculate the rotational angular velocity as a function of θ, ℓ, and g by *starting with the dynamical connection between angular momentum and torque*. Be sure to include careful, faithful sketches.

(*c*) Check your result in part (*b*) by using a more elementary method to calculate the angular velocity.

7-8 Return to the shell and four masses of section 7.5. Although ω is constant in time, the angular momentum **L** is not (because $\mathbf{e}_3$ moves through space). Calculate the torque that is being applied externally to maintain the shell and masses in uniform rotation. For which angles of inclination θ need no torque be applied? Explain with sketches.

T **7-9** *Transforming the inertia tensor.* Suppose we change from one set of body axes, the set $\mathbf{e}_i$, to another set, the set $\mathbf{e}'_i$. The origins are specified to coincide, and so the change in axes is to a set rotated relative to the original set. How is the new moment-of-inertia tensor related to the old one?

For mass m_α, the new position coordinate $x'_{i\alpha}$ is given by

$$x'_{i\alpha} = \mathbf{e}'_i \cdot \mathbf{r}_\alpha = \mathbf{e}'_i \cdot (x_{k\alpha}\mathbf{e}_k)$$

$$= \mathbf{e}'_i \cdot \mathbf{e}_k x_{k\alpha}.$$

Can you use this connection to prove that

$$I'_{ij} = (\mathbf{e}'_i \cdot \mathbf{e}_k)(\mathbf{e}'_j \cdot \mathbf{e}_\ell) I_{k\ell}?$$

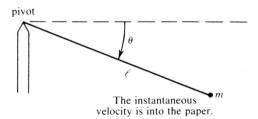

The instantaneous velocity is into the paper.

Figure P7-7

If you are familiar with matrices, can you cast this transformation law into matrix multiplication form?

For any given set of body axes fixed in a rigid body, the moment-of-inertia tensor is an array of constants. The array undergoes no dynamical change. If new axes are chosen, however, then the array changes (but the new values are dynamically constant, too).

T **7-10** *Parallel-axis theorem.* Suppose we displace the origin of the body axes from the center of mass to a location a distance ℓ away. The orientation of the axes remains unaltered. Can you show that the new moment-of-inertia tensor I_{ij} is given by

$$I_{ij} = M(\ell^2 \delta_{ij} - \ell_i \ell_j) + (I_{ij})_{\text{CM}}?$$

Here M denotes the body's total mass, and $(I_{ij})_{\text{CM}}$ is the inertia tensor calculated with the center of mass as origin. The connection between inertia tensors is called the *parallel-axis theorem*. When a body has symmetries, so that $(I_{ij})_{\text{CM}}$ is easy to calculate, but the required origin is at an awkward location, such as on the periphery of the body, the theorem is handy.

7-11 A U.S. football has the symmetry of a disk but with $I_{33} = 0.6 I_* < I_*$. We may take $I_* = 3 \times 10^{-3}$ kilogram · meter². Suppose that when the ball has been poorly thrown, the symmetry axis $\mathbf{e}_3$ describes a cone in space with a half-angle of 10°. The axis makes six circuits of the cone per second, as seen from the ground.

(*a*) Calculate the angular momentum L, the component ω_3 of the angular velocity along the symmetry axis, and the magnitude ω.

(*b*) Suppose an insect were riding along on the ball. At what rate would the insect see $\boldsymbol{\omega}$ precess about the symmetry axis? Would the precession be in the same sense as the rotation $\boldsymbol{\omega}$ itself?

(*c*) Sketch the space and body cones, analogous to those in figure 7.8-2. Why is there a qualitative difference?

7-12 The average rotational kinetic energy of a dust grain is $\tfrac{3}{2}kT$, where k is Boltzmann's constant, 1.38×10^{-23} joule/kelvin, and T is the temperature of the interstellar dust cloud. A value of $T = 80$ kelvins would be typical. Dust grains have irregular shapes; let us merely say the grain is about 5×10^{-7} meter across. The density of grains is similar to that of ice or silicate rock; take a density of 2×10^3 kilograms per cubic meter. With this information, estimate the rotational frequency of a grain (to within a factor of 3 or so).

7-13 Suppose the novel pendulum of section 7.4 and its support were inverted, so that the rod stood above the pivot. For a picture, just turn figure 7.4-1 by 180°. The vertical orientation is unstable. You balance the rod as best you can and then let it go. Calculate the angular velocity $\dot\theta$ as a function of orientation. What is its value when the rod hits the surface, having tipped over 90°? (Note that you need not solve a complicated dynamical equation for θ as a function of time.) Could you generalize your method to an inverted pendulum with a more complicated distribution of mass?

T**7-14** A disk such as the one in section 7.8 rotates about its center of mass, which is at rest. Derive an expression for the kinetic energy in terms of L^2 and L_3, the angular momentum along the the symmetry axis, together with the two constants I_* and I_{33} that characterize the moment-of-inertia tensor. (The structure carries over directly to the quantum mechanical treatment of a molecule or nucleus with the same symmetry.)

7-15 Use equation (7.11-2) to confirm conservation laws (7.11-3) and (7.11-4). Bear in mind that vector $\mathbf{e}_3$ moves as seen from the inertial frame.

T**7-16** *The fast top.* If the top in section 7.11 spins very fast around its symmetry axis, then $\mathbf{L} \simeq L_3\mathbf{e}_3$ may be a good approximation. (This amounts to ignoring—here—the contributions of nutation and precession to $\mathbf{L}$.) Given the approximation, we can write the torque in terms of $\mathbf{L}$ rather then $\mathbf{e}_3$. What do you predict for the motion of a "fast top" in this approximation? If the top is about the size of your fist and you spin it so that $\omega_3/(2\pi) = 30$ revolutions/second, estimate the precession frequency.

T**7-17** *The top more formally.* This problem outlines more formal routes to equations (7.11-6) and (7.11-11).

(a) To derive an expression for $\dot{\theta}$ in terms of $\boldsymbol{\omega}$, start with $\cos\theta = \hat{\mathbf{z}} \cdot \mathbf{e}_3$. Differentiate both sides with respect to time, and then use equation (6.1-7) on the right. Can you solve for $\dot{\theta}$ as $\dot{\theta} = \boldsymbol{\omega} \cdot$ (some unit vector)?

(b) We can compute the top's precessional angular frequency by determining the precessional motion of the unit vector $\hat{\mathbf{u}}_\perp$, equal to $\hat{\mathbf{z}} \times \mathbf{e}_3/\sin\theta$. Calculate $d\hat{\mathbf{u}}_\perp/dt$, using equation (6.1-7). That step will introduce $\boldsymbol{\omega}$; write it in terms of $\hat{\mathbf{u}}_\parallel$, $\hat{\mathbf{u}}_\perp$, and $\mathbf{e}_3$. With the aid of figure 7.11-2 and equation (7.11-6), reduce your equation to

$$\frac{d\hat{\mathbf{u}}_\perp}{dt} = \frac{\omega_\parallel}{\sin\theta}\hat{\mathbf{z}} \times \hat{\mathbf{u}}_\perp.$$

How does this result confirm equation (7.11-11)?

T**7-18** The context is section 7.11. Suppose we started the top with $\dot{\theta}(0) = 0$ but with some azimuthal motion, by giving the top a gentle sideways push. How would the symmetry axis move? Only a qualitative discussion and a sketch are asked for.

7-19 Is it possible to start the top (of section 7.11) so that it precesses only? (This means "no nutation" or $\theta =$ constant.) Can you describe the conditions on L_3 and L_z that must be met? (There is no need to work them out in detail.)

T**7-20** *The sleeping top.* Suppose the top in section 7.11 is initially vertical, so that $L_z = L_3$ and $\dot{\theta}(0) = 0$. Examine U_{eff} near $\theta = 0$, calculating its first and second derivatives at $\theta = 0$ as well as its value. How fast must the top be spinning if the vertical orientation is to be stable (against perturbation by a puff of wind that produces a tiny $\dot{\theta} \neq 0$)? When spinning sufficiently fast, the top remains —seemingly motionless—in the vertical orientation; it "sleeps" there.

7-21 A dumbbell of length D, with a mass m at each end, is suspended from a

284 EXTENDED BODIES IN ROTATION

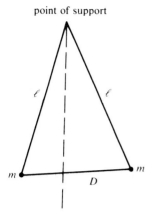

Figure P7-21

point by two strings, each of length ℓ. (See figure P7-21.) The dumbbell is in gentle oscillation in the plane of the sketch. The strings remain taut throughout the motion. Moreover, their mass and that of the rod connecting the two end masses are negligible relative to m. Compute the frequency of the small oscillations, expressing your final answer in terms of no parameters other than m, g, ℓ, and D. Does your result reduce to a familiar expression if you let D go to zero?

7-22 The nucleus of the iodine isotope $^{127}_{53}$I is slightly cigar-shaped and spins rapidly about its symmetry axis. Suppose the atomic electrons produce an electric field which, in the volume occupied by the nucleus, may be expressed as

$$\mathbf{E}(x, y, z) = C[-(x\hat{\mathbf{x}} + y\hat{\mathbf{y}}) + 2z\hat{\mathbf{z}}]$$
$$= C[-(x\hat{\mathbf{x}} + y\hat{\mathbf{y}} + z\hat{\mathbf{z}}) + 3z\hat{\mathbf{z}}],$$

where C is a positive constant. (The field has cylindrical—but not spherical—symmetry because the distribution of atomic electrons has the first symmetry but not the second.) For purposes of estimation, we may regard the nucleus as a sphere (of charge Q_s and radius R) centered on the origin plus two thin caps (of charge Q_c each) where the angular momentum axis emerges from the nucleus. That axis is inclined (initially, at least) at an angle θ relative to the $\hat{\mathbf{z}}$ axis.

Can you use symmetry to show that the net electric force on the nucleus is zero? Next, estimate the magnitude and direction of the net torque. What is the resultant nuclear motion like? Finally, for which quantities would you need numerical values in order to estimate numerically the natural frequency in the problem?

7-23 Suppose a torque is applied to a disk whose height-to-radius ratio is $\sqrt{3}$. Would the disk's response to the torque be different from that of a sphere (having the same radius and a mass 1.25 times larger)? Why or why not?

T7-24 *Intermediate-axis theorem.* Consider an object with no rotational symmetry, such as a tennis racket. When principal axes are used for the body axes, the moment-of-inertia tensor will be diagonal, but all three diagonal elements usually differ from one another in numerical value.

(a) Suppose the object is set spinning with $\omega(0)$ *precisely* aligned along one principal axis, which we may call the e_3 axis, for definiteness. No torques act. Will the object continue to rotate purely about that body axis?

(b) Suppose $\omega(0)$ is *almost* aligned with e_3. Components $\omega_1(0)$ and $\omega_2(0)$ are nonzero but small relative to $\omega_3(0)$. Is such rotation stable? That is, will ω continue to lie close to e_3? You will need to examine three situations: I_{33} is the largest, smallest, or intermediate moment of inertia.

Euler's equations provide a route to the answers. You can test your predictions by tossing a tennis racket or a book held closed with an elastic band.

T **7-25** *Angular momentum relative to the origin of the inertial frame.* In sections 7.1 and 7.3 we computed torques and the angular momentum by forming vector products with the displacement r_α from the origin of the *body* frame. What form do the equations take if you use the displacement $R + r_\alpha$ from the origin of the *inertial* frame, under the specification that the origin of the body frame is placed at the center of mass? After you have eliminated those terms that are identically zero, can you interpret the remaining terms?

7-26 In a classical picture, an electron moves around a massive nucleus with orbital angular momentum L_{orbit}. As an object with intrinsic angular momentum, the electron has angular momentum L_{spin} relative to its center of mass. When the spinning charge moves through the spherically symmetric electric field produced by the nucleus, torques arise, so that L_{spin} obeys the equation

$$\frac{dL_{spin}}{dt} = \mathcal{C}\, L_{orbit} \times L_{spin}.$$

For a nearly circular orbit, we may take $\mathcal{C}$ to be a constant.

(a) On general principles, what equation would you expect L_{orbit} to satisfy?

(b) For general initial conditions, determine the evolution of L_{spin} and L_{orbit} as far as you can.

7-27 A dumbbell-shaped satellite circles the Earth at a distance r_0. In general, what kind of tumbling or oscillatory motion should we expect? (Motion about the satellite's center of mass is meant here.) Be quantitative where you can be. (You are welcome to simplify by specifying that all motion is confined to the orbital plane.)

CHAPTER EIGHT

CROSS SECTIONS

8.1 Scattering effectiveness: The idea behind the cross section
8.2 A capture cross section
8.3 A differential cross section
8.4 Rutherford scattering
8.5 Major ideas

I had observed the scattering of alpha particles, and Dr. Geiger in my laboratory had examined it in detail. He found, in thin pieces of metal, that the scattering was usually small, of the order of one degree. One day Geiger came to me and said, "Don't you think that young Marsden, whom I am training in radioactive methods, ought to begin a small research?" Now I had thought that, too, so I said, "Why not let him see if any alpha particles can be scattered through a large angle?"

Lord Rutherford
Background to Modern Science

8.1 SCATTERING EFFECTIVENESS: THE IDEA BEHIND THE CROSS SECTION

A young hot star produces a fierce stellar wind, a stream of particles moving radially outward into space. (Particles in the stellar surface absorb light, thus acquiring momentum from the photons, and recoil outward, forming the wind.) Suppose a dust grain finds itself in this wind. Some of the hydrogen atoms in the

8.1 SCATTERING EFFECTIVENESS: THE IDEA BEHIND THE CROSS SECTION

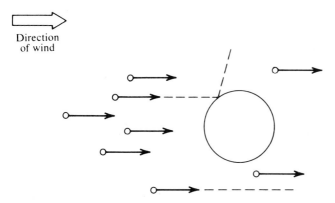

Figure 8.1-1 A dust grain in a wind of hydrogen atoms. The grain will scatter some atoms (the bent dashed line depicts the trajectory of one such atom) but leave other atoms unaffected, as indicated by the undeviated dashed line.

wind will bounce off the grain; others, though nearby, will miss the grain. The situation is sketched in figure 8.1-1.

How many scatterings per second should we expect? First we need to determine which initial trajectories lead to a collision. Let R denote the grain's radius, and let R_H denote the radius of a hydrogen atom. If an initial trajectory lies within a distance $R + R_H$ of the centerline, as sketched in figure 8.1-2, the atom will scatter off the grain.

An atom must be in a specific volume if it is to strike the grain within the next second. In time Δt, an atom moves a distance $v \Delta t$. Thus, with $\Delta t = 1$ second, the atom must be somewhere in the cylindrical volume $\pi(R + R_H)^2 \, v$ to the left of the grain. If the number of hydrogen atoms per unit volume is n_H, we can write

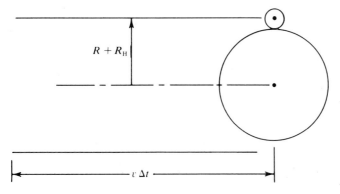

Figure 8.1-2 An initial trajectory within the perpendicular distance $R + R_H$ of the centerline will lead to a collision.

$$\text{Number of scatterings per second} = n_H v \pi (R + R_H)^2. \qquad (8.1\text{-}1)$$

If the wind becomes more intense, through an increase in n_H or v or both, the scattering rate goes up. The grain is not more effective at scattering than before; it just gets more chances to scatter. To define an honest measure of *scattering effectiveness*, we need to divide the scattering rate by some measure of the scattering opportunities. A good choice is the product $n_H v$, called the *incident flux*. It is the number of projectile atoms that flow across unit area (oriented perpendicular to the wind's direction) per second. Thus

$$\text{Measure of scattering effectiveness} = \frac{\text{number of scatterings per second}}{\text{incident flux}}$$

$$= \pi (R + R_H)^2. \qquad (8.1\text{-}2)$$

The ratio that provides the effectiveness measure has the units of an area and is called the *cross section* for scattering. Our cross section depends on the scatterer (through R) and on the incident particle (through R_H). It is not a property of either separately.

Thinking of the cross section as an area that the scatterer presents to the projectile wind is a good idea. Any particle whose incident trajectory would intersect the area will be scattered. We can recover—or compute—the scattering rate from the product of the cross section and the incident flux. Remember, however, that the scattering cross section is not the same thing as the *geometrical cross section* of the target. The distinction is evident here: $\pi(R + R_H)^2$ versus πR^2.

Later in this chapter we study the capture of incident ions by a charged dust grain. We study the deflection of alpha particles by a heavy nucleus. These are two more instances of a pervasive context: a "target" object does something to some "projectile" particles in an incident beam. The "something" can be scattering or capture or molecule formation or .... Call it, for short, an event of type X. We can always define a measure for the effectiveness of the process by a ratio:

$$\frac{\text{Effectiveness measure}}{\text{for process } X} = \frac{\text{number of events of type } X \text{ per second}}{\text{incident flux}}. \quad \bigstar$$

The ratio is then called the *cross section for events of type X*. A common symbol would be σ_X.

To see the notation in use, we can write equation (8.1-2) as

$$\sigma_{\text{scattering}} = \pi (R + R_H)^2.$$

What good is a cross section? There are at least three facets to the answer. First, a measured cross section conveys information about the interaction of projectile and target and about their individual properties. In our example, we could extract $R + R_H$, the sum of the radii, from a measurement of $\sigma_{\text{scattering}}$. Second, a theory about the behavior of the projectile and target particles can be tested by comparing the predicted and measured cross sections. This strategy has been a major theme in physics since the time of Rutherford during the first decades of this century. Third, with a known cross section we can compute the event rate to be expected for any specified incident flux.

8.2 A CAPTURE CROSS SECTION

An interstellar dust grain is often charged, usually by an excess of electrons; so let us endow our grain with an electric charge $Z(-e)$, where Z is some positive integer and $-e$ is the charge on an electron. (Plain e denotes the proton's charge.) The stellar wind contains a variety of particles, not just hydrogen atoms. Let us focus on some positive ions: charge $Z_i e$, mass m, and incident speed v_0. If such an ion strikes the surface of our grain, it will—let us suppose—be held there: it has been captured. What is the cross section for such capture?

Figure 8.2-1 sketches the situation when the ion is still far from the grain and shows also the trajectory that will be followed. The attractive electric force between ion and grain bends the initial trajectory toward the grain. The distance b is the decisive quantity. As sketched, b is small enough that capture occurs. If b were made much larger, however, the trajectory would bend in but then swing around behind the grain, and the ion would depart to infinity: no capture. This possibility is indicated by the sequence of short dashes in the figure. The crucial question for us is this: How small a value must b have if the ion is to strike the grain?

To answer the question, we can use an effective potential energy to tell us the minimum distance between the ion and the dust grain on a trajectory that is specified by b and the other parameters in the problem. Equation (5.8-4) gives the structure of such an energy as

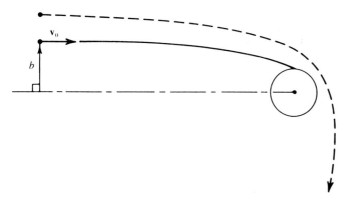

Figure 8.2-1 Two trajectories of the ion relative to the grain.

$$U_{\text{eff}}(r; L) = \frac{L^2}{2mr^2} + U(r),$$

where **r** denotes the ion's position relative to the center of the dust grain. Because the grain is so much more massive than the ion, we may regard it as permanently at rest, and the reduced mass becomes merely m, the ion's mass.

To evaluate the angular momentum L, we can use figure 8.2-2. Reasoning from the lengths and angles drawn there, we have

$$L = |m\mathbf{r}_0 \times \mathbf{v}_0| = mv_0 r_0 \sin \beta$$
$$= mv_0 b \qquad (8.2\text{-}1)$$

because $\beta = \beta'$ and $b = r_0 \sin \beta'$.

The electrical attraction varies as $1/r^2$, and so the corresponding Coulomb potential energy will vary as $1/r$:

$$U = \frac{K(-Ze)(Z_i e)}{r}.$$

Here K is the constant in Coulomb's law, dependent for its numerical value on the units used for charge; the most common choices are listed in appendix A.

The total energy E can be evaluated most easily in the initial configuration where the distance is so large that $U = 0$. Thus

$$E = \tfrac{1}{2} m v_0^2.$$

The run of U_{eff} with r and the total energy line are sketched in figure 8.2-3. The minimum separation along the trajectory occurs where the curve and line intersect. If we write L in terms of b, we have, at the minimum,

$$\frac{(mv_0 b)^2}{2m r_{\min}} + U(r_{\min}) = E.$$

The value of b for which $r_{\min}$ equals R, the grain's radius, is the critical value. Calling it b_c, we can solve for it from

$$\frac{(mv_0 b_c)^2}{2mR^2} + U(R) = \tfrac{1}{2} m v_0^2,$$

which yields

$$b_c^2 = R^2 [1 - U(R) / (\tfrac{1}{2} m v_0^2)].$$

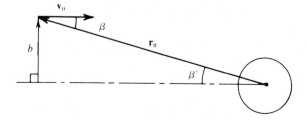

Figure 8.2-2 The initial configuration. Because the centerline is drawn parallel to $\mathbf{v}_0$, angles β and β' are equal.

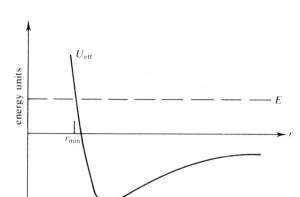

Figure 8.2-3 A graph of the effective potential energy for the ion.

All ions with $b > b_c$ miss the grain; those with $b \leq b_c$ strike it and are captured. In effect, the grain holds up a circular area πb_c^2 to the distant wind and captures all ions whose initial trajectory would take them through it. Thus we can write the capture cross section as

$$\sigma_{\text{capture}} = \pi b_c^2$$
$$= \pi R^2 \left[1 + \left(\frac{KZZ_i e^2}{R}\right) \middle/ (\tfrac{1}{2} m v_0^2)\right].$$

The cross section depends, we find, on the energy of the ions. If that energy is much larger than the potential energy (in absolute value) at the grain's surface, then σ_{capture} is approximately πR^2, the geometric cross section. We can understand that. The trajectory for an ion of very large energy will be almost a straight line, and so the ion must be headed for a direct collision, independent of the electrical attraction, if it is to be captured. If, however, the ion moves slowly when far away, then the electrical attraction can pull it in to a capture even if b is very large. Hence the cross section rises as the energy decreases.

The quantity b has been a valuable tool in our analysis and will continue in that role throughout the chapter. Its definition is best extracted from figure 8.2-2: the perpendicular distance from the initial trajectory to a parallel line running through the target's center. Its common name, the *impact parameter*, arises because b specifies where a *straight-line* trajectory would pierce a plane at the target, a plane oriented perpendicular to the centerline.

8.3 A DIFFERENTIAL CROSS SECTION

Let us return now to the neutral dust grain and hydrogen atoms. A glance at figure 8.1-1 is enough to suggest that atoms are scattered in various directions. Some atoms are deflected only a little; others go off at angles near 90°; still

292 CROSS SECTIONS

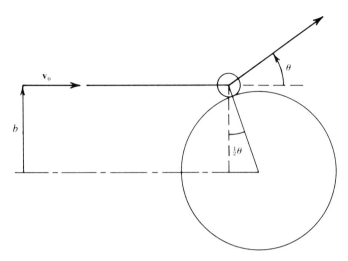

Figure 8.3-1 Relating the scattering angle to the impact parameter.

others are scattered almost dead backward. The angle of scattering depends on the impact parameter. Let us derive the exact relationship.

Figure 8.3-1 displays the essential geometry. The angle θ is the scattering angle, the angle through which the interaction deflects the initial trajectory. The distance between centers at the instant of collision is $R + R_H$. Provided the atom and grain act as hard, smooth spheres, the area of contact acts as a locally flat surface. During impact, the component of velocity along the connecting line is reduced to zero and then is reversed. The perpendicular component is not changed. Figure 8.3-2 shows how the reversal transmutes the incident velocity v_0 to the final velocity.

From the figure we can also see why the angle labeled $\frac{1}{2}\theta$ there and in figure 8.3-1 has that value. The angle between v_{final} and v_0 is θ. Thus the angle between $v_\perp$ and v_0 is $\frac{1}{2}\theta$. That angle, plus the angle between v_0 and the connecting line, equals 90°. But the latter angle, plus the angle labeled $\frac{1}{2}\theta$, also equals 90°. Thus the label indeed denotes the true value of the angle.

Now we can express the scattering angle in terms of the impact parameter. The triangle with angle $\frac{1}{2}\theta$ in figure 8.3-1 lets us write

$$b = (R + R_H) \cos \tfrac{1}{2}\theta. \tag{8.3-1}$$

Moreover, we can see how b must change to give scattering through an infinitesimally larger angle $\theta + d\theta$. Differentiating equation (8.3-1), we find

$$db = (R + R_H)(-\tfrac{1}{2} \sin \tfrac{1}{2}\theta) \, d\theta. \tag{8.3-2}$$

The change db is negative, implying a smaller impact parameter. If we look at figure 8.3-1 and mentally reduce b, we see that, indeed, the scattering angle will increase.

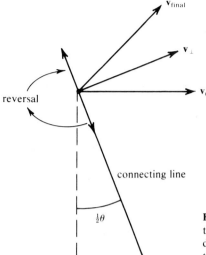

Figure 8.3-2 The reversal of the velocity component that lies along the connecting line. The vector $\mathbf{v}_\perp$ denotes the component of velocity perpendicular to the connecting line.

Figure 8.3-3 is a useful prelude to the next step. It shows a sphere of radius $\mathcal{R} \gg R$ centered on the dust grain. Two circles are drawn by rotating the radius vector, first at polar angle θ and then at $\theta + d\theta$. That generates a band. Finally a section of the band is marked off by an increment $d\varphi$ in the azimuthal angle: arcs at φ and $\varphi + d\varphi$. The patch thus defined is an infinitesimal rectangle with sides $\mathcal{R}\, d\theta$ and $\mathcal{R} \sin\theta\, d\varphi$. It defines a certain set of directions: those straight lines that emanate from the origin and pass through the patch. To characterize that bundle of directions, we specify angles θ and φ together with the ratio (patch area)/$\mathcal{R}^2$. In analogy with radian measure for angles in the plane, the ratio is called the infinitesimal *solid angle* and is denoted $d\Omega$:

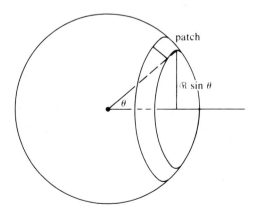

Figure 8.3-3 A sphere of radius $\mathcal{R}$ and the construction of the infinitesimal solid angle $d\Omega = \sin\theta\, d\theta\, d\phi$.

$$d\Omega \equiv \frac{\text{patch area}}{\mathcal{R}^2} = \frac{(\mathcal{R}\, d\theta)(\mathcal{R}\sin\theta\, d\varphi)}{\mathcal{R}^2}$$
$$= \sin\theta\, d\theta\, d\varphi. \qquad (8.3\text{-}3)$$

We can now inquire about the cross section for scattering into some certain $d\Omega$, meaning that the final velocity lies in the bundle of directions specified by $d\Omega$. Where must the impact parameter lie?

Figure 8.3-4 is a help here. If we think of the position vector **r** to the projectile, the polar angle changes greatly as the trajectory is followed, but the azimuthal angle φ remains constant (because there is, after all, no asymmetry in the problem to change it). As the projectile disappears toward infinity, the polar angle becomes the same as the angle between $\mathbf{v}_{\text{final}}$ and $\mathbf{v}_0$. Thus we may use the symbol θ for both the scattering angle in figure 8.3-1 and the ultimate polar angle. The upshot is that b must lie in a rectangular patch with sides $b\, d\varphi$ and $|db|$, where db is given by equation (8.3-2). The patch is the area that, in effect, is held up to the projectile wind for scattering into $d\Omega$. Thus we may write

Cross section for scattering into $d\Omega = b\, d\varphi\, |db|$ \qquad (8.3-4) ★

$$= (R + R_H)^2 \tfrac{1}{2} \sin\tfrac{1}{2}\theta \cos\tfrac{1}{2}\theta\, d\theta\, d\varphi.$$

The right-hand side has the dimensions of an area, as a cross section should, and is proportional to $d\theta\, d\varphi$ for infinitesimal angles.

If we divide the left-hand side of equation (8.3-4) by the size of $d\Omega$, we get a cross section per unit solid angle, written $d\sigma/d\Omega$ and called the *differential scattering cross section*. Thus

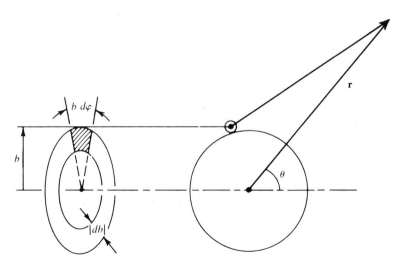

Figure 8.3-4 Range of impact parameters that yield scattering into $d\Omega$.

$$\frac{d\sigma}{d\Omega} = \frac{\text{cross section for scattering into } d\Omega}{d\Omega} \quad (8.3\text{-}5) \bigstar$$

$$= \frac{(R + R_H)^2 \tfrac{1}{2} \sin \tfrac{1}{2}\theta \cos \tfrac{1}{2}\theta \, d\theta \, d\varphi}{\sin \theta \, d\theta \, d\varphi}$$

$$= \tfrac{1}{4}(R + R_H)^2.$$

The last step follows from the trigonometric identity (A.2-5). The complete disappearance of θ is *not* typical of scattering problems; in section 8.4 a strong dependence on θ will remain. For hard spheres, however, the cross section per unit solid angle is the same for all scattering directions. The scattered particles are distributed uniformly in direction. In short, the scattering is isotropic.

From the differential cross section we can compute the total cross section by integration:

$$\sigma = \int_{\text{sphere}} \frac{d\sigma}{d\Omega} \, d\Omega$$

$$= \int_{\varphi=0}^{2\pi} \int_{\theta=0}^{\pi} \tfrac{1}{4}(R + R_H)^2 \sin \theta \, d\theta \, d\varphi$$

$$= \tfrac{1}{4}(R + R_H)^2 2\pi \, (-\cos \theta)\big|_0^\pi$$

$$= \pi(R + R_H)^2.$$

A comparison with the cross section that we derived in equation (8.1-2) gives reassuring agreement.

8.4 RUTHERFORD SCATTERING

The epigraph that introduced this chapter is worth rereading now. Rutherford set young Marsden to studying the scattering of alpha particles by gold atoms. Rutherford did not anticipate any novel results, or even much scattering. Along with the rest of his generation, he believed that the positive charge in the gold atom was spread over a volume of atomic size. The positively charged alpha particle would be repelled by all the gold's positive charge while outside that charge. Once the particle was inside the charge distribution, however, the force would diminish because only a fraction of the gold's charge would be effective—that fraction in a sphere of radius equal to the distance between the alpha particle and the center. The closer the alpha particle got to the center, the less the force. The implication was clear to Rutherford: Expect little scattering. Here is the continuation of Rutherford's story.

> I may tell you in confidence that I did not believe that they would be, since we knew that the alpha particle was a very fast, massive particle, with a great deal of energy, and you could show that if the scattering was due to the accumulated effect of a number of small scatterings, the chance of an alpha particle's being scattered backward was very small. Then I remember

two or three days later Geiger coming to me in great excitement and saying, "We have been able to get some of the alpha particles coming backward..." It was quite the most incredible event that has ever happened to me in my life. It was almost as incredible as if you fired a 15-inch shell at a piece of tissue paper and it came back and hit you. On consideration, I realized that this scattering backward must be the result of a single collision, and when I made calculations I saw that it was impossible to get anything of that order of magnitude unless you took a system in which the greater part of the mass of the atom was concentrated in a minute nucleus. It was then that I had the idea of an atom with a minute massive center carrying a charge.*

The scattering at large angles—some even backward—forced Rutherford to change his picture of how the positive charge is distributed. To generate large forces, he reasoned, the charge must be concentrated in a tiny volume. From this recognition arose the idea of the atomic nucleus. Rutherford went on to calculate the differential scattering cross section that we should expect. Let us see what it looks like.

The alpha particle has a charge Ze, with $Z = 2$; the gold nucleus, a charge $Z'e$, with $Z' = 79$. The repulsive force between the two positively charged entities generates a positive Coulomb potential energy:

$$U = \frac{K(Ze)(Z'e)}{r}$$

$$\equiv \frac{\mathcal{C}}{r},$$

where $\mathcal{C}$ is merely an abbreviation for the numerator.

A typical trajectory is sketched in figure 8.4-1. The electric force repels the alpha particle, unlike the gravitational force between the earth and the sun, which is attractive. Nonetheless, both forces are $1/r^2$ forces with $1/r$ potential energies. Thus we may carry over many results derived in section 5.4. We need merely replace $-Gm_1m_2$ by $+\mathcal{C}$ and be careful about minus signs.

The orbit of the alpha particle relative to the gold nucleus has the form displayed in equation (5.4-9):

$$\frac{1}{r} = \frac{1}{\alpha} + A \cos(\theta + \varphi_0).$$

A transcription of equation (5.4-6) gives the parameter α as

$$\alpha = \frac{L^2}{\mu(-\mathcal{C})}$$

$$= \frac{-L^2}{m\,\mathcal{C}};$$

the gold nucleus is so massive that the reduced mass μ may be replaced

---

*Lord Rutherford in *Background to Modern Science*, J. Needham and W. Pagel (eds.) (Cambridge University Press, Cambridge, England, 1938).

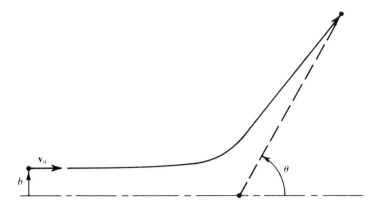

Figure 8.4-1 The repulsion between the alpha particle and the gold nucleus scatters the incident particle.

with the mass m of the alpha particle. We can work out the quantity A by proceeding step by step through the derivation in section 5.4. We can best keep the signs straight by inserting $\mathcal{C}$ and the expression for α as early as possible. The outcome is

$$A = +\frac{m\mathcal{C}}{L^2}\epsilon,$$

where

$$\epsilon = \left(1 + \frac{2EL^2}{m\mathcal{C}^2}\right)^{1/2} > 1.$$

Here E is the total energy of the alpha particle, which we can evaluate from the initial conditions:

$$E = \tfrac{1}{2}mv_0^2 + 0$$

because $U(\infty) = 0$.

Putting together the pieces, we find

$$\frac{1}{r} = \frac{m\mathcal{C}}{L^2}\left[\epsilon \cos(\theta + \varphi_0) - 1\right]. \tag{8.4-1}$$

The eccentricity ϵ is greater than 1, implying a hyperbolic orbit. Indeed, we may not allow the polar angle θ to run through a full 2π because the right-hand side would run through negative values, which is nonsensical for the inverse of a radius. What range for θ is possible?

The initial conditions, according to figure 8.4-1, correspond to $\theta = \pi$; that is, as we push the left-hand portion of the sketch farther to the left (at constant b), the polar angle to the initial location descends toward π. We can insert the initial configuration into equation (8.4-1) and thereby determine the constant φ_0:

$$\frac{1}{\infty} = \frac{m\mathcal{C}}{L^2}\left[\epsilon \cos(\pi + \varphi_0) - 1\right],$$

which implies

$$\varphi_0 = -\pi + \arccos\frac{1}{\epsilon},$$

where we may take the inverse cosine to have a positive value.

With φ_0 in hand, we may return to equation (8.4-1) and write it as

$$\frac{1}{r} = \frac{m\mathcal{C}}{L^2}\left[\epsilon \cos\left(\theta - \pi + \arccos\frac{1}{\epsilon}\right) - 1\right]. \quad (8.4\text{-}2)$$

Let us start with $\theta = \pi$ and decrease θ, following the trajectory. At first the argument of the cosine decreases from a net value of $\arccos(1/\epsilon)$ to zero. The value zero gives $\cos 0 = 1$ and therefore the largest value of $1/r$. We are at the angular location where the alpha particle is closest to the nucleus. As θ decreases further, the argument of the cosine becomes negative. The cosine is an even function of its argument, however, and so some negative values of the argument are acceptable. The full argument may decrease until it reaches $-\arccos(1/\epsilon)$. Then the right-hand side of equation (8.4-2) vanishes, implying that r has gone to infinity. The value of θ that sends r to infinity is the scattering angle; so we may compute it from the relation

$$\theta - \pi + \arccos\frac{1}{\epsilon} = -\arccos\frac{1}{\epsilon}.$$

Thus the scattering angle is given by

$$\theta = \pi - 2\arccos\frac{1}{\epsilon}. \quad (8.4\text{-}3)$$

The dependence of the scattering angle on the impact parameter is hidden in ϵ. We can reveal it by recalling, from equation (8.2-1), that $L = mv_0 b$, so that ϵ may be expressed as

$$\epsilon = \left[1 + \frac{2E(mv_0 b)^2}{m\mathcal{C}^2}\right]^{1/2}$$

$$= \left(1 + \frac{4E^2 b^2}{\mathcal{C}^2}\right)^{1/2}. \quad (8.4\text{-}4)$$

If we solve equation (8.4-3) for ϵ, we find, as a first step,

$$\arccos\frac{1}{\epsilon} = \frac{1}{2}\pi - \frac{1}{2}\theta.$$

Then, taking the cosine of each side, we get

$$\frac{1}{\epsilon} = \cos(\tfrac{1}{2}\pi - \tfrac{1}{2}\theta)$$
$$= \cos\tfrac{1}{2}\pi \cos(-\tfrac{1}{2}\theta) - \sin\tfrac{1}{2}\pi \sin(-\tfrac{1}{2}\theta)$$
$$= +\sin\tfrac{1}{2}\theta.$$

Combining this expression for ϵ with equation (8.4-4) gives

$$\left(1 + \frac{4E^2 b^2}{\mathcal{C}^2}\right)^{1/2} = \frac{1}{\sin\tfrac{1}{2}\theta}. \tag{8.4-5}$$

To increase θ, it looks as if we need to decrease b. Let us check by differentiating:

$$\frac{1}{2}\left(1 + \frac{4E^2 b^2}{\mathcal{C}^2}\right)^{-1/2} \frac{4E^2}{\mathcal{C}^2} 2b \, db = -\frac{\cos\tfrac{1}{2}\theta}{\sin^2\tfrac{1}{2}\theta} \frac{1}{2} d\theta. \tag{8.4-6}$$

When θ lies in the range $0 < \theta < \pi$, the angle $\tfrac{1}{2}\theta$ lies in the first quadrant, where the cosine is positive. Thus positive $d\theta$ requires negative db, as we expected. It makes sense: to increase the scattering angle, we need to aim the initial trajectory closer to the gold nucleus.

In section 8.3 we learned that the cross section for scattering into solid angle $d\Omega$ can be computed as $b \, d\varphi \, |db|$; that was formulated as equation (8.3-4). Therefore we can use equations (8.4-5) and (8.4-6) to write

$$\text{Cross section for scattering into } d\Omega = \frac{\mathcal{C}^2}{8E^2} \frac{\cos\tfrac{1}{2}\theta}{\sin^3\tfrac{1}{2}\theta} d\theta \, d\varphi.$$

To get the differential cross section, we divide by $d\Omega$ and use the identity $\sin\theta = 2\sin\tfrac{1}{2}\theta \cos\tfrac{1}{2}\theta$:

$$\frac{d\sigma}{d\Omega} = \frac{\mathcal{C}^2}{8E^2} \frac{\cos\tfrac{1}{2}\theta}{\sin^3\tfrac{1}{2}\theta} \frac{d\theta \, d\varphi}{\sin\theta \, d\theta \, d\varphi}$$
$$= \frac{(KZZ'e^2)^2}{16E^2} \frac{1}{\sin^4\tfrac{1}{2}\theta}. \tag{8.4-7}\bigstar$$

What emerges is a differential cross section that depends on both energy and scattering angle. The dependence on angle is suggested by table 8.4-1. It is very strong—indeed, violent. The coulomb force diminishes slowly with distance—only as $1/r^2$—and thus extends to infinity. It affects particles that pass far away from the force center. To be sure, it does not deflect those distant particles much, but it affects a huge number, so many that an integral for the total cross section diverges. We could try to define a total cross section σ by

$$\sigma \equiv \lim_{\theta_{\text{least}} \to 0} \int_{\varphi=0}^{2\pi} \int_{\theta=\theta_{\text{least}}}^{\pi} \frac{d\sigma}{d\Omega} d\Omega,$$

where θ_{least} is some least scattering angle that we are considering. If we look back to the first line of equation (8.4-7), we can infer that the double integral that

Table 8.4-1 How the differential cross section varies with scattering angle

The last angle, 1 milliradian, corresponds to the angle subtended by 1 millimeter at the distance of 1 meter. It is a small angle, but readily measured.

θ	$\dfrac{d\sigma/d\Omega}{(KZZ'e^2)^2/(16E^2)}$
π	1
$\dfrac{\pi}{2}$	4
$\dfrac{\pi}{4}$	47
0.1	1.6×10^5
0.01	1.6×10^9
0.001	1.6×10^{13}

defines σ is proportional to

$$\int_{\theta_{\text{least}}}^{\pi} \frac{\cos\tfrac{1}{2}\theta}{\sin^3\tfrac{1}{2}\theta}\,d\theta = -\frac{1}{\sin^2\tfrac{1}{2}\theta}\bigg|_{\theta_{\text{least}}}^{\pi}$$

$$= \frac{1}{\sin^2\tfrac{1}{2}\theta_{\text{least}}} - 1.$$

This diverges as θ_{least} goes to zero.

We calculated $d\sigma/d\Omega$ with classical mechanics. What emerges if the analysis is done with quantum theory? Remarkably, the very same expression. The exact agreement between quantum and classical cross sections, however, is a peculiarity of the $1/r^2$ force law. Typically, the two theories give different results, although the classical result can be found as a limit of the quantum mechanical expression. Nowhere is the distinction more striking than in the total cross section σ. If the force is radial and proportional to $1/r^n$ at large distances, then σ_{quantum} is finite provided $n > 3$. The classical cross section, however, diverges for all positive n. (A way to derive that classical conclusion is outlined in problem 8-5.)

8.5 MAJOR IDEAS

A summary of the major ideas is in order. A cross section is a measure of effectiveness, an event rate for some process divided by the incident flux. The incident flux is itself a measure of the opportunities for the process to occur. The cross section is the ratio of the event rate to the opportunity rate.

Thinking of the cross section as an area that the target presents to the projectile wind is a good idea. Any particle whose incident trajectory would intersect that area will undergo the process, will "become an event." The larger the cross section, the larger the area held up and hence the larger the event rate. We can recover—or compute—the event rate from the product of the cross section and the incident flux.

In classical physics, once the force law or potential energy is specified, the trajectory is determined by the energy E and the impact parameter b. To calculate a cross section (at given E), we need to determine the range of b's that lead to that kind of occurrence.

For a capture process, the range of b's is typically from $b = 0$ to some critical b, denoted b_c. Then

$$\sigma_{\text{capture}} = \pi b_c^2, \qquad (8.5\text{-}1)$$

and we need merely to determine b_c by studying the trajectories in detail.

For scattering into a narrow bundle of directions characterized by θ, φ, and $d\Omega$, we go to figure 8.3-4. A patch of area $b\, d\varphi\, |db|$ is, in effect, held up to the projectile wind for scattering into $d\Omega$. Thus equation (8.3-4) follows:

$$\text{Cross section for scattering into } d\Omega = b\, d\varphi\, |db|. \qquad (8.5\text{-}2)$$

In each specific context we need to determine θ as a function of b and E. Then we can compute $|db|$ by differentiating the relation $b = $ function of θ.

If we want the cross section per unit solid angle, we divide by $d\Omega$ and write

$$\frac{d\sigma}{d\Omega} = \frac{b\, d\varphi\, |db|}{\sin\theta\, d\varphi\, d\theta}. \qquad (8.5\text{-}3)$$

The $d\varphi$'s cancel, of course, and $|db|$ is proportional to $d\theta$; their ratio comes from differentiating the relation $b = $ function of θ. Thus the right-hand side is a well-defined ratio of infinitesimals.

What good is a cross section? The question was raised in the last paragraph of section 8.1 and was answered there. Here is a succinct recapitulation. (1) A measured cross section provides information about the interaction of projectile and target and about their individual properties. (2) A theory about the behavior of the projectile and target particles can be tested: we compare the predicted and measured cross sections. (3) With a known cross section we can compute the event rate to be expected for any specified incident flux.

Without cross sections, modern physics would be unthinkable. Indeed, their role is preeminent: the particle accelerators of today—grown to dimensions measured in kilometers—attest to that.

WORKED PROBLEMS

WP8-1 An air molecule is roughly spherical, with a radius R_{air} of about 2×10^{-10} meter. The number of such molecules per unit volume is about $n = 3 \times 10^{25}$ molecules per cubic meter.

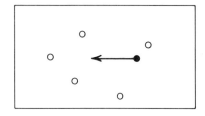

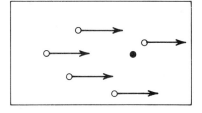

Figure WP8-1 The chosen air molecule, and some others, as seen from the laboratory and from the molecule itself. (The random velocities of the other molecules may be ignored because they average to zero.)

Problem. On the average, how far does an air molecule travel between collisions with other such molecules?

Suppose the chosen air molecule has velocity **v**, as in figure WP8-1. To turn the problem into a typical scattering context, let us ride with the chosen air molecule. (It provides an inertial reference frame between collisions.) Then our molecule is at rest but subject to a flux of magnitude nv. If the scattering cross section is σ, then incident molecules are scattered at a rate of

$$\text{Flux} \times \sigma = nv\sigma$$

events per second. Let τ be the time we must wait (on the average) to get one scattering event. Then

$$nv\sigma\tau = 1,$$

and so

$$\text{mfp} \equiv v\tau = \frac{1}{n\sigma},$$

where mfp denotes the *mean free path,* the average distance traveled between collisions.

The cross section, based on a hard-sphere model and section 8.1, is

$$\sigma = \pi (R_{\text{air}} + R_{\text{air}})^2$$
$$= 5 \times 10^{-19} \text{ meter}^2.$$

Thus

$$\text{mfp} = 7 \times 10^{-8} \text{ meter}.$$

An aside. How does the mfp compare with the average separation between air molecules? The cube of the average separation times the number density n

must yield one molecule. Thus the average separation is $n^{-1/3} = 3 \times 10^{-9}$ meter. Comparing this with the mfp, we find that the typical molecule goes about 20 times the average molecular separation between collisions.

WP8-2 *Ion-molecule reactions.* If a neutral molecule comes near a positive ion, the ion's electric field distorts the molecule: the molecule's electrons are pulled closer while the nuclei are pushed away. The near end of the molecule becomes a bit negative; the far end, a bit positive, by default. This is sketched in figure WP8-2. Because the negative end of the molecule is closer to the ion and hence is in a stronger field, the electrical attraction there dominates the repulsion at the distant end. There is a net attractive force.

We can work out the associated potential energy. The distorted molecule acts as though it consisted of charges $\pm q$ separated by a small distance s. Thus

$$U = \frac{-KQ_i q}{r - \tfrac{1}{2}s} + \frac{KQ_i s}{r + \tfrac{1}{2}s}$$

$$\simeq \frac{-KQ_i qs}{r^2},$$

after we use the binomial expansion on $[1 \pm s/(2r)]^{-1}$, given that $s/r \ll 1$. Here Q_i denotes the ion's charge. The distortion is produced by the ion's electric field, and so the product qs must be proportional to that field, i.e., proportional to $1/r^2$. In short, we can write

$$U = -\frac{\mathcal{K}}{r^4}$$

for some positive constant $\mathcal{K}$.

If the molecule and ion come into contact, let us suppose they react chemically.

Problem. Estimate the cross section for reaction, given $\mathcal{K}$ and energy $E = \tfrac{1}{2}mv_0^2$ of the molecule. (The ion is relatively massive and remains at rest.)

We can pattern the calculation after the work in section 8.2. Picking one molecule from a wind of molecules incident on our ion would lead to a picture like figure 8.2-2. The molecule has angular momentum $L = mbv_0$. How close to

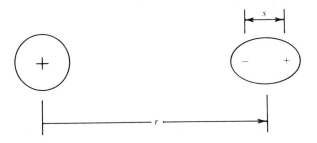

Figure WP8-2 Distortion of the neutral molecule by the ion.

the ion will it get? The effective potential energy is

$$U_{\text{eff}}(r;L) = \frac{L^2}{2mr^2} - \frac{\mathcal{K}}{r^4}. \tag{1}$$

A graph of this function already exists: the right-hand side of figure 5.6-1. If the energy E lies above the maximum in U_{eff}, the molecule will pass over the hump, plummet down the well, and react. If E lies below the maximum, there will be a turning point; we can suppose the molecule leaves without having reacted.

We need to determine the impact parameter for which the maximum in U_{eff} equals E. The maximum occurs where

$$\frac{dU_{\text{eff}}}{dr} = -\frac{L^2}{mr^3} + \frac{4\mathcal{K}}{r^5} = 0,$$

that is, at

$$r = \left(\frac{4\mathcal{K}m}{L^2}\right)^{1/2}.$$

Inserting this into equation (1) gives U_{eff} at its maximum:

$$U_{\text{eff}}\bigg|_{\text{max}} = \frac{1}{16}\frac{L^2}{\mathcal{K}m^2}.$$

After we set this equal to E and write $L = mb_c v_0$, we can solve for the critical value b_c:

$$b_c = \left(\frac{4\mathcal{K}}{E}\right)^{1/4}.$$

All molecules with $b < b_c$ will go over their respective humps and react. Thus the reaction cross section is πb_c^2:

$$\sigma_{\text{reaction}} = \pi\left(\frac{4\mathcal{K}}{E}\right)^{1/2}.$$

PROBLEMS

8-1 Another gas, helium, is added to the air in WP8-1. A helium atom has a radius R_{He} of about 1.5×10^{-10} meter. The helium number density is $n_{\text{He}} = 10^{26}$ atoms per cubic meter; the number density n_{air} of air molecules is kept the same as earlier. Calculate the new mean free path for an air molecule.

8-2 A beam of xenon atoms strikes a target volume with 10^8 target xenon atoms. The incident flux is 5×10^8 atoms/(meter^2 · second). In an hour of observing time, 117 scattering events are detected.
  (a) What is the event rate per second and per target atom?
  (b) Determine the scattering cross section.
  (c) What value do you infer for the radius of a xenon atom?

8-3 The context is section 8.2, except that now we want the cross section for capturing an additional electron when the grain already has an excess of Z electrons. Develop an expression for that capture cross section.

Suppose the grain's radius R is 3×10^{-7} meter. What is the capture cross section for electrons whose energy equals $2KZe^2/R$? What is it for those whose energy equals $\frac{1}{3}KZe^2/R$?

8-4 A gold atom is bombarded by alpha particles, each of which has an energy of 4 million electronvolts, equivalent to 6.4×10^{-13} joule.

(a) Compute the cross section for scattering through an angle of $\pi/2$ or more, that is, for scattering into a direction in the backward hemisphere.

(b) The gold foils used by Rutherford had a thickness of about 5×10^{-7} meter. An area of 1 square centimeter, times that thickness, yields a volume with about 10^{18} gold atoms. If this area is bombarded with an incident flux of 10^8 alpha particles per meter² · second, how many scatterings into the backward hemisphere would you expect to see in a minute of observing?

T **8-5** *Classical small-angle scattering.* If the impact parameter is large, the projectile will be deflected only a little. The trajectory will be almost a straight line. This is sketched in figure P8-5. To get a good approximation to the scattering angle, we can divide the y momentum that the particle acquires by the original x momentum. And we can compute the final y momentum (to adequate accuracy) by approximating the trajectory with a straight line. In short,

$$\theta_{\text{scattering}} \simeq \frac{1}{mv_0} \int_{-\infty}^{\infty} \mathbf{F}(\mathbf{r}) \cdot \hat{\mathbf{y}} \, dt,$$

and we replace dt by dx/v_0, integrating over $-\infty \leq x \leq +\infty$ at $y = b$.

(a) Suppose the force is a radial power law: $\mathbf{F} = (\mathcal{C}/r^n)\hat{\mathbf{r}}$. Here n is a constant greater than 1 but not necessarily an integer, and $\mathcal{C}$ is some other constant. Can you show that

$$\theta_{\text{scattering}} = \frac{\mathcal{C}}{E} b^{1-n} \times \begin{pmatrix} \text{integral dependent} \\ \text{on } n \text{ but independent of } b \end{pmatrix}?$$

You need only develop this form, not evaluate the integral fully (which, for general n, is impossible to do in terms of elementary functions).

(b) Compute the differential scattering cross section $d\sigma/d\Omega$ as a function of the scattering angle. (A constant, known in principle but not in practice, will

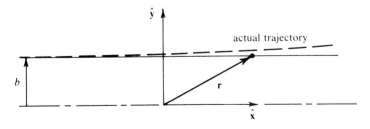

Figure P8-5

carry over from the equation above.) Does your result agree with section 8.4 if you set $n = 2$?

(c) Does $d\sigma/d\Omega$ diverge as $\theta \to 0$? Does $\int (d\sigma/d\Omega)\, d\Omega$ diverge? Can you relate the behavior of the integral over θ for small θ to a corresponding integral of $b\, d\varphi\, |db|$? Consequently, can you see why divergence is inevitable?

8-6 The radius of a gold nucleus is about 6×10^{-15} meter. When an alpha particle gets that close, the specifically nuclear forces come into play, in addition to the electric force. The differential scattering cross section will differ somewhat from what we calculated.

(a) For which direction of scattering would you expect $d\sigma/d\Omega$ to change first as you increased the energy of the incident alpha particles?

(b) Estimate the energy at which a change would first appear.

T **8-7** *Beam attenuation.* The metal cadmium is remarkably good at absorbing slow neutrons. For the isotope $^{113}_{48}\text{Cd}$, the capture cross section is $\sigma = 2 \times 10^{-24}$ meter^2; that is larger than the geometric cross section of the cadmium nucleus by 4 orders of magnitude. Suppose a slab of cadmium is placed in front of a beam of slow neutrons. How will the cadmium attenuate the beam?

There are, let us say, N atoms of $^{113}_{48}\text{Cd}$ per cubic meter in the slab. The number density n of neutrons in the beam will vary with position because neutrons are removed from the beam by absorption. Figure P8-7 shows a portion of the entire slab and draws attention to a thin "subslab" in the interior. Write symbolic expressions for the number of neutrons that enter the subslab in Δt, for the number that leave, and for the number absorbed. Can you justify the expression

$$n(x + \Delta x) = n(x) - n(x)\sigma N\, \Delta x?$$

Convert the above equation to a differential equation, and solve for $n(x)$, given the initial number density $n(0)$.

If $N = 6 \times 10^{27}$ meters^{-3} how thin a slab will suffice to attenuate the beam by a factor of 100? How much attenuation would 2 millimeters provide? (These numbers provide some sense of why the control rods in a nuclear reactor are made of cadmium.)

8-8 A sphere of radius R finds itself in a hail of tiny particles with velocity $\mathbf{v}$, number density n, mass m, and radius R_0. Calculate the rate at which the sphere acquires momentum from the hail. First, estimate the rate by qualitative

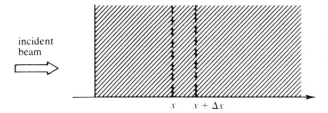

Figure P8-7

reasoning, asking (in part), What can the rate depend on and how? (Be sure your answer has the correct dimensions.) Then calculate the rate exactly.

8-9 Atoms, though overall neutral, will attract each other. The attraction is adequately described by a $1/r^6$ potential energy:

$$U = -U_0 \frac{r_0^6}{r^6},$$

where r_0 is a fixed length of atomic size and U_0 is a positive constant with the dimensions of an energy.

Suppose a beam of hydrogen atoms passes into some cesium vapor. A cesium hydride molecule is formed if, let us assume, the distance between a hydrogen and a cesium atom is ever as small as r_0.

The hydrogen atoms in the beam have an energy E. The mass ratio is $m_{hydrogen}/m_{cesium} = 1/133 \ll 1$. Calculate the cross section for forming a cesium hydride molecule.

Why must we distinguish a "low energy" domain, $E < 2U_0$, from a "high energy" domain, $E > 2U_0$?

8-10 Our calculation of $d\sigma/d\Omega$ in section 8.4 ignored the electrons that surround the gold nucleus. With them in mind, estimate the *finite* size of the total scattering cross section: alpha particles on gold atoms. (You may find it useful to note that in bulk gold the nuclei are separated by about 2.6×10^{-10} meter.) An order-of-magnitude estimate will suffice.

Although the electrons greatly influence the total cross section, they have little effect on the large-angle scattering, $\theta_{scattering} > 90°$, say. Why?

APPENDIX A

EXPANSIONS, IDENTITIES, AND MISCELLANY

A.1 EXPANSIONS

Taylor Series

$$f(x) = f(x_0) + f'(x_0)(x - x_0) + \tfrac{1}{2}f''(x_0)(x - x_0)^2 + \cdots$$
$$= \sum_{n=0}^{\infty} \frac{1}{n!} \left.\frac{d^n f}{dx^n}\right|_{x=x_0} (x - x_0)^n. \tag{A.1-1}$$

With the Taylor expansion, we can confirm—or derive—the series given below.

Binomial

$$(1 + \alpha)^n = 1 + n\alpha + \frac{n(n-1)}{2}\alpha^2 + \cdots \quad \text{if } |\alpha| < 1. \tag{A.1-2}$$

Common specific instances of the binomial expansion:

$$(1 + \alpha)^{1/2} = 1 + \tfrac{1}{2}\alpha - \tfrac{1}{8}\alpha^2 + \cdots, \tag{A.1-3}$$

$$(1 + \alpha)^{-1/2} = 1 - \tfrac{1}{2}\alpha + \tfrac{3}{8}\alpha^2 + \cdots. \tag{A.1-4}$$

Exponential

$$e^\alpha = 1 + \alpha + \frac{1}{2}\alpha^2 + \cdots + \frac{1}{n!}\alpha^n + \cdots. \qquad (A.1\text{-}5)$$

The exponential series converges for all finite α but is useful numerically only if $|\alpha| \ll 1$.

Logarithm

$$\ln(1+\alpha) = \alpha - \tfrac{1}{2}\alpha^2 + \cdots \qquad \text{if } |\alpha| < 1. \qquad (A.1\text{-}6)$$

Sine and Cosine

$$\sin \alpha = \frac{e^{i\alpha} - e^{-i\alpha}}{2i} = \alpha - \tfrac{1}{6}\alpha^3 + \cdots, \qquad (A.1\text{-}7)$$

$$\cos \alpha = \frac{e^{i\alpha} + e^{-i\alpha}}{2} = 1 - \tfrac{1}{2}\alpha^2 + \cdots. \qquad (A.1\text{-}8)$$

A.2 TRIGONOMETRIC IDENTITIES

All the identities given here can be proved algebraically by using equations (A.1-7) and (A.1-8), where sine and cosine are expressed in terms of complex exponentials.

$$\sin^2 a + \cos^2 a = 1. \qquad (A.2\text{-}1)$$

$$\cos(a+b) = \cos a \cos b - \sin a \sin b. \qquad (A.2\text{-}2)$$

$$\sin(a+b) = \sin a \cos b + \cos a \sin b. \qquad (A.2\text{-}3)$$

$$\cos 2a = \cos^2 a - \sin^2 a = 1 - 2\sin^2 a. \qquad (A.2\text{-}4)$$

$$\sin 2a = 2 \sin a \cos a. \qquad (A.2\text{-}5)$$

$$\cos^3 a = \tfrac{1}{4}\cos 3a + \tfrac{3}{4}\cos a. \qquad (A.2\text{-}6)$$

$$\sin^3 a = -\tfrac{1}{4}\sin 3a + \tfrac{3}{4}\sin a. \qquad (A.2\text{-}7)$$

A.3 MISCELLANY

$$\int \frac{dx}{(a^2 - x^2)^{1/2}} = \arcsin \frac{x}{a} \qquad \text{if } |x| \leq a.$$

$$\int \frac{x^2\, dx}{(a^2 - x^2)^{1/2}} = \frac{-x(a^2 - x^2)^{1/2}}{2} + \frac{a^2}{2} \arcsin \frac{x}{a}.$$

Earth's mass: $m_\oplus = 5.98 \times 10^{24}$ kilograms
Earth's (equatorial) radius: $R_\oplus = 6.38 \times 10^6$ meters
Mean earth-sun distance: "$r_\oplus$" $= 1.50 \times 10^{11}$ meters

Sun's mass: $m_\odot = 1.99 \times 10^{30}$ kilograms
Solar radius: $R_\odot = 6.96 \times 10^8$ meters

Moon's mass: $m_{\mathbb{C}} = 7.35 \times 10^{22}$ kilograms
Mean earth-moon distance: $r_{\oplus\mathbb{C}} = 3.84 \times 10^8$ meters

Gravitational constant: $G = 6.67 \times 10^{-11}$ newton · meter²/kilogram²
Standard acceleration of free fall: $g = 9.81$ meters/second² or newtons/kilogram

Proton's mass: $m_p = 1.67 \times 10^{-27}$ kilogram
Electron's mass: $m_e = 9.11 \times 10^{-31}$ kilogram

Coulomb's law: the electric force between point charges q_1 and q_2 has the magnitude $K|q_1 q_2|/(\text{separation})^2$. The constant K depends on the units used for charge:

Proton charge e	Constant K
1.60×10^{-19} coulomb	$1/(4\pi\epsilon_0) = 9.00 \times 10^9$ newton · meter²/coulomb²
4.80×10^{-10} esu	1 dimensionless

APPENDIX
B

VECTOR PRODUCT

In this appendix we review the vector product and then develop several computational aids, based on components and cartesian unit vectors.

To define the vector product **A** × **B** of two vectors **A** and **B**, we need to specify the magnitude and direction of the newly constructed vector. Here is a geometric prescription:

Magnitude: $\quad |\mathbf{A} \times \mathbf{B}| = |\mathbf{A}| |\mathbf{B}| |\sin(\mathbf{A}, \mathbf{B})|,$ (B.1)

Direction: $\quad$ Perpendicular to plane
defined by **A** and **B**. (B.2)

The prescription is not quite complete: which of the two directions perpendicular to the plane should be used? Let us adopt the *right-hand rule:* Rotate the fingers of your right hand from the first vector in the product to the second, going through the smaller of the two possible angles; the orientation of your thumb then gives the correct perpendicular. When we adopt this geometric definition—equations (B.1) and (B.2) and the right-hand rule—the product **A** × **B** is independent of any coordinate axes and is properly called a vector.

An illustration is provided in figure B.1.

The vector product is anti-commutative:

$$\mathbf{B} \times \mathbf{A} = -\mathbf{A} \times \mathbf{B}.$$

312 VECTOR PRODUCT

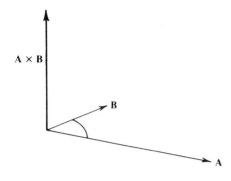

Figure B 1 The vector product. Given $|\mathbf{A}| = 2.3$, $|\mathbf{B}| = 0.8$, and $60°$ for the angle between $\mathbf{A}$ and $\mathbf{B}$, the magnitude of $\mathbf{A} \times \mathbf{B}$ is $(2.3)(0.8)(\sqrt{3}/2) = 1.59$.

If **A** and **B** are parallel (or antiparallel), they cease to define a plane, and the direction of $\mathbf{A} \times \mathbf{B}$ might be indeterminate. The magnitude, however, which is proportional to the sine of the angle between **A** and **B**, vanishes and thus saves the day. Moreover, the equation above is consistent with this because $0 = -0$ both for scalars and for vectors.

Distributivity holds for the vector product:

$$\mathbf{A} \times (\mathbf{B} + \mathbf{C}) = \mathbf{A} \times \mathbf{B} + \mathbf{A} \times \mathbf{C},$$

but a proof based on the geometric definition of the product is cumbersome.

Components and Orthogonal Unit Vectors

There is an advantage, sometimes, to using a set of orthogonal unit vectors and representing a general vector **A** in terms of them and components. The representation looks like this:

$$\mathbf{A} = \sum_{i=1}^{3} A_i \mathbf{e}_i. \tag{B.3}$$

Because the unit vectors $\mathbf{e}_i$ are specified to be orthogonal, their scalar products can be written

$$\mathbf{e}_i \cdot \mathbf{e}_j = \delta_{ij}, \tag{B.4}$$

where δ_{ij}, called the *Kronecker delta,* is defined by the prescription

$$\delta_{ij} = \begin{cases} 1 & \text{if } i = j \\ 0 & \text{if } i \neq j. \end{cases} \tag{B.5}$$

With this device, we can extract from equation (B.3) an expression for the jth component A_j. Taking the scalar product of both sides with $\mathbf{e}_j$ gives

$$\mathbf{e}_j \cdot \mathbf{A} = \sum_{i=1}^{3} A_i \mathbf{e}_j \cdot \mathbf{e}_i$$

$$= \sum_{i=1}^{3} A_i \delta_{ij} = A_j.$$

The component A_j is just the scalar product of $\mathbf{e}_j$ and $\mathbf{A}$, sometimes called *the projection of* $\mathbf{A}$ *onto the unit vector* $\mathbf{e}_j$. That, indeed, is a fruitful way to think of components.

Some computations entail several summations. To indicate all the summations with summation signs Σ and with explicit limits would hopelessly clutter the equations. We adopt the convention that *the indices in any pair of repeated indices are to be summed over automatically*. With this convention, known as the *Einstein summation convention*, we can write equation (B.3) tersely as

$$\mathbf{A} = A_i \mathbf{e}_i.$$

In the notation introduced here, the scalar product of vectors $\mathbf{A}$ and $\mathbf{B}$ takes the form

$$\mathbf{A} \cdot \mathbf{B} = (A_i \mathbf{e}_i) \cdot (B_j \mathbf{e}_j)$$
$$= A_i B_j \mathbf{e}_i \cdot \mathbf{e}_j = A_i B_j \delta_{ij}$$
$$= A_i B_i.$$

The summation over the index j in $B_j \delta_{ij}$ gives a nonzero value only when $j = i$, and then it yields B_i, whence the last line follows, which itself is a sum over three products:

$$A_i B_i = A_1 B_1 + A_2 B_2 + A_3 B_3.$$

Going back now to the vector product, we can write

$$\mathbf{A} \times \mathbf{B} = (A_i \mathbf{e}_i) \times (B_j \mathbf{e}_j)$$
$$= A_i B_j \mathbf{e}_i \times \mathbf{e}_j. \tag{B.6}$$

By examining figure B.2, we can confirm that

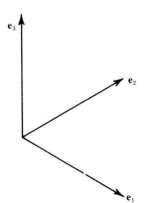

Figure B.2 Orthogonal unit vectors in a right-handed system. (The perspective intends $\mathbf{e}_2$ to be in the background.)

$$\mathbf{e}_i \times \mathbf{e}_j = \epsilon_{ijk}\mathbf{e}_k \tag{B.7}$$

in a right-handed system. The permutation symbol ϵ_{ijk} is defined by the following prescription:

1. If the sequence of numbers ijk is the sequence 123 or can be generated from 123 by an even number of transpositions, then $\epsilon_{ijk} = +1$.
2. If the sequence ijk results from 123 by an odd number of transpositions, then $\epsilon_{ijk} = -1$.
3. If any index is equal to another index, then $\epsilon_{ijk} = 0$.

Explicitly, the nonzero elements are

$$\epsilon_{123} = \epsilon_{312} = \epsilon_{231} = +1,$$
$$\epsilon_{132} = \epsilon_{213} = \epsilon_{321} = -1.$$

If we combine equations (B.6) and (B.7), we arrive at

$$\mathbf{A} \times \mathbf{B} = \epsilon_{ijk} A_i B_j \mathbf{e}_k. \tag{B.8}\bigstar$$

Equation (B.8) is the basis for a mnemonic. We can express the vector product symbolically as a determinant:

$$\mathbf{A} \times \mathbf{B} = \begin{vmatrix} \mathbf{e}_1 & \mathbf{e}_2 & \mathbf{e}_3 \\ A_1 & A_2 & A_3 \\ B_1 & B_2 & B_3 \end{vmatrix} \tag{B.9}$$

Some Identities

The triple product may be written

$$\begin{aligned} \mathbf{C} \cdot (\mathbf{A} \times \mathbf{B}) &= C_m \mathbf{e}_m \cdot (\epsilon_{ijk} A_i B_j \mathbf{e}_k) \\ &= \epsilon_{ijk} A_i B_j C_m \delta_{mk} \\ &= \epsilon_{ijm} A_i B_j C_m. \end{aligned} \tag{B.10}$$

The structure of the right-hand side is particularly simple. Any permutation of vectors **A, B,** and **C** on the left, with the $\cdot$ and $\times$ operations left in place, will generate a similar right-hand side. Therefore, the magnitude of the triple product is unchanged by permutations; only the sign may change. For example,

$$\begin{aligned} \mathbf{A} \cdot (\mathbf{C} \times \mathbf{B}) &= A_i \mathbf{e}_i \cdot (\epsilon_{mjk} C_m B_j \mathbf{e}_k) \\ &= \epsilon_{mji} A_i B_j C_m \\ &= -\epsilon_{ijm} A_i B_j C_m \\ &= -\mathbf{C} \cdot (\mathbf{A} \times \mathbf{B}). \end{aligned}$$

The step to the third line follows because ϵ_{mji} can be generated from ϵ_{ijm} by three transpositions, an odd number:

$$\epsilon_{mji} = -\epsilon_{mij} = -(-)\,\epsilon_{imj} = -(-)(-)\,\epsilon_{ijm}.$$

After that, comparision with equation (B.10) completes the demonstration.

When the triple product appears in a calculation, often we can make it easier to evaluate or make its implications easier to see if we permute the vectors. On several occasions, we will permute the vectors cyclically: each vector moves to the right, and the last goes around to the front, as in

$$\mathbf{C} \cdot (\mathbf{A} \times \mathbf{B}) \quad \text{becomes} \quad \mathbf{B} \cdot (\mathbf{C} \times \mathbf{A});$$

or each vector moves to the left, and the first goes around to the end, as in

$$\mathbf{C} \cdot (\mathbf{A} \times \mathbf{B}) \quad \text{becomes} \quad \mathbf{A} \cdot (\mathbf{B} \times \mathbf{C}).$$

We can check that under such cyclic permutation the triple product does not change in either magnitude or sign. For example,

$$\mathbf{B} \cdot (\mathbf{C} \times \mathbf{A}) = B_j \mathbf{e}_j \cdot (\epsilon_{mik} C_m A_i \mathbf{e}_k)$$
$$= \epsilon_{mij} A_i B_j C_m.$$

Two transpositions will convert ϵ_{mij} through $-\epsilon_{imj}$ to $+\epsilon_{ijm}$, and then comparison with equation (B.10) completes the proof.

Whenever two vector products arise in a problem, two permutation symbols will appear, and frequently they will have an index in common, over which we are to sum, for example, $\epsilon_{ink}\epsilon_{jmk}$. Such a sum can be evaluated in terms of Kronecker deltas. Imagine summing over the common index: $k = 1, 2, 3$. Each product will be zero unless ($i = j$ and $n = m$) or ($i = m$ and $n = j$). Because the second alternative incorporates a transposition, the full result is

$$\epsilon_{ink}\epsilon_{jmk} = \delta_{ij}\delta_{nm} - \delta_{im}\delta_{nj}. \tag{B.11}\bigstar$$

There is no substitute for testing this proposition with specific numbers.

With identity (B.11), we can reduce a double vector product to simpler form, as follows:

$$\mathbf{A} \times (\mathbf{B} \times \mathbf{C}) = (A_i \mathbf{e}_i) \times (\epsilon_{jmk} B_j C_m \mathbf{e}_k)$$
$$= \epsilon_{ikn}\epsilon_{jmk} A_i B_j C_m \mathbf{e}_n$$
$$= -(\delta_{ij}\delta_{nm} - \delta_{im}\delta_{nj})\,A_i B_j C_m \mathbf{e}_n$$
$$= (A_i C_i B_n - A_i B_i C_n)\mathbf{e}_n$$
$$= (\mathbf{A} \cdot \mathbf{C})\mathbf{B} - (\mathbf{A} \cdot \mathbf{B})\mathbf{C}. \tag{B.12}$$

A single transposition in the second line, namely, $\epsilon_{ikn} = -\epsilon_{ink}$, puts that line into a form where the identity (B.11) can be invoked.

A × B as an Axial Vector

Some people take equation (B.8) as the *definition* of the vector product:

$$\mathbf{A} \times \mathbf{B} \equiv \epsilon_{ijk} A_i B_j \mathbf{e}_k. \tag{B.13}$$

316 VECTOR PRODUCT

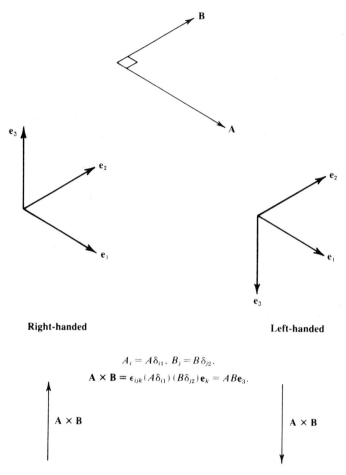

Figure B.3 Right- versus left-handed. Vectors **A** and **B** are sketched at the top, as they exist in space. (For convenience only, the vectors are perpendicular to each other.) If equation (B.13) is taken as the *definition* of the vector product, then the product has the *form* ABe_3 in both right- and left-handed systems. The direction of the product in space, however, is not the same.

Such a definition makes the direction of **A** × **B** depend on whether the unit vectors form a right- or left-handed system. Figure B.3 shows what this means.

A "vector" that reverses direction on the change from a right-handed representation to a left-handed one is called an *axial vector*. In distinction, a vector such as a position or velocity vector, which never has such a reversal property, is called a *polar vector*. Whenever we employ a set e_i of unit vectors, a right-handed system is specified, though perhaps only implicitly. The question of which definition to adopt—equation (B.1), equation (B.2), and the right-hand rule or equation (B.13)—then becomes irrelevant. We do not need to distinguish axial from polar vectors, and we will not.

APPENDIX
C

THE AVERAGING METHOD

Problems in mechanics and elsewhere in science often lead to equations of the form

$$\ddot{x} = -\omega_0^2 x + f(x, \dot{x}) \tag{C.1}$$

where $f(x, \dot{x})$ incorporates all the complicated terms. If f were always zero, the general solution would be

$$x = \text{const} \times \cos\left[\omega_0 t + \text{const}'\right]. \tag{C.2}$$

When f is not zero, it may be that f has only a small influence on the motion during any single time interval of order $2\pi/\omega_0$. (We may not know this beforehand, but we may always try assuming it and see whether it is self-consistent.) If the influence is small, we may be able to get a good approximation by letting the amplitude and phase in equation (C.2) change with time.

In short, we try the form

$$x = \mathcal{A}(t) \cos[\omega_0 t + \varphi(t)]. \tag{C.3}$$

We are to let the functions $\mathcal{A}$ and φ vary in such a way that equation (C.1) is satisfied or at least approximately so.

To find equations for $\mathcal{A}$ and φ, we differentiate equation (C.3) and insert into equation (C.1), as follows:

$$\dot{x} = -\omega_0 \alpha \sin[\cdots] + \{\dot{\alpha}\cos[\cdots] - \dot{\varphi}\alpha\sin[\cdots]\}; \quad (C.4)$$

$$\ddot{x} = -\omega_0^2 \alpha \cos[\cdots] - \omega_0 \dot{\alpha}\sin[\cdots] - \omega_0 \dot{\varphi}\alpha\cos[\cdots] + \frac{d}{dt}\{\cdots\};$$

and so insertion produces

$$-\omega_0 \dot{\alpha}\sin[\cdots] - \omega_0 \dot{\varphi}\alpha\cos[\cdots] + \frac{d}{dt}\{\cdots\} = f(x, \dot{x}). \quad (C.5)$$

(It is easier to use dots than to write out the arguments or to remember what a new symbol means.) We have here one equation in two unknowns, α and φ. We need a second equation.

We can get a second equation—and can simplify the equation we already have—by insisting that the quantity in the curly brackets be zero; that is, we impose the equation

$$\dot{\alpha}\cos[\cdots] - \dot{\varphi}\alpha\sin[\cdots] = 0. \quad (C.6)$$

We can still choose $\alpha(0)$ and $\varphi(0)$ to meet specified initial conditions, and the expression for $\dot{x}$ becomes much simpler, indeed, just what it would be if α and φ were actually constants.

Why do we have the right to impose equation (C.6)? Because we replaced a single variable, namely x, by two variables, α and φ. Through that increase in the number of variables we introduced some freedom into the mathematics; here we use the freedom to simplify the calculation.

The derivatives $\dot{\alpha}$ and $\dot{\varphi}$ appear in equations (C.5) and (C.6). We can solve for them algebraically, as follows:

$$\frac{-\sin[\cdots]}{\omega_0} \times (C.5) + \cos[\cdots] \times (C.6) \Rightarrow \dot{\alpha} = -\frac{1}{\omega_0} f \sin[\cdots], \quad (C.7a)$$

$$\frac{-\cos[\cdots]}{\omega_0 \alpha} \times (C.5) - \frac{\sin[\cdots]}{\alpha} \times (C.6) \Rightarrow \dot{\varphi} = -\frac{1}{\omega_0 \alpha} f \cos[\cdots]. \quad (C.7b)$$

Equations (C.7a and b) are still miserably difficult. At this point we make our single, but fundamental, approximation. If α and φ change only slowly and if we are satisfied with only the long-term behavior, we may approximate the right-hand sides by using their average over one period $2\pi/\omega_0$, with the average calculated as though α and φ were literally constant. Spelled out for $\dot{\alpha}$, this means

$$\dot{\alpha}(t) \simeq -\frac{1}{\omega_0} \frac{1}{2\pi/\omega_0} \int_t^{t+2\pi/\omega_0} dt' \, f \sin[\omega_0 t' + \varphi(t)],$$

where f is to be expressed in terms of α and φ via equations (C.3) and (C.4), but then α and φ are to be regarded as constant over the relatively short integration interval. The succinct expressions are

$$\dot{\mathcal{A}} = -\frac{1}{\omega_0} \langle f \sin[\cdots] \rangle, \qquad (C.8a) \bigstar$$

$$\dot{\varphi} = -\frac{1}{\omega_0} \left\langle \frac{f \cos[\cdots]}{\mathcal{A}} \right\rangle. \qquad (C.8b) \bigstar$$

Equality signs are used because what we do subsequently with the equations will follow exactly; as long as we remember that we made one approximation in getting to equations (C.8a and b), the equality signs will prevent ambiguity in the future.

The equations for $\mathcal{A}$ and φ depend on certain averages of the complicated quantity f. This characteristic generates the name of the approximation: the *averaging method*.

We can make sense out of the structure that equations (C.8a and b) have. The velocity $\dot{x}$ is $-\omega_0 \mathcal{A} \sin[\cdots]$. Thus the right-hand side of equation (C.8a) is proportional to $+\langle f\dot{x} \rangle$; the product $f\dot{x}$ is the rate at which the "force" f does work. A positive value for the average work rate should increase the oscillation amplitude, and that is just what equation (C.8a) says.

For equation (C.8b), we should recall that $x = \mathcal{A} \cos[\cdots]$, and so on the right-hand side we have an average proportional to $\langle fx \rangle$. If $\langle fx \rangle > 0$, say, so that f and x are positively correlated, then the function f (or part of it) is acting like $+x$. The effect of f in the differential equation (C.1) will be like a decrement in a spring constant, and so the full angular argument $\omega_0 t + \varphi(t)$ should change with time less rapidly than just $\omega_0 t$ would. The negative value for $\dot{\varphi}$ will produce precisely this behavior.

A last comment. If $\langle f \rangle$ itself is nonzero, we should not expect $\langle x \rangle$ to be zero. Rather, the latter will depart from zero so that the average of the "restoring force" $\langle -\omega_0^2 x \rangle$ will cancel with $\langle f \rangle$. In this circumstance we must augment the trial form displayed in equation (C.3) by adding the constant term $\langle f \rangle/\omega_0^2$. [This conclusion follows formally by averaging equation (C.1) term by term and noting that the acceleration, averaged over a period of the motion, must be zero because the velocity returns to its original value:

$$\langle \ddot{x} \rangle = -\omega_0^2 \langle x \rangle + \langle f \rangle,$$

together with $\langle \ddot{x} \rangle = 0$, implies

$$\langle x \rangle = \frac{\langle f \rangle}{\omega_0^2},$$

as reasoned above.]

APPENDIX
D

THE CRAFT OF THE PHYSICIST

Most topics and problems in an undergraduate physics curriculum surrender to assault by a smallish number of techniques. And much the same is true in professional research. To display some of those techniques—and to provide practice with them—is a major aim of this book.

The techniques could be left implicit, left as they arise in the text. But there is merit in making them explicit, in gathering them together and thus bringing them more strongly into your consciousness.

This appendix is an attempt to be more explicit. In table D.1 are listed two dozen analytical tricks of the trade, phrased as suggestions to the problem solver. Some instances of their use in the text are cited, usually the earlier ones. (Many more instances occur in the homework problems, sometimes with hints about which route you might try, at other times without.) As for the whole of undergraduate physics, no list of problem-solving techniques is ever complete, even for major items; differences of opinion alone ensure that judgment. Still, the present list will help you assimilate the techniques that this book seeks to teach, and table D.1 will serve as a checklist for a stymied problem solver.

Two reservations must be emphasized. The list is not a system for solving homework problems, in this book or in others. Much less is the list a prescription for discovering the relativity of simultaneity or for inventing a geometric theory of gravitation. The great steps in physics come from a deeper source. But much of physics does proceed at a level where the items listed are the methods of the physicist's craft.

Table D.1 Analytical tricks of the trade
The section numbers in the right column cite examples in the text.

1. General strategy
   a. Ask, What could it depend on? (Thus, select appropriate parameters and variables.) — 2.5, 7.2
   b. Make a sketch or a rough graph. (It aids the intuition.) — 1.1, 4.5, 5.2
   c. Use vectors heavily, for they display the essentials succinctly. — 1.3, 5.1
   d. Choose the simplest model. — 5.5, 7.8
   e. Decide, Is it better to work with a vector equation or with components? — 1.3, 6.3
   f. Choose coordinates advantageously.
      (1) Use coordinates that display the symmetry or preferred directions. — 1.3, 1.6
      (2) Make a coordinate transformation (to get a better point of view, as in center-of-mass or rotating coordinates). — 5.1, 6.3
   g. Use the conservation laws (or first integrals). — 3.1, 5.2, 7.11
   h. Reason from symmetry. — 7.1, 7.6
   i. Estimate the order of magnitude. — 1.1, WP6-1, 7.12
   j. Superpose: split the problem into known or simpler parts; recombine them later. — 1.6, 2.3, 7.9
   k. Solve a simpler version first; then, with new insight, return to the actual problem. — 5.4 & 5.5
   l. Test your answer to see whether it makes physical sense (when you vary parameters or take limits). — 1.1, 7.4

2. Some tactics
   a. Check dimensions (e.g., terms in the same equation, the final answer).
   b. Examine the behavior in the limit (as you vary parameters or variables). — 3.3, 4.5
   c. Map onto a solved problem. (Sometimes this means "adopt a tractable model, one from among a limited number." At other times, it means "change variables to get a well-studied equation." Or "just recognize that you have a previously studied equation.") — 5.4, 5.5, WP6-1
   d. Approximate:
      (1) Drop small terms. — 1.6, 6.4
      (2) Ignore slow variations (at least at first). — 2.1
      (3) Institute successive approximations. — 6.3
      (4) Linearize equations. — 3.1, 5.6
   e. Expand:
      (1) Binomial expansion. — 1.6, 3.1
      (2) Taylor series. — 1.1, 3.1
      (3) More general series (e.g., trigonometric functions). — 3.5
   f. Look for average behavior. — 3.3, 5.5
   g. Exploit an extremum: locate one or expand about one. — 2.4, 3.1
   h. Guess a solution (with adjustable parameters). — 2.4, 3.2

i. Try dimensional analysis.*	3.1
j. Use a variational principle.	4.1, 4.5
k. Integrate by parts.	4.2, 4.3
l. Replace a vector field by a (usually scalar) potential and some differential operator.	1.4, 6.5

---

*Here is the essence of the method:
1. Pose the question, What quantities could the answer depend on? Then respond, using your best judgment.
2. From the selected quantities, form an expression with the answer's dimensions.
3. That is your answer—except for a factor that is an unknown numerical constant (*presumed* to be of order 1) or, unfortunately, is an unknown function of whatever dimensionless products can be formed from the quantities.

INDEX

Amplitude jumps, 81–87
Analytical tricks, summary of, 321–323
Angular frequency, defined, 48
Angular momentum:
  defined, 29, 238
  in plane polar coordinates, 160–161
  and relation to angular velocity, 244–245, 249–251
  relative to origin of inertial frame, 285
  response to torque of, 238–239
Angular velocity vector:
  defined, 204
  instantaneous, 205–207
Anti-damping, 88
Astronomical unit, 174
Averaging method:
  as applied to: Duffing's equation, 93–94
    Foucault pendulum, 221–222
    Newton's law of damping, 95–97
    planetary orbits, 175–181
    van der Pol's equation, 89–93
  conditions for validity of, 104–105
  derived, 317–319
  understanding the, 319
  when $\langle f \rangle \neq 0$, 319

Bifurcation, 136
Brachistochrone problem, 149–150

Calculus of variations, 118–123
  Euler's equations and, 122
Center of mass:
  as defined for more than two masses, 237
  motion of, 237–238
Centrifuge, 229–231
Chandler wobble, 260–264
Cometary orbits, 174–175

Cone:
  body, 259
  space, 259
Constants, physical, 310
Coriolis effect and wind, 216–217
Coriolis term, 209
Coupled pendula, 154–155
Cross section:
  beam attenuation and, 306
  for capture, 289–291, 301
  defined, 288
  differential, 291–295, 301
  mean free path and, 301–303
  for reaction, 303–304
  Rutherford, 295–300
  for small-angle scattering, 305–306
Cylindrical coordinates:
  acceleration in, 150
  unit vectors in, 125–126

Dimensional analysis, 323n.
Duffing's equation, 81–87
Dust grain:
  bodily precession of, 242–244
  precession of orbital plane of, 276–277

Earth's bulge, mass of, 227–228
Eccentricity, 170, 173, 191
Effective potential energy, 161, 190
Electric units, 310
Energy:
  conservation of, 17, 29–30
  kinetic, 15–17
  potential (*see* Potential energy)
  in rotational motion, 264–265
  work and, 15–17
Euler-Lagrange equation, 125

Euler's equations for a rigid body, 254–255
Expansions, mathematical series, 308–309
Extremal principle (*see* Variational principle)

Fermat's principle, 116–118
Figure of the earth, 223–228
Fixed-axis rotation, 249
Foucault pendulum, 217–223

g as defined on rotating earth, 211–212
Generalized coordinate, 131
Generalized momentum, 137, 140
Generalized velocity, 136
Geopotential, 226
Gradient:
  in cartesian coordinates, 19–20
  defined, 17–18
  in plane polar coordinates, 42
Gravitational potential energy, 20–27

Hamiltonian:
  as constant of the motion, 154
  defined, 139
  energy and, 154
  Liouville's theorem and, 140–144
Hamilton's equations, 136–140
Harmonic oscillator:
  absorption profile of, 59–61
  critically damped, 67
  damped, 45–51
  energy of damped, 49–51
  heavily damped, 67
  Lissajous figures and, 68
  Lorentz atom and, 58–61, 68–69
  phase relations and, 57, 70
  quality factor of, 61–62
  sinusoidally driven, 56–61
  in two dimensions, 54–56
Hysteresis, 86

Impact parameter, 291
Incident flux, 288
Inertial frame, 31–32
Instability (*see* Stability)
Intermediate axis theorem, 284–285

Invariance:
  implications of, 27–30
  Lagrangian and, 153
  rotational, 28–29
  space translational, 27–28
  time translation, 29–30

Keplerian ellipse, 171–173
Kepler's laws, 197
Kinetic energy in rotational motion, 264–265
Kronecker delta, 312

Lagrange's equations, 123–125
  in cylindrical coordinates, 125–128
  summarized, 146–147
Lagrangian:
  constraints and, 128–136, 147–149
  defined, 125, 128
  quantum mechanics and, 144–146
Limit cycle:
  defined, 92
  examples of, 112
  in van der Pol's equation, 88–93
Linearize, 184–185
Liouville's theorem, 140–144

Massive limit, 191
Moment-of-inertia tensor:
  for cross, 251
  defined, 245
  diagonal form for, 253
  for diatomic molecule, 278
  for disk, 252
  for ring, 246
  for sphere, 252
  transformation of, 281–282

Newton and shape of the earth, 223–225
Newton's laws, 30–34
Normal modes, 154–155

Oblate sun, 175–181
Orbit:
  in attractive $1/r^2$ field, 168–175, 191

Orbit (*Cont.*):
  data on objects in solar, 174
  period of elliptical, 187–189
  time-independent equation for shape of, 164–165, 190
Orientation, equation for changes in, 238–239

Parallel axis theorem, 282
Perihelion:
  defined, 174
  precession of, 175–181
Permutation symbol, 314
Phase space, 51–54, 88–93, 140–144
  harmonic oscillator in, 51–54
  Liouville's theorem and, 140–144
Physical constants, 310
Plane polar coordinates and Newton's second law, 182–183, 190
Planetary orbits, 168–181
Potential energy, 15–20
  curl **F** and, 42
  defined, 16–17, 29–30
  gravitational, 20–27
  of massive ring, 23–27
  ways to compute, 35
Precession:
  of the equinoxes, 272–275
  of perihelion, 175–181
  simple, 240–241
  of supported top, 265–272
  of torque-free top, 255–264
Pressure gradient, 224, 229–230
Principal axes, 253

Quality factor, 61–62

Rayleigh's equation, 111
Reduced mass:
  angular momentum and, 195
  defined, 158
  kinetic energy and, 195
Rotating frame, 202–209
  velocity and acceleration as seen in, 208–209
Rotation:
  general, 205–207
  uniform, 204

Satellite paradox, 195
Secular term, defined, 101
Series expansions, 97–104
  with secular terms precluded, 100–103
  synopsis, 103–104
Small oscillations theory:
  development, 73–74
  summary, 80–81
Solar oblateness, 175–181
Solid angle, 293–294
Stability:
  of circular orbits, 181–187
  criteria for, 187
  potential energy and, 23–27
  restoring forces and, 26–27, 80, 187
Summation convention, 313

Top:
  axisymmetric and supported, 265–272
  axisymmetric and torque-free, 255–264
  fast, 283
  sleeping, 283
Torque, internal forces and, 239
Transfer orbit, 200–201
Travel time, integral for, 77
Trigonometric identities, 309
Two-body problem, 156–191

Units, electric, 310

Van der Pol's equation, 88–93
Variational principle, 118–123
  defined, 122
  Newton's second law as a, 123–128
  quantum mechanics and, 144–146
Vectors:
  axial, 315–316
  defined, 7
  double vector product of, 315
  orthogonal, 8
  polar, 316
  scalar product of, 7–8
  triple product of, 314–315
    cyclic permutation and, 315
  unit, 8
  vector product of, 311–316
Velocity, derivative definition of, 9–10, 34
Virial theorem, 43